HUMAN CYTOSOLIC SULFOTRANSFERASES

HUMAN CYTOSOLIC SULFOTRANSFERASES

EDITED BY

Gian Maria Pacifici
Michael W.H. Coughtrie

CRC Press
Taylor & Francis Group
Boca Raton London New York

CRC Press is an imprint of the
Taylor & Francis Group, an **informa** business
A TAYLOR & FRANCIS BOOK

CRC Press
Taylor & Francis Group
6000 Broken Sound Parkway NW, Suite 300
Boca Raton, FL 33487-2742

First issued in paperback 2019

© 2005 by Taylor & Francis Group, LLC
CRC Press is an imprint of Taylor & Francis Group, an Informa business

No claim to original U.S. Government works

ISBN-13: 978-0-415-28462-2 (hbk)
ISBN-13: 978-0-367-39256-7 (pbk)
Library of Congress Card Number 2004030011

Library of Congress Cataloging-in-Publication Data

Human cytosolic sulfotransferases / edited by Gian Maria Pacifici and Michael W.H. Coughtrie.
     p. cm.
Includes bibliographical references and index.
ISBN 0-415-28462-7 (alk. paper)
[DNLM: 1. Sulfotransferases.] I. Pacifici, Gian Maria. II. Coughtrie, Michael W.H.

QP606.S8H85 2005
612'.015192--dc22

2004030011

**Visit the Taylor & Francis Web site at
http://www.taylorandfrancis.com**

**and the CRC Press Web site at
http://www.crcpress.com**

Preface

The early studies in drug metabolism, conducted in the first half of the nineteenth century by Ure, Woehler, and others, were of conjugation and the idea that conjugation reactions generally form the true detoxification step of xenobiotic metabolism. Today's research in human cytosolic sulfotransferases has arisen from these pioneering studies. The role of sulfation in the biotransformation of xenobiotics was demonstrated in the 1870s, when Baumann isolated a sulfate conjugate of phenol from a patient treated with carbolic acid — therefore sulfation was one of the first biotransformation reactions to be identified. The requirement for an "active" form of sulfate (i.e., 3'-phosphoadenosine 5'-phosphosulfate [PAPS]) in the sulfation reaction was not confirmed until the late 1950s, however, when Lipmann described the structure of PAPS and showed that it acted as the sulfuryl donor for liver sulfotransferase (or sulfokinase) activity. As early as 1958, it was recognized that multiple sulfotransferases (SULTs) existed, as phenol and steroid sulfation activities could be resolved from liver cytosol preparations. The study of the sulfation of xenobiotics and endogenous compounds in man has therefore had a long and distinguished history.

To date, there has been only one book dedicated to sulfation and the sulfotransferases — the seminal volume *Sulfation of Drugs and Related Compounds* from 1981 edited by Gerard Mulder. The current volume brings the sulfotransferase field up-to-date. Within its covers, the reader will find a series of chapters summarizing the most exciting findings of sulfotransferases in *Homo sapiens* — the most biologically relevant species for human metabolism! The chapters, contributed by leading experts in the field, outline the sulfotransferase field from a wide perspective and cover the latest advances in enzymology, structural biology, genetics, and toxicology.

We are extremely grateful to all the authors for their excellent contributions. We hope this book will provide a useful resource for scientists and students working in the field of sulfation and drug metabolism in general and will serve to inspire the next generation of sulfotransferase researchers.

G.M. Pacifici

M.W.H. Coughtrie

Editors

Gian Maria Pacifici achieved the Laurea (highest degree from the Italian University) in Biological Sciences at the University of Pisa (Italy) in 1971. In 1983, he achieved the Ph.D. in clinical pharmacology at Karolinska Institutet (Stockholm, Sweden). He holds the permanent position of Associate Professor in pharmacology, Medical School, University of Pisa, awarded in 1983.

From 1971 until the present, Dr. Pacifici has worked uninterruptedly in the metabolism of drugs in humans. In 1980, he established a bank of human tissues in his laboratory and it has continuously been developed and expanded. His training and scientific work is confined to the area of clinical pharmacology.

Dr. Pacifici's training consists of 3 years at Mario Negri Institute (Laboratory of Clinical Pharmacology), Milan, Italy (1971–1974), followed by several experiences abroad: Department of Clinical Pharmacology, Royal Postgraduate Medical School, Hammersmith Hospital, London, United Kingdom (1978–1979); Department of Clinical Pharmacology, Karolinska Institutet, Stockholm, Sweden (1979–1980 and for several periods from 1981 to 1984); Department of Clinical Pharmacology, Uppsala Universitet, Uppsala, Sweden (two periods in 1985 and 1986); Department of Pharmacology and Therapeutics, Liverpool, United Kingdom (Summer 1987); Unit of Clinical Pharmacology, The Mayo Clinic, The Mayo Foundation, Rochester, Minnesota, U.S. (Summer 1988); Unit of Clinical Pharmacology, Oxford University, Oxford, United Kingdom (Summer 1989); and finally he has visited the laboratory of Professor Klaus Brendel, Department of Pharmacology, University of Tucson, Arizona (October 1993).

Dr. Pacifici has published over 150 articles in international peer-reviewed scientific journals and over 20 chapters for books. Most of his work deals with the sulfation of drugs in tissues obtained from human fetuses and adult subjects.

Michael W.H. Coughtrie, B.Sc., Ph.D., received his Ph.D. in Biochemistry from the University of Dundee, Scotland, in 1986 for studies on UDP-glucuronosyltransferases carried out while at Dundee and at the National Institute of Environmental Health Sciences, Research Triangle Park, North Carolina. He is currently professor of Biochemical Pharmacology and Head of the Division of Pathology and Neuroscience in the Medical School, University of Dundee, Scotland, where he has been a faculty member since 1988. His major research activity is in sulfation, sulfotransferases, and sulfatases although he has recently rekindled his long-held interest in glucuronidation and the UDP-glucuronosyltransferases.

Dr. Coughtrie has published extensively in the sulfation and glucuronidation fields, with more than 80 papers in peer-reviewed journals in addition to many review articles and book chapters. He has delivered numerous invited lectures at national and international meetings and has organized several international meetings and workshops in the field of drug metabolism and conjugation. He is a member of the Biochemical Society (for which he was a member of the Cell and Molecular Pharmacology Group Committee), the International Society for the Study of Xenobiotics, the European Society for Biochemical Pharmacology, and the Society for Endocrinology. He is also an editorial board member for the journals *Xenobiotica* and *Pharmacogenetics*.

Dr. Coughtrie has received more than £5 million in research grant funding, including the first Caledonian Research Foundation Personal Research Fellowship in 1990, and major awards from the Commission of the European Communities, the Medical Research Council, the Biotechnology and Biological Sciences Research Council, the Wellcome Trust, and a number of major pharmaceutical companies.

Contributors

Araba A. Adjei
Department of Molecular
  Pharmacology and Experimental
  Therapeutics and Medicine
Mayo Medical School, Mayo Clinic,
  Mayo Foundation
Rochester, Minnesota

Amanda Barnett
School of Biomedical Sciences
University of Queensland
Queensland, Australia

Rebecca L. Blanchard
Department of Pharmacology
Fox Chase Cancer Center
Philadelphia, Pennsylvania

Gerard S. Chetrite
Hormones and Cancer Research
  Unit
Institut de Puériculture
Paris, France

Michael W.H. Coughtrie
Division of Pathology and
  Neuroscience
Ninewells Hospital and
  Medical School
University of Dundee
Dundee, Scotland

Ronald G. Duggleby
School of Molecular and Microbial
  Sciences
University of Queensland
Queensland, Australia

Charles N. Falany
Department of Pharmacology and
  Toxicology
University of Alabama
Birmingham, Alabama

Josie L. Falany
Department of Pharmacology and
  Toxicology
University of Alabama
Birmingham, Alabama

Niranjali Gamage
School of Biomedical Sciences
University of Queensland
Queensland, Australia

Hansruedi Glatt
German Institute of Human Nutrition
  (DIfE)
Potsdam-Rehbruecke, Germany

Dongning He
Department of Pharmacology and
  Toxicology
University of Alabama
Birmingham, Alabama

Nadine Hempel
School of Biomedical Sciences
University of Queensland
Queensland, Australia

Robert Hume
Tayside Institute of Child Health
University of Dundee
Dundee, Scotland

Monique H.A. Kester
Department of Internal Medicine,
  Erasmus Medical Center
Rotterdam, The Netherlands

Jennifer L. Martin
Institute for Molecular Biosciences
University of Queensland
Queensland, Australia

Michael E. McManus
School of Biomedical Sciences
University of Queensland
Queensland, Australia

Connie A. Meloche
Department of Pharmacology and
  Toxicology
University of Alabama
Birmingham, Alabama

Kiyoshi Nagata
Department of Drug Metabolism and
  Molecular Toxicology
Graduate School of Pharmaceutical
  Sciences
Tohoku University
Sendai, Japan

Gian Maria Pacifici
Department of Neurosciences
Section of Pharmacology, Medical
  School
Pisa, Italy

Jorge R. Pasqualini
Hormones and Cancer Research Unit
Institut de Puériculture
Paris, France

Nancy B. Schwartz
Departments of Biochemistry and
  Molecular Biology
Pediatrics and Committee on
  Developmental Biology
University of Chicago
Chicago, Illinois

Miki Shimada
Department of Drug Metabolism and
  Molecular Toxicology
Graduate School of Pharmaceutical
  Sciences
Tohoku University
Sendai, Japan

Emma L. Stanley
Ninewells Hospital and
  Medical School
University of Dundee
Dundee, Scotland

Charles A. Strott
Section on Steroid Regulation
Endocrinology and Reproductive
  Research Branch
National Institute of Child Health and
  Human Development
National Institutes of Health
Bethesda, Maryland

Jyrki Taskinen
Department of Pharmacology
Viikki Drug Discovery Technology
  Center
University of Helsinki
Helsinki, Finland

Theo J. Visser
Department of Internal Medicine,
  Erasmus Medical Center
Rotterdam, The Netherlands

Richard M. Weinshilboum
Department of Molecular
  Pharmacology and Experimental
  Therapeutics and Medicine
Mayo Medical School, Mayo Clinic,
  Mayo Foundation
Rochester, Minnesota

Kelly F. Windmill
School of Biomedical Sciences
University of Queensland
Queensland, Australia

Yasushi Yamazoe
Department of Drug Metabolism and
  Molecular Toxicology
Graduate School of Pharmaceutical
  Sciences
Tohoku University
Sendai, Japan

Table of Contents

1 Nomenclature and Molecular Biology of the Human Sulfotransferase Family

Rebecca L. Blanchard

CONTENTS

INTRODUCTION

As evidenced throughout this book, the cytosolic sulfotransferase (*SULT*) gene superfamily encodes enzymes catalyzing the sulfate conjugation of a wide variety of endogenous and exogenous substrates. Substrates include steroid hormones such as 17β-estradiol and dehydroepiandrosterone, thyroid hormones, catecholamines, xenobiotics such as *N*-hydroxy-2-acetylaminofluorene and isoflavones, and many drugs, including acetaminophen and minoxidil. At least 65 distinct eukaryotic SULT proteins (33 isoforms) have been identified and functionally characterized. The 1990s and early 2000s witnessed an explosion in the cloning of novel SULT cDNAs and genes, and we are still witnessing identification of allelic variants of SULT genes.

1

This rapid pace resulted in confusion with regard to SULT nomenclature, with many laboratories using their own versions. The need for a widely accepted nomenclature system was widely recognized, and the SULT community responded by developing and publishing guidelines for a unified nomenclature system (Blanchard et al., 2004). These guidelines do not extend to the membrane bound SULTs that are localized to the Golgi and catalyze the sulfation of macromolecules, such as tyrosylproteins and glucosaminoglycans. The membrane bound SULTs exhibit a low degree of amino acid sequence identity with the SULTs and have historically been treated as a separate superfamily.

GENE AND ENZYME DESIGNATION OF KNOWN SULTS

Attendees at the Third International Sulfation Workshop held in Drymen, Scotland, 1996, and the Fourth International ISSX meeting in Seattle, WA, 1995, first agreed that "SULT" would be adopted as the abbreviation for cytosolic sulfotransferase enzymes, and a symbol was accepted by the HUGO Gene Nomenclature Committee (http://www.gene.ucl.ac.uk/nomenclature/). It was agreed that members of the SULT superfamily would be assigned family and subfamily designations on the basis of their amino acid sequence identity. SULTs sharing at least 45% amino acid sequence identity should be considered members of the same family, and subfamily members should share at least 60% identity. SULT families were designated by an Arabic numeral immediately following the name (e.g., SULT4), and subfamilies were iden-tified by alphabetic categories, with the first-identified subfamily of a new family labeled "A" (e.g., SULT4A). Unique isoforms within subfamilies were identified using an Arabic numeral following the subfamily designation (e.g., SULT4A1). Generally, the first published sequence in a subfamily was designated as isoform 1 and subsequent isoforms within that subfamily have been assigned on the basis of percentage amino acid identity relative to the "1" isoform. In some cases, as with SULT2A1, the desire to maintain historical use of a name necessitated exceptions. All gene names follow the enzyme designation except that the gene name will be in italics (e.g., *SULT2B1*, *Sult1a1*) according to widely accepted conventions. Gene naming conventions for different organisms were followed as closely as possible; e.g., for human, rat, and bovine genes, gene names should be capitalized; for mouse and Drosophila genes, the first letter only should be capitalized, etc. The letters "I," "O," "P," "X," and "Y" were not and shall not be used to avoid confusion with other genetic designations.

Orthology is a source of confusion for researchers within and outside the field. In light of experiences with nomenclature systems for other drug metabolizing enzyme families, the developed SULT nomenclature included reference to the spe-cies. Species names were designated by a three- to five-letter code, in parentheses, prefixed to the SULT name (e.g., (MOUSE)Sult2b1 for the 2b1 isoform from mouse, *Mus musculus*, and *(BOVIN)SULT1A1* for the *1A1* gene from cattle, *Bos taurus*). It was recommended that the three- to five-letter code already established by the Committee on Standardization in Human Cytogenetics be used for different species (http://www.gene.ucl.ac.uk/nomenclature/guidelines.html), primarily to facilitate the identification of orthologous SULT isoforms in different species. A recent article by Fitch (2000) covers these issues thoroughly.

At least one *SULT* gene, *(HUMAN)SULT2B1*, encodes two SULT isoforms that differ in their N-terminal amino acid sequences as a result of alternative transcription initiation and alternative splicing (Her et al., 1998). Isoforms with different amino acid sequences that are encoded by the same gene were differentiated by "_vx" at the end of the SULT name, where "x" designates the sequential variant number. The proteins initially referred to as SULT2B1a and SULT2B1b were renamed (HUMAN)SULT2B1_v1 and (HUMAN)SULT2B1_v2, respectively, under the proposed nomenclature guidelines. One SULT pseudogene, *(HUMAN)SULT1D1*, has been identified to date (Meinl and Glatt, 2001). Pseudogenes were designated by a "*P*" following the gene name (e.g., *(HUMAN)SULT1D1P*). If more than one paralogous pseudogene is identified, the name will be followed by an Arabic numeral. For example, the second pseudogene of the hypothetical human SULT8G3 gene would be named *(HUMAN)SULT8G3P2*.

Figure 1.1 depicts the dissection of the SULT nomenclature as just described. The SULTs analyzed here represent mammalian, insect, and plant sequences because SULT proteins have been characterized from those groups of organisms. In recognition that the genomes of other members of the animal kingdom as well as prokaryotes likely encode SULT proteins, SULT families 1–200 were reserved for members of the animal kingdom, 201–400 for plant SULTs, and 401 and higher for prokaryotes. Table 1.1 lists the collection of SULT cDNAs for which these nomenclature guidelines were applied. The proposed SULT nomenclature, previously used names for each isoform, all GenBank accession numbers (including redundant entries), and references for those sequences that have been published are listed.

Several *SULT* genes have also been cloned and sequenced. Table 1.2 depicts their gene designations, GenBank accession numbers, chromosomal localization (where available), and references. SULT gene structures are remarkably well conserved (Blanchard et al., 2004; Weinshilboum et al., 1997). Figure 1.2 illustrates the gene structure of (HUMAN)SULT1A1 as representative of the superfamily. Exons

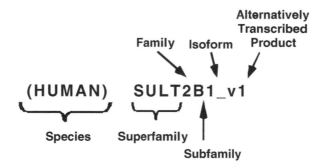

FIGURE 1.1 Dissection diagram of a representative SULT name. A complete SULT protein name contains species, superfamily, family, isoform, and alternatively transcribed product designations as shown by applying the guidelines described in this manuscript.

TABLE 1.1
cDNAs Comprising the SULT Superfamily

cDNA	Reported cDNA Designations	Accession Number	Reference
	SULT1 Family		
SULT1A Subfamily			
(MOUSE)SULT1A1	mST$_{p1}$	L02331	Kong et al., 1993a
(RAT)SULT1A1	PST-1	X52883	Ozawa et al., 1990
		S42994	Ozawa et al., 1990
	Mx-STb	L19998	Hirshey et al., 1992
	AST IV	X68640	Yerokun et al., 1992
	Tyrosine-ester ST	—	Chen et al., 1992
	PST-1	—	Cruickshank et al., 1993
(BOVIN)SULT1A1	PST	U35253	Schauss et al., 1995
(PIG)SULT1A1		AY193893	Lin et al., 2004
(CANFA)SULT1A1	dPST-1	D29807	
cSULT1A1		AY069922	Tsoi et al., 2002
(MACFA)SULT1A1	monPST-1	D85514	
(ORNAN)SULT1A1	SULT1A	AY044182	
(RABIT)SULT1A1	SULT1A	AF360872	Riley et al., 2002
(HUMAN)SULT1A1	P-PST-1	L19999	Wilborn et al., 1993
*2	HAST1	L10819	Zhu et al., 1993b
*2	HAST2	L19955	Zhu et al., 1993a
		U09031	Zhu et al., 1993b
	P-PST	X84654	Jones et al., 1995
*1	ST1A3	X78283	Ozawa et al., 1995
	H-PST	U26309	Hwang et al., 1995
*1	SULT1A1	AJ007418	Dajani et al., 1998
(HUMAN)SULT1A2			
*3	ST1A2	X78282	Ozawa et al., 1995
*1	HAST4V	U28169	Zhu et al., 1996
*2	HAST4	U28170	Zhu et al., 1996
(HUMAN)SULT1A3	HAST3	L19956	Zhu et al., 1993a
	TL PST	U08032	Wood et al., 1994
	hEST	L25275	Bernier et al., 1994b
	m-PST	X84653	Jones et al., 1995
	hM-PST	—	Ganguly et al., 1995
SULT1B Subfamily			
(MOUSE)SULT1B1	mSULT1B1	U92076	Saeki et al., 1998
(RAT)SULT1B1	rT/DST	U38419	Sakakibara et al., 1995
	ST1B1	D89375	Fujita et al., 1997
(CHICK)SULT1B1	SULT1B	AJ494980	
(CANFA)SULT1B1	cSULT1B1	AY004332	Tsoi et al., 2001b
(TRIVU)SULT1B1	SULT1B	AY044181	
(HUMAN)SULT1B1	ST1B2	D89479	Fujita et al., 1997
		U95726	Wang et al., 1998

(continued)

TABLE 1.1 (CONTINUED)
cDNAs Comprising the SULT Superfamily

cDNA	Reported cDNA Designations	Accession Number	Reference
SULT1C Subfamily			
(MOUSE)SULT1C1	mOlfST	AF033653	Tamura et al., 1998
(RAT)SULT1C1	ST1C1,	HAST-1 L22339	Nagata et al., 1993
(CHICK)SULT1C1	SULT1C	AJ416889	
(MOUSE)SULT1C2		AY005469	
(RAT)SULT1C2	rSULT1C2	AJ238391	Xiangrong et al., 2000
(RABIT)SULT1C2	SULT1C2	AF026304	Hehonah et al., 1999
(HUMAN)SULT1C2	SULT1C1	U66036	Her et al., 1997
	ST1C2	AB008164	Yoshinari et al., 1998b
	HAST5	AF026303	Yoshinari et al., 1998b
(RAT)SULT1C3	rSULT1C2A	AJ238392	Xiangrong et al., 2000
(HUMAN)SULT1C4	hSULT1C	AF055584	Sakakibara et al., 1998a
SULT1D Subfamily			
(MOUSE)SULT1D1	Clone 679153	U32371	Sakakibara et al., 1998b
	SULT-N	AF026073	Sakakibara et al., 1998b
(RAT)SULT1D1	Tyrosine-ester sulfotransferase	U32372	
(CANFA)SULT1D1	c-SULT1D1	AY004331	Tsoi et al., 2001a
SULT1E Subfamily			
(MOUSE)SULT1E1	mEST	S78182	Song et al., 1995
(RAT)SULT1E1	ESTr	M86758	Demyan et al., 1992
	rEST-3	S76489	Falany et al., 1995b
	Ste1	U50204	Rikke and Roy, 1996
	Ste2	U50205	Rikke and Roy, 1996
(CAVPO)SULT1E1	EST$_r$	S45979	Oeda et al., 1992
		U09552	
(BOVIN)SULT1E1	OST	X56395	Nash et al., 1988
		M54942	Nash et al., 1988
		M38672	Nash et al., 1988
(PIG)SULT1E1		AF389855	Kim et al., 2002
(HUMAN)SULT1E1	hEST	U08098	Aksoy et al., 1994b
	hEST-1	S77383	Falany et al., 1995a
	SULT1E1	Y11195	Rubin et al., 1999
(RAT)SULT1E2	rEST-6	S76490	Falany et al., 1995b
SULT1F Subfamily			
(BRARE)SULT1F1	ZFSULT1ST#1,ZFST #2	AY181064	Sugahara et al., 2003b
(BRARE)SULT1F2	ZFSULT1ST#2,ZFST #3	AY181065	Sugahara et al., 2003b

(continued)

TABLE 1.1 (CONTINUED)
cDNAs Comprising the SULT Superfamily

cDNA	Reported cDNA Designations	Accession Number	Reference
	SULT2 Family		
SULT2A Subfamily			
(MOUSE)SULT2A1	mST$_{a1}$	L02335	Kong et al., 1993b
(RAT)SULT2A1	ST-20	M31363	Ogura et al., 1989
	ST-21a	D14987	Watabe et al., 1994
	ST-21b	D14988	Watabe et al., 1994
(RABIT)SULT2A1	AST-RB2, ST2A8	AB006053	Yoshinari et al., 1998c
(MACFA)SULT2A1	monHST-1	D85521	
(HUMAN)SULT2A1			
*1	DHEA ST	U08024	Otterness et al., 1992
*1		U08025	Otterness et al., 1992
	hST$_a$	S43859	Kong et al., 1992
		L02337	Kong et al., 1992
	DHEA-ST8	X70222	Comer et al., 1993
		S53620	Comer et al., 1993
		L20000	Comer et al., 1993
*1	HST-hfa	X84816	Forbes et al., 1995
(MOUSE)SULT2A2	mST$_{a2}$	L27121	Kong and Fei, 1994
(RAT)SULT2A2	SMP-2	J02643	Chatterjee et al., 1987
(CAVPO)SULT2A2	gpHST	U06871	Lee et al., 1994
(RAT)SULT2A3	ST-40	M33329	Ogura et al., 1990
	ST-41	X63410	Ogura et al., 1990
(CAVPO)SULT2A3	gpHST2	U35115	Luu et al., 1995
	gpPreg-ST	55944	Dufort et al., 1996
(RAT)SULT2A4	ST-60	D14989	Watabe et al., 1994
SULT2B Subfamily			
(MOUSE)SULT2B1	Clone 445155	AF026072	Sakakibara et al., 1998b
(HUMAN)SULT2B1_v1	hSULT2B1a	U92314	Her et al., 1998
(HUMAN)SULT2B1b_v2	hSULT2B1b	U92315	Her et al., 1998
SULT2C Subfamily			
(BRARE)SULT2C1	ZF SULT2ST, ZFDHEAST	AY181063	Sugahara et al., 2003c
	SULT3 Family		
(MOUSE)SULT3A1	ST3A1	AF026075	
(RABIT)SULT3A1	AST-RB1, ST3A1	D86219	Yoshinari et al., 1998a
	SULT4 Family		
(MOUSE)SULT4A1	SULTx3	AF059257	
(RAT)SULT4A1	rBR-STL	AF188699	Falany et al., 2000

(continued)

TABLE 1.1 (CONTINUED)
cDNAs Comprising the SULT Superfamily

cDNA	Reported cDNA Designations	Accession Number	Reference
	SULT4 Family (continued)		
(HUMAN)SULT4A1	hBR-STL	AF188698	Falany et al., 2000
	hSULT4A1	AF251263	Walther et al., 1999
	SULTX3	AF115311	
	SULT5 Family		
(MOUSE)SULT5A1	SULT-X1	AF026074	
	SULT6 Family		
(CHICK)SULT6A1		AF033189	Cao et al., 1999
	SULT7 Family		
(BRARE)SULT7A1	zebrafish ST,ZFST#1	AY180110	Sugahara et al., 2003a
	SULT101 Family		
(SPOFR)SULT101A1		U28654	Grun et al., 1996
	SULT201 Family		
(FLACH)SULT201A1	pFST3	M84135	Varin et al., 1992
(FLABI)SULT201A1	pBFST3	U10275	Ananvoranich et al., 1994
(FLACH)SULT201A2	pFST4′	M84136	Varin et al., 1992
(FLABI)SULT201A3	pBFSTX	U10277	Ananvoranich et al., 1995
	SULT202 Family		
SULT202A Subfamily			
(ARATH)SULT202A1	RaRo47	Z46823	Lacomme et al., 1996
(BRANA)SULT202A1	BnST3	AF000307	Rouleau et al., 1999
(BRANA)SULT202A2	BnST1	AF000305	Rouleau et al., 1999
(BRANA)SULT202A3	BnST2	AF000306	Rouleau et al., 1999
SULT202B Subfamily			
(ARATH)SULT202B1	AtST2a	T43254	Gidda et al., 2003

Note: *1, *2, and *3 are allelic designations. ARATH, *Arabidopsis thaliana*; BOVIN, *Bos taurus*; BRANA, *Brassica napus*; BRARE, *Brachydanio rerio*; CANFA, *Canis familiaris*; CAVPO, *Cavia porcellus*; FLABI, *Flaveria bidentis*; FLACH, *Flaveria chloraefolia*; CHICK, *Gallus gallus*; HUMAN, *Homo sapiens*; MACFA, *Macaca fascicularis*; MOUSE, *Mus musculus*; ORNAN, *Ornithorhynchus anatinus*; RABIT, *Oryctolagus caniculus*; RAT, *Rattus norvegicus*; SPOFR, *Spodoptera frugiperda*; PIG, *Sus scrofa*; TRIVU, *Trichosurus vulpecula*.

TABLE 1.2
SULT Gene Nomenclature and Chromosomal Localization

New Name	Old Name	Chromosomal Localization	GenBank Accession Number	Reference
(MOUSE)Sult1a1	Stp, St1a4	7 (distal)	AB029487	Dooley et al., 1993
(RAT)SULT1A1	Ast-IV	1q35-q37	L16241	Khan et al., 1993; Nagai et al., 1999
(BOVIN)SULT1A1	PST		U34753	Henry et al., 1996
(HUMAN)SULT1A3	STM, HAST	16p11.2	L34160	Bernier et al., 1994a
			U20499	Aksoy et al., 1994a; Aksoy and Weinshilboum, 1995
			U37686	Dooley et al., 1994
(HUMAN)SULT1A2	STP2	16p12.1-11.2	U34804	Her et al., 1996
(HUMAN)SULT1A1	STP, STP1	16p12.1-11.2	U52852	Raftogianis et al., 1996
			U54701	Bernier et al., 1996
			U71086	Dooley et al., 1993; Dooley and Huang, 1996
(HUMAN)SULT1B1	SULT1B2	4q13.1	AF184894[b]	Meinl and Glatt, 2001
(RAT)Sult1c1[a]	St1c1	9q11.2-q12.1		Nagai et al., 1999
(HUMAN)SULT1C2	SULT1C1	2q11.2	AF186257-62	Freimuth et al., 2000
(HUMAN)SULT1C4	SULT1C2	2q11.2	AF186263	Freimuth et al., 2000
(CAVPO)SULT1E1	EST		L11117	Chiba et al., 1995; Komatsu et al., 1993
(RAT)Sult1e1	Ste1		AJ131835[b]	Unpublished
(HUMAN)SULT1E1	STE	4q13.1	U20514-21	Her et al., 1995
(RAT)Sult2a1[a]	St1	1q21.3-q22.1		Nagai et al., 1996
(HUMAN)SULT2A1	STD	19q13.3	U13056-61	Luu-The et al., 1995
			L36191-196	Otterness et al., 1995
(RAT)Sult2a3a	St2	1q21.3-q22.1		Nagai et al., 1996
(HUMAN)SULT2B1	SULT2B1	19q13.3	U92316-22	Her et al., 1998
(HUMAN)SULT4A1	SULT4A1	22q13.1-13.2	Z97055	Dunham et al., 1999

[a] Gene has been mapped but not cloned.
[b] Partial gene sequence.

and introns are numbered with Arabic numerals. A number of *SULT* genes, including *(HUMAN)SULT1A1*, contain alternatively transcribed first exons that represent untranslated mRNA sequences (Weinshilboum et al., 1997). Untranslated first exons are designated exon "1" Alphabetic characters (such as "1A" and "1B") differentiate alternative exons 1, with A the most 3′ exon 1 sequence. Therefore, the first coding exon for any *SULT* gene is designated exon 2 — even if no exon 1 has yet been identified. The advantages of this are twofold. First, upstream exons can be identified in the future, and second, *SULT* gene structures can be compared directly. Note that SULTs whose cDNA sequences had been assigned to the same SULT family were

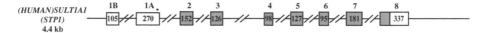

FIGURE 1.2   The structure for *(HUMAN)SULT1A1*, representative of the SULT superfamily of genes. Suggested gene nomenclature is presented at the left with previous gene designation indicated beneath (in parentheses). The number beneath the gene name represents approximate length, in kb, of the gene from the initial through the final exon. Black rectangles represent coding regions, and open rectangles represent noncoding exons. Numbers above exons represent their lengths in bp, and above that is the exon number.

ultimately discovered to share very similar gene structures. Also, *SULT* gene subfamilies were found to be located within subfamily-specific chromosomal clusters. For example, in humans the *SULT1A* subfamily is found at chromosome 16p11.2-12, the *SULT1C* subfamily at chromosome 2q11.2, and the *SULT2A* and *SULT2B* genes localize to human chromosome 19q13.3 (Table 1.2). Human *SULT1B1* and *SULT1E1* genes are localized to 4q11-13.1. This suggests that the emergence of subfamilies might have resulted from independent gene duplication events.

ORTHOLOGS

The ability to identify orthologous SULT isoforms in different species is desirable. Orthology has been defined as "the relationship of any two homologous characters whose common ancestor lies in the cenancestor of the taxa from which the two species were obtained," whereas paralogy has been defined as "the relationship of any two homologous characters arising from a duplication of the gene for that character" (Fitch, 2000). We have attempted to assign SULT orthologs where possible, recognizing that identification of orthologs would greatly facilitate the interpretation of gene regulation and function studies. For example, the ability to recognize mouse and human orthologs would facilitate the extrapolation of results from *SULT* gene knockout experiments in mice to understand the function of that isoform in humans. The assignment of orthologs was made on the basis of amino acid sequence identity, isoform function (where known), and gene structure (where known). As more SULTs are identified, an update of SULT ortholog assignments may be necessary.

NOMENCLATURE FOR HUMAN *SULT* ALLELES AND ALLOZYMES

The discovery of nucleotide polymorphisms and haplotypes in human genes is progressing at a rapid pace. This creates considerable difficulties in relation to gene nomenclature, therefore a recommended nomenclature for *SULT* alleles was also developed. The proposed nomenclature for *SULT* alleles was based on the work of authors describing nomenclature for allelic variants of other drug metabolizing enzyme genes (Daly et al., 1996; Hein et al., 2000; Ingelman-Sundberg et al., 2000; Nebert, 2000).

Alleles are recognized as genomic haplotypes that are defined by permutations of often several genetic variations. Those variations include nucleotide insertions, deletions, variant length repeats, and, predominantly, single nucleotide polymorphisms (SNPs). SNPs can result in alleles that encode altered amino acid sequences (nonsynonymous SNPs) or those that encode identical amino acid sequences (synonymous SNPs). The protein products of alleles that are defined by nonsynonymous SNPs are referred to as allozymes. Common synonymous and nonsynonymous SNPs have been identified within several SULT genes, including *SULT1A1*, *SULT1A2*, *SULT1C2*, and *SULT2A1* (Coughtrie et al., 1999; Freimuth et al., 2001; Glatt et al., 2000; Ozawa et al., 1998, 1999; Raftogianis et al., 1997, 1999; Thomae et al., 2002; Wood et al., 1996), and others certainly will be discovered and characterized. Different permutations of those SNPs have given rise to several SULT haplotypes that encode alleles and allozymes with functional differences in level of protein expression and enzymatic properties (Coughtrie et al., 1999; Freimuth et al., 2001; Glatt et al., 2000; Ozawa et al., 1998, 1999; Raftogianis et al., 1997, 1999; Thomae et al., 2002; Wood et al., 1996). Given the widespread use of allele names for the *SULT* variants already identified, they were not renamed and the reader should consult the relevant publications for the nomenclature of these alleles. The following guidelines were proposed for *SULT* alleles yet to be discovered.

Alleles will be named based on available haplotype. Investigators are strongly encouraged to, whenever possible, report *SULT* haplotypes and haplotype frequency rather than discrete SNPs. We propose that *SULT* alleles be named with an asterisk and Arabic numeral following the gene isoform number. Normally, the reference amino acid sequence, i.e., the first sequence established, as proposed by Nebert (2000, 2002), would be assigned *1. Alleles defined by synonymous SNPs, and thus encoding the same amino acid sequence, would be named *1A, *1B, etc., while those defined by nonsynonymous SNPs, and thus encoding different allozymes, would be named *2, *3, etc. Additional nucleotide sequence variants related to the founding allele of each class (i.e., sequences defining individual haplotypes but encoding the same protein) would be indicated by an additional letter, e.g., *2B, *2C, etc., with the founding allele *2 renamed *2A upon discovery of such related suballeles (Figure 1.3). In cases where several alleles are identified simultaneously

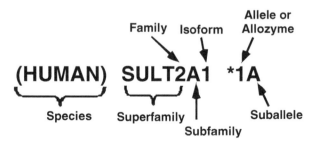

FIGURE 1.3   Dissection diagram of a representative SULT allele name. A complete SULT allele name contains species, superfamily, family, isoform, allele (allozyme), and suballele designations.

and where allele frequencies are determined, alleles should be named in the rank order of most frequent to least. If differences in allele frequencies among ethnic groups are known at the time of allele naming, the rank order should be determined based on the mean frequency in a combined population. Should the number of alleles discovered exceed 22 (avoid use of "I," "O," "P," "X," and "Y"), 2 letters should be used, e.g., *2AA*, *2AB*, *2AS*, etc. The use of additional subclassifications (e.g., *1A1*, *1Ba*, etc.) is discouraged. The first genomic sequence of the *SULT1B1* gene recorded would be given the reference sequence designation *SULT1B1*1*. The first nonsynonymous allelic variant of this gene to be discovered and confirmed would be *SULT1B1*2*, and a synonymous nucleotide variant of the *2 allele would be *SULT1B1*2B* (Figure 1.3). The null allele, where the whole gene or the major part of it is deleted, should be designated with a zero, e.g., *SULT1B1*0*. The SULT enzymes (allozymes) derived from these variants should be named according to the same rules; however, because suballelic variants (SULT1B1*2A, SULT1B1*2B, etc.) do not result in the substitution of amino acids, the allozyme encoded by each of those suballeles would be the same (SULT1B1*2). Therefore, by definition, no suballozyme designations are necessary.

In agreement with the suggestions made by Nebert (2000, 2002), the term "wild-type allele" should not be used in reference to human gene variants and a gene extends from at least 3 kb 5′ of the transcription start site to at least 150 bp 3′ of the last exon. The reader is encouraged to consult the above referenced papers for additional guidance on the naming of alleles.

SULT ISOFORM FUNCTION

Sulfation is an important reaction for the biotransformation of numerous endogenous compounds, drugs, and environmental chemicals. Although the criteria for naming SULT isoforms described here are based on amino acid sequence identity, application of those criteria, to a certain extent, also resulted in the segregation of SULTs by function (Table 1.3). Some SULT isoforms were well characterized biochemically long before any SULT cDNAs were cloned. Others have only recently been identified by the application of reverse genetic techniques, and their functions are not well understood.

SULT1 FAMILY

The SULT1 family comprises at least 11 proteins segregated into 3 subfamilies (Table 1.1). The SULT1A subfamily members are also known as the "phenol SULTs" because each of the three 1A isoforms catalyzes the sulfation of phenolic molecules (Table 1.3). (HUMAN)SULT1A1 has been studied most often within the context of conjugation of phenolic drugs, such as acetaminophen, minoxidil, and 17α-ethinylestradiol. This isoform also sulfates endogenous compounds, such as 17β-estradiol and iodothyronines, as well as environmental xenobiotics, such as the isoflavones. It is highly expressed in liver tissue and is expressed to a lesser degree in brain, breast, intestine, endometrium, kidney, lung, and platelets (Falany et al., 1998; Falany and Falany,

TABLE 1.3
Common SULT Substrates

SULT Isoform	Prototypic Substrates
SULT1A1	small phenols: p-nitrophenol, 1-naphthol, acetaminophen, minoxidil, 17α-ethinylestradiol, 17β-estradiol, isoflavones, 4′-OH-PhIP
SULT1A2	p-nitrophenol (low affinity), N-OH-2-acetylaminoflorene
SULT1A3	monoamines: dopamine, norepinephrine
SULT1B1	thyroid hormones, 2-naphthol, dopamine
SULT1C1	N-OH-2-acetylaminoflorene, p-nitrophenol
SULT1C4	N-OH-2-acetylaminoflorene, p-nitrophenol
SULT1D1	p-nitrophenol, dopamine, 4′-OH-PhIP, eicosanoids
SULT1E1	endogenous and exogenous estrogens (high affinity)
SULT1E2	endogenous and exogenous estrogens (high affinity)
SULT1F1	hydroxychlorobiphenyls: 3,3′,5,5′-tetrachloro-4,4′-biphenyldiol, catechin, and n-propyl gallate
SULT1F2	estrone, 17β-estradiol, quercetin, and genistein, hydroxychlorobiphenyls: 3,3′,5,5′-tetrachloro-4,4′-biphenyldiol, catechin, and n-propyl gallate
SULT2A1	dehydroepiandrosterone, polycyclic aromatic hydrocarbons
SULT2B1_v1	dehydroepiandrosterone, pregnenolone
SULT2B1_v2	dehydroepiandrosterone, cholesterol, pregnenolone
SULT2C1	dehydroepiandrosterone, pregnenolone
SULT3A1	desiprimine, aniline, 2-naphthylamine
SULT6A1	estradiol, corticosterone
SULT7A1	flavanoids, T_3, T_4, dopamine, p-nitrophenol, ß-naphthylamine, daidaein, caffeic acid
SULT101A1	retinol, dopamine, 4-nitrophenol, hydroxybenzylhydrazine, serotonin, vanillin, ethanol
SULT201A1	flavanoids
SULT201A2	flavanoids
SULT201A3	flavanoids
SULT202A1	brassinosteroids
SULT202A2	brassinosteroids
SULT202A3	brassinosteroids
SULT202B1	jasmonic acid

1996; Hume et al., 1996; Kester et al., 1999; Weinshilboum and Otterness, 1994). (HUMAN)SULT1A2 is very closely related to (HUMAN) SULT1A1 at the amino acid sequence level yet exhibits very different affinity for phenolic compounds (Her et al., 1996; Ozawa et al., 1995; Raftogianis et al., 1999; Wood et al., 1994). Although SULT1A2 cDNAs have been isolated from human liver and colon tissue, it is not clear that this isoform is highly expressed in any tissue. Northern- and western-blot analyses of human SULT1A1 and SULT1A2 are complicated by their high degree of identity. However, (HUMAN)SULT1A1 activity can be measured in tissues by virtue of the high affinity (low K_m) of this enzyme for p-nitrophenol compared with other human SULTs, such as SULT1A2 and SULT1A3. (HUMAN)SULT1A3 has a major role in the sulfation of catecholamines, such as dopamine and norepinephrine, a function that is probably specific to humans and possibly other primates and that is reflected in the fact that an enzyme with similar specificity has not been found in any other species so far (Falany, 1997; Weinshilboum et al., 1997). (HUMAN)SULT1A3 is highly expressed in intestinal tissue and brain but is also present in other tissues, such as platelets (Weinshilboum and Otterness, 1994). Expression in adult liver is negligible. The crystal structure of (HUMAN)SULT1A3 has recently been solved (Bidwell et al., 1999; Dajani et al., 1999), providing important information about the structure of this protein and a framework from which other SULT structures might be modeled.

The SULT1B subfamily has been identified recently and may be involved in the sulfation of thyroid hormones (Fujita et al., 1997, 1999; Saeki et al., 1998; Sakakibara et al., 1995; Tamura et al., 1999; Wang et al., 1998). This family also catalyzes the sulfation of prototypic phenolic substrates, such as 2-naphthol and dopamine, but not steroid hormones. (HUMAN)SULT1B1 mRNA has been detected in liver, colon, small intestine, and blood leukocytes.

The SULT1C subfamily is also a recently identified gene family that encodes at least four isoforms (Hehonah et al., 1999; Her et al., 1997; Nagata et al., 1993; Sakakibara et al., 1998a; Tamura et al., 1998; Xiangrong et al., 2000; Yoshinari et al., 1998b). The function of these enzymes is not yet fully understood but (RAT)SULT1C1 and (HUMAN)SULT1C4 catalyze the sulfation of the procarcinogen N-OH-2-acetylaminoflorene as well as p-nitrophenol, the prototypic phenolic SULT substrate. (RAT)SULT1C1 is liver specific, and (HUMAN)SULT1C2 is expressed primarily in thyroid, stomach, and kidney, whereas (HUMAN)SULT1C4 is expressed predominantly in fetal kidney and lung but also in adult ovary and brain.

Very little is known about the SULT1D subfamily. It has been identified in rat, mouse, and dog and appears to be most active toward small phenolic- and amine-containing molecules, including p-nitrophenol, dopamine, and 4'-OH-PhIP; to be moderately active toward 2-methylbenzyl alcohol; to exhibit low activity toward triiodothyronine, estradiol, and minoxidil; and to exhibit no activity toward other steroids or thyroid hormones (Sakakibara et al., 1998b; Shimada et al., 2004; Tsoi et al., 2001a). Further, (MOUSE)Sult1d1 catalyzes the *in vitro* sulfation of several eicosanoids, including prostaglandins, thromboxane B_2, and leukotriene E_4 (Liu et al., 1999). The biological function of those reactions is not yet understood. In mouse, Sult1d1 is expressed in kidney and uterus but apparently not in liver, brain, lung, intestinal, testicular, or glandular tissue (Sakakibara et al., 1998b; Shimada et al., 2004).

The SULT1E subfamily is among the most widely studied of the SULT enzymes, and the first SULT protein for which a crystal structure was solved was (MOUSE)Sult1e1 (Kakuta et al., 1997, 1998). Those studies, along with the human SULT1A3 and SULT2A1 crystal structures (Bidwell et al., 1999; Dajani et al., 1999; Pedersen et al., 2000), have facilitated the prediction of additional SULT structures. SULT1E family members catalyze the sulfation of the 3α-hydroxy group of endogenous (e.g., estrone and β-estradiol) and xenobiotic (e.g., 17α-ethinylestradiol) estrogens, generally with very high affinity (Aksoy et al., 1994b; Coughtrie et al., 1994, 1998; Falany, 1997; Falany et al., 1994). Although SULT1A1 and SULT2A members also sulfate estrogens, the affinity of those enzymes for endogenous estrogens is typically several orders of magnitude lower than is the affinity of SULT1E1 for those substrates. SULT1E1 and 1E2 isoforms are expressed in several steroid hormone responsive tissues, such as endometrium, testes, breast, adrenal gland, and placenta as well as in the liver and small intestine, depending on species.

Recently, two SULT cDNAs with homology to the SULT1 family were characterized from zebrafish (Sugahara et al., 2003b). These sequences shared greater than 45% but less than 60% amino acid sequence homology to other SULT1 family members and shared 82% amino acid sequence identity. Hence, those two sequences comprised a novel SULT1 subfamily, SULT1F, and were named (BRARE)SULT1F1 and (BRARE)SULT1F2. The proteins encoded by these sequences were functionally characterized and found to exhibit broad substrate specificity, including highest activity toward hydroxychlorobiphenyls, such as 3,3',5,5'-tetrachloro-4,4'-biphenyl-diol, catechin, and n-propyl gallate (Sugahara et al., 2003b). Further, SULT1F2 was shown to exhibit estrogen-sulfating activity, with substrates including estrone, 17ß-estradiol, quercetin, and genistein (Ohkimoto et al., 2004).

SULT2 FAMILY

The SULT2 family comprises two subfamilies, SULT2A and SULT2B (Table 1.1). Isoforms within these subfamilies catalyze sulfation of the 3β-hydroxy groups of steroids with unsaturated "A" rings, such as dehydroepiandrosterone (DHEA), as well as 3β-hydroxy groups in steroids, such as androsterone and allopregnanolone (Falany, 1997; Geese and Raftogianis, 2001; Her et al., 1998; Meloche and Falany, 2001). A major difference between the activities of the SULT2A and SULT2B enzymes appears to be that only the SULT2A isoforms are capable of catalyzing the sulfation of the phenolic hydroxy group at the 3 position of estrogens, while the SULT2B1 isoforms are more selective for 3β-hydroxysteroids. The two isoforms differ also in their pattern of tissue expression. Members of the SULT2A family also sulfate a range of xeno-biotics, including the bioactivation of benzylic alcohols of polycyclic aromatic hydro-carbons (Glatt et al., 2000). (HUMAN)SULT2A is expressed in adrenal cortex, liver, brain, and intestine (Falany, 1997; Weinshilboum and Otterness, 1994). Recently, the crystal structure of human SULT2A1 was solved (Pedersen et al., 2000). SULT2B1 isoforms are expressed in humans in prostate, placenta, skin, platelets, and trachea and in mouse in intestine, epididymis, and uterus (Her et al., 1998; Higashi et al., 2004; Sakakibara et al., 1998b; Yanai et al., 2004). Interestingly, alternative transcription initiation results in the expression of two distinct SULT2B1 variants (SULT2B1_v1 and SULT2B1_v2). The two enzymes differ in the amino terminal residues, and these differences impart apparent biochemical distinction such that SULT2B1_v1 preferentially catalyzes the sulfation of pregnenolone while SULT2B1_v2 sulfates cholesterol and pregnenolone (Fuda et al., 2002).

Recently, a member of the SULT2 family was cloned from zebrafish, whose amino acid sequence exhibited homology with the SULT2 family and represented a novel subfamily, SULT2C (Sugahara et al., 2003c). This sequence has been named (BRARE)SULT2C1. Recombinant SULT2C1 exhibited sulfation activity toward dehydroepiandrosterone, pregnenolone, and several pregnenolone derivatives.

SULTS 3, 4, 5, 6, AND 7

Several recently cloned SULT cDNAs define novel SULT families. SULT3A1 has been identified in mouse and rabbit and catalyzes the sulfation of heterocyclic amines, such as desiprimine, aniline, and 2-naphthylamine (Yoshinari et al., 1998a). SULT4A1 has been identified in mouse, rat, and humans (Falany et al., 2000; Walther et al., 1999). Little is known about the function of this isoform, but it is expressed in brain and apparently has little activity toward any prototypic SULT substrate (Falany et al., 2000; Walther et al., 1999) though it may exhibit marginal activity toward L-triiodothyronine, thyroxine, estrone, p-nitrophenol, 2-naphthylamine, and 2-naphthol (Sakakibara et al., 2002). The SULT4A1 enzymes share a remarkable degree of cross-species similarity, with the human, mouse, and rat isoforms sharing at least 97% amino acid sequence identity, suggesting an important, conserved endogenous function. The sequence identified here as SULT4A1 has also been referred to as SULT5A1 (Nagata and Yamazoe, 2000); however, SULT4A1 has been in more common use in the literature and this nomenclature is retained here.

SULT5A1 has been cloned from mouse (GenBank accession number AF026074, unpublished), but no information on its function is available. SULT6A1 has been cloned from chicken liver and was differentially downregulated in GH-receptor deficient dwarf chickens (Cao et al., 1999). Recombinant (CHICK)SULT6A1 catalyzed the sulfation of estradiol and corticosterone. (BRARE)SULT7A1 was cloned from zebrafish and the recombinant protein exhibited sulfation activity toward a broad range of substrates, including 3,5,3'-Triiodothyronine (T_3), 3,5,3',5'-Tetraiodothyronine (T_4), dopamine, p-nitrophenol, ß-naphthylamine, daidaein, and caffeic acid and particularly high activity toward flavanoids, such as n-propyl gallate (Sugahara et al., 2003a).

SULT101

One insect SULT has been cloned and characterized (Grun et al., 1996). That protein is unique among members of the SULT superfamily in that it apparently functions as a retinol dehydratase with sulfation an intermediary catalytic reaction. Thus, the final product is not sulfated but rather dehydrated. (SPOFR)SULT101A1 was cloned from the moth *Spodoptera frugiperda* and shown to catalyze the conversion of retinol to anhydroretinol, probably via a retinyl sulfate intermediate (Grun et al., 1996). The SULT101A1 protein contains Region I and IV SULT signature domains and requires 3'-phosphoadenosine 5'-phosphosulfate (PAPS) for activity. This protein also exhibited SULT activity toward dopamine, 4-nitrophenol, hydroxybenzylhydrazine, serotonin, vanillin, and ethanol (Vakiani et al., 1998). The crystal structure of SULT101A1 has been solved (Pakhomova et al., 2000). Structure- and site-directed mutation studies demonstrate that a helical lid not found in other SULTs is necessary for anhydroretinol production but not for sulfation (Pakhomova et al., 2001). SULT101A1 is 20–26% identical to other SULT isoforms.

SULTS 201 AND 202

In addition to animals, several plant species express cytosolic SULTs. The plant SULTs fall into two separate SULT families. We have arbitrarily named those plant families SULT201 and SULT202 to allow for the identification and naming of future SULT families within the animal kingdom (up to SULT200). The SULT201 family members catalyze the sulfation of flavanoids, while the SULT202 isoforms sulfate plant steroids, such as the brassinosteroids, and signaling molecules, such as jasmonic acid (Gidda et al., 2003; Rouleau et al., 1999; Varin et al., 1997).

SULT WEBSITE

A cytosolic SULT nomenclature website has been developed and is located at http://www.fccc.edu/research/labs/blanchard/sult/. This website will maintain updated information about SULT isoforms, orthologs, GenBank accession numbers, and publications. Links to GenBank entries and PubMed citations will be included. The site will provide a forum for SULT researchers to exchange information and submit new SULT sequences. Before a new SULT sequence or allelic variant is published, authors should confer with a member of the SULT Nomenclature Committee for nomenclature assignment. Committee members have been selected to reflect several areas of SULT research, and their names and e-mail addresses will be posted on the SULT website. Discussion with one or more of the committee members regarding the novel SULT amino acid sequence identity to other SULT family members should allow for assignment of nomenclature. Contact information for committee members is available on the website.

SUMMARY

A unified nomenclature system for the cytosolic SULT superfamily was developed and is now in common use (Blanchard et al., 2004). These guidelines suggest that SULT isoforms with at least 45% amino acid sequence identity be assigned to the same SULT family and those with at least 60% amino acid sequence identity share the same subfamily. Table 1.1 depicts the new nomenclature for the set of known eukaryotic SULTs and cross-references those names with previous nomenclature and GenBank accession numbers. Recommendations for the naming of allelic variants of SULT genes and allozymes were developed. A SULT Nomenclature Committee has formed that should be consulted for nomenclature assignment of novel SULT isoforms. Furthermore, a cytosolic SULT website has been developed as a resource for SULT nomenclature and a contact point for the SULT Nomenclature Committee.

ACKNOWLEDGMENTS

Dr. Blanchard acknowledges that these nomenclature guidelines were developed with the help of numerous members of the SULT community of investigators. Drs. Robert Freimuth, Jochen Buck, Richard Weinshilboum, and Michael Coughtrie were key members of these efforts. The following investigators are also acknowledged for important advice regarding nomenclature assignments: Drs. Joe Beckman, Bandana Chatterjee, Thomas Dooley, Charles Falany, Hansruedi Glatt, Fred Kadlubar, A.-N. Tony Kong, Kenichiro Ogura, Shogo Ozawa, Michael McManus, Melissa Runge-Morris, Charles Strott, Luc Varin, and Yasushi Yamazoe.

REFERENCES

Aksoy, I.A., Callen, D.F., Apostolou, S., Her, C., and Weinshilboum, R.M., (1994a) Thermolabile phenol sulfotransferase gene (STM): localization to human chromosome 16p11.2. *Genomics* 23: 275–277.

Aksoy, I.A. and Weinshilboum, R.M., (1995) Human thermolabile phenol sulfotransferase gene (STM): molecular cloning and structural characterization. *Biochem Biophys Res Commun* 208: 786–795.

Aksoy, I.A., Wood, T.C., and Weinshilboum, R., (1994b) Human liver estrogen sulfotransferase: identification by cDNA cloning and expression. *Biochem Biophys Res Commun* 200: 1621–1629.

Ananvoranich, S., Gulick, P., and Ibrahim, R.K., (1995) Flavonol sulfotransferase-like cDNA clone from *Flaveria bidentis*. *Plant Physiol* 107: 1019–1020.

Ananvoranich, S., Varin, L., Gulick, P., and Ibrahim, R., (1994) Cloning and regulation of flavonol 3-sulfotransferase in cell-suspension cultures of *Flaveria bidentis*. *Plant Physiol* 106: 485–491.

Bernier, F., Leblanc, G., Labrie, F., and Luu-The, V., (1994a) Structure of human estrogen and aryl sulfotransferase gene: two mRNA species issued from a single gene. *J Biol Chem* 269: 28200–28205.

Bernier, F., Lopez Solache, I., Labrie, F., and Luu-The, V., (1994b) Cloning and expression of cDNA encoding human placental estrogen sulfotransferase. *Mol Cell Endocrinol* 99: R11–R15.

Bernier, F., Soucy, P., and Luu-The, V., (1996) Human phenol sulfotransferase gene contains two alternative promoters: structure and expression of the gene. *DNA Cell Biol* 15: 367–375.

Bidwell, L.M., McManus, M.E., Gaedigk, A., Kakuta, Y., Negishi, M., Pedersen, L., and Martin, J.L., (1999) Crystal structure of human catecholamine sulfotransferase. *J Mol Biol* 293: 521–530.

Blanchard, R.L., Freimuth, R.R., Buck, J., Weinshilboum, R.M., and Coughtrie, M.W., (2004) A proposed nomenclature system for the cytosolic sulfotransferase (SULT) superfamily. *Pharmacogenetics* 14: 199–211.

Cao, H., Agarwal, S.K., and Burnside, J., (1999) Cloning and expression of a novel chicken sulfotransferase cDNA regulated by GH. *J Endocrinol* 160: 491–500.

Chatterjee, B., Majumdar, D., Ozbilen, O., Murty, C.V., and Roy, A.K., (1987) Molecular cloning and characterization of cDNA for androgen-repressible rat liver protein, SMP-2. *J Biol Chem* 262: 822–825.

Chen, X., Yang, Y.S., Zheng, Y., Martin, B.M., Duffel, M.W., and Jakoby, W.B., (1992) Tyrosine-ester sulfotransferase from rat liver: bacterial expression and identification. *Protein Expr Purif* 3: 421–426.

Chiba, H., Komatsu, K., Lee, Y.C., Tomizuka, T., and Strott, C.A., (1995) The 3′-terminal exon of the family of steroid and phenol sulfotransferase genes is spliced at the N-terminal glycine of the universally conserved GXXGXXK motif that forms the sulfonate donor binding site. *Proc Natl Acad Sci U S A* 92: 8176–8179.

Comer, K.A., Falany, J.L., and Falany, C.N., (1993) Cloning and expression of human liver dehydroepiandrosterone sulphotransferase. *Biochem J* 289: 233–240.

Coughtrie, M.W., Bamforth, K.J., Sharp, S., Jones, A.L., Borthwick, E.B., Barker, E.V., Roberts, R.C., Hume, R., and Burchell, A., (1994) Sulfation of endogenous compounds and xenobiotics — interactions and function in health and disease. *Chem Biol Interact* 92: 247–256.

Coughtrie, M.W., Gilissen, R.A., Shek, B., Strange, R.C., Fryer, A.A., Jones, P.W., and Bamber, D.E., (1999) Phenol sulphotransferase SULT1A1 polymorphism: molecular diagnosis and allele frequencies in Caucasian and African populations. *Biochem J* 337: 45–49.

Coughtrie, M.W., Sharp, S., Maxwell, K., and Innes, N.P., (1998) Biology and function of the reversible sulfation pathway catalysed by human sulfotransferases and sulfatases. *Chem Biol Interact* 109: 3–27.

Cruickshank, D., Sansom, L.N., Veronese, M.E., Mojarrabi, B., McManus, M.E., and Zhu, X., (1993) cDNA expression studies of rat liver aryl sulphotransferase. *Biochem Biophys Res Commun* 191: 295–301.

Dajani, R., Cleasby, A., Neu, M., Wonacott, A.J., Jhoti, H., Hood, A.M., Modi, S., Hersey, A., Taskinen, J., Cooke, R.M., Manchee, G.R., and Coughtrie, M.W., (1999) X-ray crystal structure of human dopamine sulfotransferase, SULT1A3: molecular modeling and quantitative structure-activity relationship analysis demonstrate a molecular basis for sulfotransferase substrate specificity. *J Biol Chem* 274: 37862–37868.

Dajani, R., Hood, A.M., and Coughtrie, M.W., (1998) A single amino acid, glu146, governs the substrate specificity of a human dopamine sulfotransferase, SULT1A3. *Mol Pharmacol* 54: 942–948.

Daly, A.K., Brockmoller, J., Broly, F., Eichelbaum, M., Evans, W.E., Gonzalez, F.J., Huang, J.D., Idle, J.R., Ingelman-Sundberg, M., Ishizaki, T., Jacqz-Aigrain, E., Meyer, U.A., Nebert, D.W., Steen, V.M., Wolf, C.R., and Zanger, U.M., (1996) Nomenclature for human CYP2D6 alleles. *Pharmacogenetics* 6: 193–201.

Demyan, W.F., Song, C.S., Kim, D.S., Her, S., Gallwitz, W., Rao, T.R., Slomczynska, M., Chatterjee, B., and Roy, A.K., (1992) Estrogen sulfotransferase of the rat liver: complementary DNA cloning and age- and sex-specific regulation of messenger RNA. *Mol Endocrinol* 6: 589–597.

Dooley, T.P. and Huang, Z., (1996) Genomic organization and DNA sequences of two human phenol sulfotransferase genes (STP1 and STP2) on the short arm of chromosome 16. *Biochem Biophys Res Commun* 228: 134–140.

Dooley, T.P., Mitchison, H.M., Munroe, P.B., Probst, P., Neal, M., Siciliano, M.J., Deng, Z., Doggett, N.A., Callen, D.F., Gardiner, R.M., et al., (1994) Mapping of two phenol sulphotransferase genes, STP and STM, to 16p: candidate genes for Batten disease. *Biochem Biophys Res Commun* 205: 482–489.

Dooley, T.P., Obermoeller, R.D., Leiter, E.H., Chapman, H.D., Falany, C.N., Deng, Z., and Siciliano, M.J., (1993) Mapping of the phenol sulfotransferase gene (STP) to human chromosome 16p12.1-p11.2 and to mouse chromosome 7. *Genomics* 18: 440–443.

Dufort, I., Tremblay, Y., Belanger, A., Labrie, F., and Luu-The, V., (1996) Isolation and characterization of a stereospecific 3β-hydroxysteriod sulfotransferase (pregnenolone sulfotransferase) cDNA. *DNA Cell Biol* 15: 481–487.

Dunham, I., Shimizu, N., Roe, B.A., Chissoe, S., Hunt, A.R., Collins, J.E., Bruskiewich, R., Beare, D.M., Clamp, M., Smink, L.J., Ainscough, R., Almeida, J.P., Babbage, A., Bagguley, C., Bailey, J., Barlow, K., Bates, K.N., Beasley, O., Bird, C.P., Blakey, S., Bridgeman, A.M., Buck, D., Burgess, J., Burrill, W.D., O'Brien, K.P., et al., (1999) The DNA sequence of human chromosome 22 [see comments]. *Nature* 402: 489–495.

Falany, C.N., (1997) Enzymology of human cytosolic sulfotransferases. *Faseb J* 11: 206–216.

Falany, C.N., Krasnykh, V., and Falany, J.L., (1995a) Bacterial expression and characterization of a cDNA for human liver estrogen sulfotransferase. *J Steroid Biochem Mol Biol* 52: 529–539.

Falany, C.N., Wheeler, J., Oh, T.S., and Falany, J.L., (1994) Steroid sulfation by expressed human cytosolic sulfotransferases. *J Steroid Biochem Mol Biol* 48: 369–375.

Falany, C.N., Xie, X., Wang, J., Ferrer, J., and Falany, J.L., (2000) Molecular cloning and expression of novel sulphotransferase-like cDNAs from human and rat brain. *Biochem J* 346 (Pt 3): 857–864.

Falany, J.L., Azziz, R., and Falany, C.N., (1998) Identification and characterization of cytosolic sulfotransferases in normal human endometrium. *Chem Biol Interact* 109: 329–339.

Falany, J.L. and Falany, C.N., (1996) Expression of cytosolic sulfotransferases in normal mammary epithelial cells and breast cancer cell lines. *Cancer Res* 56: 1551–1555.

Falany, J.L., Krasnykh, V., Mikheeva, G., and Falany, C.N., (1995b) Isolation and expression of an isoform of rat estrogen sulfotransferase. *J Steroid Biochem Mol Biol* 52: 35–44.

Fitch, W.M., (2000) Homology: a personal view on some of the problems. *Trends Genet* 16: 227–331.

Forbes, K.J., Hagen, M., Glatt, H., Hume, R., and Coughtrie, M.W., (1995) Human fetal adrenal hydroxysteroid sulphotransferase: cDNA cloning, stable expression in V79 cells and functional characterisation of the expressed enzyme. *Mol Cell Endocrinol* 112: 53–60.

Freimuth, R.R., Eckloff, B., Wieben, E.D., and Weinshilboum, R.M., (2001) Human sulfotransferase SULT1C1 pharmacogenetics: gene resequencing and functional genomic studies. *Pharmacogenetics* 11: 747–756.

Freimuth, R.R., Raftogianis, R.B., Wood, T.C., Moon, E., Kim, U.J., Xu, J., Siciliano, M.J., and Weinshilboum, R.M., (2000) Human sulfotransferases SULT1C1 and SULT1C2: cDNA characterization, gene cloning, and chromosomal localization. *Genomics* 65: 157–165.

Fuda, H., Lee, Y.C., Shimizu, C., Javitt, N.B., and Strott, C.A., (2002) Mutational analysis of human hydroxysteroid sulfotransferase SULT2B1 isoforms reveals that exon 1B of the SULT2B1 gene produces cholesterol sulfotransferase, whereas exon 1A yields pregnenolone sulfotransferase. *J Biol Chem* 277: 36161–36166.

Fujita, K., Nagata, K., Ozawa, S., Sasano, H., and Yamazoe, Y., (1997) Molecular cloning and characterization of rat ST1B1 and human ST1B2 cDNAs, encoding thyroid hormone sulfotransferases. *J Biochem (Tokyo)* 122: 1052–1061.

Fujita, K., Nagata, K., Yamazaki, T., Watanabe, E., Shimada, M., and Yamazoe, Y., (1999) Enzymatic characterization of human cytosolic sulfotransferases: identification of ST1B2 as a thyroid hormone sulfotransferase. *Biol Pharm Bull* 22: 446–452.

Ganguly, T.C., Krasnykh, V., and Falany, C.N., (1995) Bacterial expression and kinetic characterization of the human monoamine-sulfating form of phenol sulfotransferase. *Drug Metab Dispos* 23: 945–950.

Geese, W.J. and Raftogianis, R.B., (2001) Biochemical characterization and tissue distribution of human SULT2B1. *Biochem Biophys Res Commun* 288: 280–289.

Gidda, S.K., Miersch, O., Levitin, A., Schmidt, J., Wasternack, C., and Varin, L., (2003) Biochemical and molecular characterization of a hydroxyjasmonate sulfotransferase from Arabidopsis thaliana. *J Biol Chem* 278: 17895–17900.

Glatt, H., Engelke, C.E., Pabel, U., Teubner, W., Jones, A.L., Coughtrie, M.W., Andrae, U., Falany, C.N., and Meinl, W., (2000) Sulfotransferases: genetics and role in toxicology. *Toxicol Lett* 112–113: 341–348.

Grun, F., Noy, N., Hammerling, U., and Buck, J., (1996) Purification, cloning, and bacterial expression of retinol dehydratase from *Spodoptera frugiperda*. *J Biol Chem* 271: 16135–16138.

Hehonah, N., Zhu, X., Brix, L., Bolton-Grob, R., Barnett, A., Windmill, K., and McManus, M., (1999) Molecular cloning, expression, localisation and functional characterisation of a rabbit SULT1C2 sulfotransferase. *Int J Biochem Cell Biol* 31: 869–882.

Hein, D.W., Grant, D.M., and Sim, E., (2000) Update on consensus arylamine N-acetyltransferase gene nomenclature. *Pharmacogenetics* 10: 291–292.

Henry, T., Kliewer, B., Palmatier, R., Ulphani, J.S., and Beckmann, J.D., (1996) Isolation and characterization of a bovine gene encoding phenol sulfotransferase. *Gene* 174: 221–224.

Her, C., Aksoy, I.A., Kimura, S., Brandriff, B.F., Wasmuth, J.J., and Weinshilboum, R.M., (1995) Human estrogen sulfotransferase gene (STE): cloning, structure, and chromosomal localization. *Genomics* 29: 16–23.

Her, C., Kaur, G.P., Athwal, R.S., and Weinshilboum, R.M., (1997) Human sulfotransferase SULT1C1: cDNA cloning, tissue-specific expression, and chromosomal localization. *Genomics* 41: 467–470.

Her, C., Raftogianis, R., and Weinshilboum, R.M., (1996) Human phenol sulfotransferase STP2 gene: molecular cloning, structural characterization, and chromosomal localization. *Genomics* 33: 409–420.

Her, C., Wood, T.C., Eichler, E.E., Mohrenweiser, H.W., Ramagli, L.S., Siciliano, M.J., and Weinshilboum, R.M., (1998) Human hydroxysteroid sulfotransferase SULT2B1: two enzymes encoded by a single chromosome 19 gene. *Genomics* 53: 284–295.

Higashi, Y., Fuda, H., Yanai, H., Lee, Y., Fukushige, T., Kanzaki, T., and Strott, C.A., (2004) Expression of cholesterol sulfotransferase (SULT2B1b) in human skin and primary cultures of human epidermal keratinocytes. *J Invest Dermatol* 122: 1207–1213.

Hirshey, S.J., Dooley, T.P., Reardon, I.M., Heinrikson, R.L., and Falany, C.N., (1992) Sequence analysis, *in vitro* translation, and expression of the cDNA for rat liver minoxidil sulfotransferase. *Mol Pharmacol* 42: 257–264.

Hume, R., Barker, E.V., and Coughtrie, M.W., (1996) Differential expression and immunohistochemical localisation of the phenol and hydroxysteroid sulphotransferase enzyme families in the developing lung. *Histochem Cell Biol* 105: 147–152.

Hwang, S.R., Kohn, A.B., and Hook, V.Y., (1995) Molecular cloning of an isoform of phenol sulfotransferase from human brain hippocampus. *Biochem Biophys Res Commun* 207: 701–707.

Ingelman-Sundberg, M., Daly, A.K., Oscarson, M., and Nebert, D.W., (2000) Human cytochrome P450 (CYP) genes: recommendations for the nomenclature of alleles. *Pharmacogenetics* 10: 91–93.

Jones, A.L., Hagen, M., Coughtrie, M.W., Roberts, R.C., and Glatt, H., (1995) Human platelet phenolsulfotransferases: cDNA cloning, stable expression in V79 cells and identification of a novel allelic variant of the phenol-sulfating form. *Biochem Biophys Res Commun* 208: 855–862.

Kakuta, Y., Pedersen, L.G., Carter, C.W., Negishi, M., and Pedersen, L.C., (1997) Crystal structure of estrogen sulphotransferase. *Nat Struct Biol* 4: 904–908.

Kakuta, Y., Petrotchenko, E.V., Pedersen, L.C., and Negishi, M., (1998) The sulfuryl transfer mechanism: crystal structure of a vanadate complex of estrogen sulfotransferase and mutational analysis. *J Biol Chem* 273: 27325–27330.

Kester, M.H., Kaptein, E., Roest, T.J., van Dijk, C.H., Tibboel, D., Meinl, W., Glatt, H., Coughtrie, M.W., and Visser, T.J., (1999) Characterization of human iodothyronine sulfotransferases. *J Clin Endocrinol Metab* 84: 1357–1364.

Khan, A.S., Taylor, B.R., Chung, K., Etheredge, J., Gonzales, R., and Ringer, D.P., (1993) Genomic structure of rat liver aryl sulfotransferase IV-encoding gene. *Gene* 137: 321–326.

Kim, J.G., Vallet, J.L., Rohrer, G.A., and Christenson, R.K., (2002) Characterization of porcine uterine estrogen sulfotransferase. *Domest Anim Endocrinol* 23: 493–506.

Komatsu, K., Oeda, T., and Strott, C.A., (1993) Cloning and sequence analysis of the 5'-flanking region of the estrogen sulfotransferase gene: steroid response elements and cell-specific nuclear DNA-binding proteins. *Biochem Biophys Res Commun* 194: 1297–1304.

Kong, A.N. and Fei, P., (1994) Molecular cloning of three sulfotransferase cDNAs from mouse liver. *Chem Biol Interact* 92: 161–168.

Kong, A.N., Ma, M., Tao, D., and Yang, L., (1993a) Molecular cloning of cDNA encoding the phenol/aryl form of sulfotransferase (mSTp1) from mouse liver. *Biochim Biophys Acta* 1171: 315–318.

Kong, A.N., Yang, L., Ma, M., Tao, D., and Bjornsson, T.D., (1992) Molecular cloning of the alcohol/hydroxysteroid form (hSTa) of sulfotransferase from human liver. *Biochem Biophys Res Commun* 187: 448–454.

Kong, A.T., Tao, D., Ma, M., and Yang, L., (1993b) Molecular cloning of the alcohol/hydroxy-steroid form (mSTa1) of sulfotransferase from mouse liver. *Pharm Res* 10: 627–630.

Lacomme, C. and Roby, D., (1996) Molecular cloning of a sulfotransferase in Arabidopsis thaliana and regulation during development and in response to infection with pathogenic bacteria. *Plant Mol Biol* 30: 995–1008.

Lee, Y.C., Park, C.S., and Strott, C.A., (1994) Molecular cloning of a chiral-specific 3 alpha-hydroxysteroid sulfotransferase. *J Biol Chem* 269: 15838–15845.

Lin, Z., Lou, Y., and Squires, J.E., (2004) Molecular cloning and functional analysis of porcine SULT1A1 gene and its variant: a single mutation SULT1A1 causes a significant decrease in sulfation activity. *Mamm Genome* 15: 218–226.

Liu, M.C., Sakakibara, Y., and Liu, CC., (1999) Bacterial expression, purification, and characterization of a novel mouse sulfotransferase that catalyzes the sulfation of eicosanoids. *Biochem Biophys Res Commun* 254: 65–69.

Luu, N.X., Driscoll, W.J., Martin, B.M., and Strott, C.A., (1995) Molecular cloning and expression of a guinea pig 3-hydroxysteroid sulfotransferase distinct from chiral-specific 3 alpha-hydroxysteroid sulfotransferase. *Biochem Biophys Res Commun* 217: 1078–1086.

Luu-The, V., Dufort, I., Paquet, N., Reimnitz, G., and Labrie, F., (1995) Structural characterization and expression of the human dehydroepiandrosterone sulfotransferase gene. *DNA Cell Biol* 14: 511–518.

Meinl, W., and Glatt, H., (2001) Structure and localization of the human SULT1B1 gene: neighborhood to SULT1E1 and a SULT1D pseudogene. *Biochem Biophys Res Commun* 288: 855–862.

Meloche, C.A. and Falany, C.N., (2001) Expression and characterization of the human 3β-hydroxysteroid sulfotransferases (SULT2B1a and SULT2B1b). *J Steroid Biochem Mol Biol* 77: 261–269.

Nagai, F., Satoh, H., Hirota, M., Ogawa, K., Homma, H., and Matsui, M., (1996) Assignment of two hydroxysteroid sulfotransferase genes (St1 and St2) to rat chromosome bands 1q21.3 — >q22.1 by *in situ* hybridization. *Cytogenet Cell Genet* 74: 111–112.

Nagai, F., Satoh, H., Usui, T., Tamura, H., and Matsui, M., (1999) Chromosome assignments of rat phenol sulfotransferase ST1A1 and ST1C1 genes (Sult1a1 and Sult1c1) by fluorescence *in situ* hybridization. *Cytogenet Cell Genet* 84: 145–147.

Nagata, K., Ozawa, S., Miyata, M., Shimada, M., Gong, D.W., Yamazoe, Y., and Kato, R., (1993) Isolation and expression of a cDNA encoding a male-specific rat sulfotransferase that catalyzes activation of N-hydroxy-2-acetylaminofluorene. *J Biol Chem* 268: 24720–24725.

Nagata, K. and Yamazoe, Y., (2000) Pharmacogenetics of sulfotransferase. *Annu Rev Pharmacol Toxicol* 40: 159–176.

Nash, A.R., Glenn, W.K., Moore, S.S., Kerr, J., Thompson, A.R., and Thompson, E.O., (1988) Oestrogen sulfotransferase: molecular cloning and sequencing of cDNA for the bovine placental enzyme. *Aust J Biol Sci* 41: 507–516.

Nebert, D.W., (2000) Suggestions for the nomenclature of human alleles: relevance to ecogenetics, pharmacogenetics, and molecular epidemiology. *Pharmacogenetics* 10: 279–290.

Nebert, D.W., (2002) Proposal for an allele nomenclature system based on the evolutionary divergence of haplotypes. *Hum Mutat* 20: 463–472.

Oeda, T., Lee, Y.C., Driscoll, W.J., Chen, H.C., and Strott, C.A., (1992) Molecular cloning and expression of a full-length complementary DNA encoding the guinea pig adrenocortical estrogen sulfotransferase. *Mol Endocrinol* 6: 1216–1226.

Ogura, K., Kajita, J., Narihata, H., Watabe, T., Ozawa, S., Nagata, K., Yamazoe, Y., and Kato, R., (1989) Cloning and sequence analysis of a rat liver cDNA encoding hydroxysteroid sulfotransferase. *Biochem Biophys Res Commun* 165: 168–174.

Ogura, K., Kajita, J., Narihata, H., Watabe, T., Ozawa, S., Nagata, K., Yamazoe, Y., and Kato, R., (1990) cDNA cloning of the hydroxysteroid sulfotransferase STa sharing a strong homology in amino acid sequence with the senescence marker protein SMP-2 in rat livers. *Biochem Biophys Res Commun* 166: 1494–1500.

Ohkimoto, K., Liu, M.Y., Suiko, M., Sakakibara, Y., and Liu, M.C., (2004) Characterization of a zebrafish estrogen-sulfating cytosolic sulfotransferase: inhibitory effects and mechanism of action of phytoestrogens. *Chem Biol Interact* 147: 1–7.

Otterness, D.M., Her, C., Aksoy, S., Kimura, S., Wieben, E.D., and Weinshilboum, R.M., (1995) Human dehydroepiandrosterone sulfotransferase gene: molecular cloning and structural characterization. *DNA Cell Biol* 14: 331–341.

Otterness, D.M., Wieben, E.D., Wood, T.C., Watson, W.G., Madden, B.J., McCormick, D.J., and Weinshilboum, R.M., (1992) Human liver dehydroepiandrosterone sulfotransferase: molecular cloning and expression of cDNA. *Mol Pharmacol* 41: 865–872.

Ozawa, S., Nagata, K., Gong, D.W., Yamazoe, Y., and Kato, R., (1990) Nucleotide sequence of a full-length cDNA (PST-1) for aryl sulfotransferase from rat liver. *Nucleic Acids Res* 18: 4001.

Ozawa, S., Nagata, K., Shimada, M., Ueda, M., Tsuzuki, T., Yamazoe, Y., and Kato, R., (1995) Primary structures and properties of two related forms of aryl sulfotransferases in human liver. *Pharmacogenetics* 5: S135–S140.

Ozawa, S., Shimizu, M., Katoh, T., Miyajima, A., Ohno, Y., Matsumoto, Y., Fukuoka, M., Tang, Y.M., Lang, N.P., and Kadlubar, F.F., (1999) Sulfating-activity and stability of cDNA-expressed allozymes of human phenol sulfotransferase, ST1A3*1 ((213)Arg) and ST1A3*2 ((213)His), both of which exist in Japanese as well as Caucasians. *J Biochem (Tokyo)* 126: 271–277.

Ozawa, S., Tang, Y.M., Yamazoe, Y., Kato, R., Lang, N.P., and Kadlubar, F.F., (1998) Genetic polymorphisms in human liver phenol sulfotransferases involved in the bioactivation of N-hydroxy derivatives of carcinogenic arylamines and heterocyclic amines. *Chem Biol Interact* 109: 237–248.

Pakhomova, S., Kobayashi, M., Buck, J., and Newcomer, M.E., (2001) A helical lid converts a sulfotransferase to a dehydratase. *Nat Struct Biol* 8: 447–451.

Pakhomova, S., Luz, J.G., Kobayashi, M., Mellman, D., Buck, J., and Newcomer, M.E., (2000) Crystallization of retinol dehydratase from *Spodoptera frugiperda*: improvement of crystal quality by modification by ethylmercurythiosalicylate. *Acta Crystallogr D Biol Crystallogr* 56: 1641–1643.

Pedersen, L.C., Petrotchenko, E.V., and Negishi, M., (2000) Crystal structure of SULT2A3, human hydroxysteroid sulfotransferase. *FEBS Lett* 475: 61–64.

Raftogianis, R.B., Her, C., and Weinshilboum, R.M., (1996) Human phenol sulfotransferase pharmacogenetics: STP1 gene cloning and structural characterization. *Pharmacogenetics* 6: 473–487.

Raftogianis, R.B., Wood, T.C., Otterness, D.M., Van Loon, J.A., and Weinshilboum, R.M., (1997) Phenol sulfotransferase pharmacogenetics in humans: association of common SULT1A1 alleles with TS PST phenotype. *Biochem Biophys Res Commun* 239: 298–304.

Raftogianis, R.B., Wood, T.C., and Weinshilboum, R.M., (1999) Human phenol sulfotransferases SULT1A2 and SULT1A1: genetic polymorphisms, allozyme properties, and human liver genotype-phenotype correlations. *Biochem Pharmacol* 58: 605–616.

Rikke, B.A. and Roy, A.K., (1996) Structural relationships among members of the mammalian sulfotransferase gene family. *Biochim Biophys Acta* 1307: 331–338.

Riley, E., Bolton-Grob, R., Liyou, N., Wong, C., Tresillian, M., and McManus, M.E., (2002) Isolation and characterisation of a novel rabbit sulfotransferase isoform belonging to the SULT1A subfamily. *Int J Biochem Cell Biol* 34: 958–969.

Rouleau, M., Marsolais, F., Richard, M., Nicolle, L., Voigt, B., Adam, G., and Varin, L., (1999) Inactivation of brassinosteroid biological activity by a salicylate-inducible steroid sulfotransferase from *Brassica napus*. *J Biol Chem* 274: 20925–20930.

Rubin, G.L., Harrold, A.J., Mills, J.A., Falany, C.N., and Coughtrie, M.W., (1999) Regulation of sulphotransferase expression in the endometrium during the menstrual cycle, by oral contraceptives and during early pregnancy. *Mol Hum Reprod* 5: 995–1002.

Saeki, Y., Sakakibara, Y., Araki, Y., Yanagisawa, K., Suiko, M., Nakajima, H., and Liu, M.C., (1998) Molecular cloning, expression, and characterization of a novel mouse liver SULT1B1 sulfotransferase. *J Biochem (Tokyo)* 124: 55–64.

Sakakibara, Y., Suiko, M., Pai, T.G., Nakayama, T., Takami, Y., Katafuchi, J., and Liu, M.C., (2002) Highly conserved mouse and human brain sulfotransferases: molecular cloning, expression, and functional characterization. *Gene* 285: 39–47.

Sakakibara, Y., Takami, Y., Zwieb, C., Nakayama, T., Suiko, M., Nakajima, H., and Liu, M.C., (1995) Purification, characterization, and molecular cloning of a novel rat liver dopa/tyrosine sulfotransferase. *J Biol Chem* 270: 30470–30478.

Sakakibara, Y., Yanagisawa, K., Katafuchi, J., Ringer, D.P., Takami, Y., Nakayama, T., Suiko, M., and Liu, M.C., (1998a) Molecular cloning, expression, and characterization of novel human SULT1C sulfotransferases that catalyze the sulfonation of N-hydroxy-2-acetylaminofluorene. *J Biol Chem* 273: 33929–33935.

Sakakibara, Y., Yanagisawa, K., Takami, Y., Nakayama, T., Suiko, M., and Liu, M.C., (1998b) Molecular cloning, expression, and functional characterization of novel mouse sulfotransferases. *Biochem Biophys Res Commun* 247: 681–686.

Schauss, S.J., Henry, T., Palmatier, R., Halvorson, L., Dannenbring, R., and Beckmann, J.D., (1995) Characterization of bovine tracheobronchial phenol sulphotransferase cDNA and detection of mRNA regulation by cortisol. *Biochem J* 311 (Pt 1): 209–217.

Shimada, M., Terazawa, R., Kamiyama, Y., Honma, W., Nagata, K., and Yamazoe, Y., (2004) Unique properties of a renal sulfotransferase, st1d1, in dopamine metabolism. *J Pharmacol Exp Ther* 310: 808–814.

Song, W.C., Moore, R., McLachlan, J.A., and Negishi, M., (1995) Molecular characterization of a testis-specific estrogen sulfotransferase and aberrant liver expression in obese and diabetogenic C57BL/KsJ-db/db mice. *Endocrinology* 136: 2477–2484.

Sugahara, T., Liu, C.C., Govind Pai, T., and Liu, M.C., (2003a) Molecular cloning, expression, and functional characterization of a novel zebrafish cytosolic sulfotransferase. *Biochem Biophys Res Commun* 300: 725–730.

Sugahara, T., Liu, C.C., Pai, T.G., Collodi, P., Suiko, M., Sakakibara, Y., Nishiyama, K., and Liu, M.C., (2003b) Sulfation of hydroxychlorobiphenyls: molecular cloning, expression, and functional characterization of zebrafish SULT1 sulfotransferases. *Eur J Biochem* 270: 2404–2411.

Sugahara, T., Yang, Y.S., Liu, C.C., Pai, T.G., and Liu, M.C., (2003c) Sulphonation of dehydroepiandrosterone and neurosteroids: molecular cloning, expression, and functional characterization of a novel zebrafish SULT2 cytosolic sulphotransferase. *Biochem J* 375 (Pt 3): 785–791.

Tamura, H., Miyawaki, A., Yoneshima, H., Mikoshiba, K., and Matsui, M., (1999) Molecular cloning, expression and characterization of a phenol sulfotransferase cDNA from mouse intestine. *Biol Pharm Bull* 22: 234–239.

Tamura, H.O., Harada, Y., Miyawaki, A., Mikoshiba, K., and Matsui, M., (1998) Molecular cloning and expression of a cDNA encoding an olfactory-specific mouse phenol sulphotransferase. *Biochem J* 331 (Pt 3): 953–958.

Thomae, B.A., Eckloff, B.W., Freimuth, R.R., Wieben, E.D., and Weinshilboum, R.M., (2002) Human sulfotransferase SULT2A1 pharmacogenetics: genotype-to-phenotype studies. *Pharmacogenomics J* 2: 48–56.

Tsoi, C., Falany, C.N., Morgenstern, R., and Swedmark, S., (2001a) Identification of a new subfamily of sulphotransferases: cloning and characterization of canine SULT1D1. *Biochem J* 356: 891–897.

Tsoi, C., Falany, C.N., Morgenstern, R., and Swedmark, S., (2001b) Molecular cloning, expression, and characterization of a canine sulfotransferase that is a human ST1B2 ortholog. *Arch Biochem Biophys* 390: 87–92.

Tsoi, C., Morgenstern, R., and Swedmark, S., (2002) Canine sulfotransferase SULT1A1: molecular cloning, expression, and characterization. *Arch Biochem Biophys* 401: 125–133.

Vakiani, E., Luz, J.G., and Buck, J., (1998) Substrate specificity and kinetic mechanism of the insect sulfotransferase, retinol dehydratase. *J Biol Chem* 273: 35381–35387.

Varin, L., DeLuca, V., Ibrahim, R.K., and Brisson, N., (1992) Molecular characterization of two plant flavonol sulfotransferases. *Proc Natl Acad Sci U S A* 89: 1286–1290.

Varin, L., Marsolais, F., Richard, M., and Rouleau, M., (1997) Sulfation and sulfotransferases 6: biochemistry and molecular biology of plant sulfotransferases. *Faseb J* 11: 517–525.

Walther, S., Dunbrack, R., and Raftogianis, R., (1999) Cloning, expression and characterization of a human sulfotransferase, SULT4A1, that represents a novel SULT family. *ISSX Proceedings* 15: 195.

Wang, J., Falany, J.L., and Falany, C.N., (1998) Expression and characterization of a novel thyroid hormone-sulfating form of cytosolic sulfotransferase from human liver. *Mol Pharmacol* 53: 274–282.

Watabe, T., Ogura, K., Satsukawa, M., Okuda, H., and Hiratsuka, A., (1994) Molecular cloning and functions of rat liver hydroxysteroid sulfotransferases catalysing covalent binding of carcinogenic polycyclic arylmethanols to DNA. *Chem Biol Interact* 92: 87–105.

Weinshilboum, R., and Otterness, D.M., Sulfotransferase enzymes, in *Conjugation-Deconjugation Reactions in Drug Metabolism and Toxicity*, Kauffman, F., Ed., Vol. 112, Springer-Verlag, Berlin, 1994, pp. 45–78.

Weinshilboum, R.M., Otterness, D.M., Aksoy, I.A., Wood, T.C., Her, C., and Raftogianis, R.B., (1997) Sulfation and sulfotransferases 1: sulfotransferase molecular biology: cDNAs and genes. *Faseb J* 11: 3–14.

Wilborn, T.W., Comer, K.A., Dooley, T.P., Reardon, I.M., Heinrikson, R.L., and Falany, C.N., (1993) Sequence analysis and expression of the cDNA for the phenol-sulfating form of human liver phenol sulfotransferase. *Mol Pharmacol* 43: 70–77.

Wood, T.C., Aksoy, I.A., Aksoy, S., and Weinshilboum, R.M., (1994) Human liver thermolabile phenol sulfotransferase: cDNA cloning, expression and characterization. *Biochem Biophys Res Commun* 198: 1119–1127.

Wood, T.C., Her, C., Aksoy, I., Otterness, D.M., and Weinshilboum, R.M., (1996) Human dehydroepiandrosterone sulfotransferase pharmacogenetics: quantitative western analysis and gene sequence polymorphisms. *J Steroid Biochem Mol Biol* 59: 467–478.

Xiangrong, L., Johnk, C., Hartmann, D., Schestag, F., Kromer, W., and Gieselmann, V., (2000) Enzymatic properties, tissue-specific expression, and lysosomal location of two highly homologous rat SULT1C2 sulfotransferases, in process citation. *Biochem Biophys Res Commun* 272: 242–250.

Yanai, H., Javitt, N.B., Higashi, Y., Fuda, H., and Strott, C.A., (2004) Expression of cholesterol sulfotransferase (SULT2B1b) in human platelets. *Circulation* 109: 92–96.

Yerokun, T., Etheredge, J.L., Norton, T.R., Carter, H.A., Chung, K.H., Birckbichler, P.J., and Ringer, D.P., (1992) Characterization of a complementary DNA for rat liver aryl sulfotransferase IV and use in evaluating the hepatic gene transcript levels of rats at various stages of 2-acetylaminofluorene-induced hepatocarcinogenesis. *Cancer Res* 52: 4779–4786.

Yoshinari, K., Nagata, K., Ogino, M., Fujita, K., Shiraga, T., Iwasaki, K., Hata, T., and Yamazoe, Y., (1998a) Molecular cloning and expression of an amine sulfotransferase cDNA: a new gene family of cytosolic sulfotransferases in mammals. *J Biochem (Tokyo)* 123: 479–486.

Yoshinari, K., Nagata, K., Shimada, M., and Yamazoe, Y., (1998b) Molecular characterization of ST1C1-related human sulfotransferase. *Carcinogenesis* 19: 951–953.

Yoshinari, K., Nagata, K., Shiraga, T., Iwasaki, K., Hata, T., Ogino, M., Ueda, R., Fujita, K., Shimada, M., and Yamazoe, Y., (1998c) Molecular cloning, expression, and enzymatic characterization of rabbit hydroxysteroid sulfotransferase AST-RB2. *J Biochem (Tokyo)* 123: 740–746.

Zhu, X., Veronese, M.E., Bernard, C.C., Sansom, L.N., and McManus, M.E., (1993a) Identification of two human brain aryl sulfotransferase cDNAs. *Biochem Biophys Res Commun* 195: 120–127.

Zhu, X., Veronese, M.E., Iocco, P., and McManus, M.E., (1996) cDNA cloning and expression of a new form of human aryl sulfotransferase. *Int J Biochem Cell Biol* 28: 565–571.

Zhu, X., Veronese, M.E., Sansom, L.N., and McManus, M.E., (1993b) Molecular characterisation of a human aryl sulfotransferase cDNA. *Biochem Biophys Res Commun* 192: 671–676.

2 Structure and Function of Sulfotransferases

Jyrki Taskinen and Michael W.H. Coughtrie

CONTENTS

INTRODUCTION

The cytosolic SULTs catalyze the transfer of the sulfuryl group from 3′-phospho-adenosine 5′-phosphosulfate (PAPS) to a nucleophilic acceptor group of a small molecule substrate, giving as products the sulfated substrate and 3′-phosphoadenos-ine 5′-phosphate (PAP; Figure 2.1). The SULT enzymes, which are derived from a superfamily of genes, have been divided into six families based on amino acid sequence analysis. Eleven human isoforms have been identified representing three families, SULT1, SULT2, and SULT4. The human SULT1 family comprises seven isoforms (1A1, 1A2, 1A3, 1B1, 1C2, 1C4, and 1E1), which typically sulfate phenolic hydroxyl groups. The three enzymes of the SULT2 family (2A1, 2B1a, and 2B1b) sulfate typically alcoholic hydroxyls of steroid substrates. No substrates are presently known for SULT4A1, which is expressed exclusively in the brain. Several previously

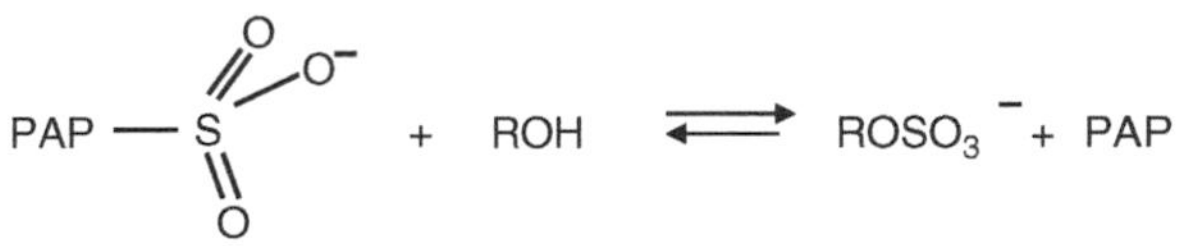

$$PAP - \overset{\overset{\displaystyle O}{\|}}{\underset{\underset{\displaystyle O}{\|}}{S}} - O^- \quad + \quad ROH \quad \rightleftharpoons \quad ROSO_3^- + PAP$$

FIGURE 2.1 The reaction catalyzed by SULTs.

unknown SULT genes (and pseudogenes) were revealed on completion of the human genome sequence; however, the function and even expression of these genes is unknown.

The turnover numbers of SULTs reported for their prototypical substrates are rather low: 1–5 min^{-1} (Petrotchenko et al., 1999; Zhang et al., 1998), whereas the catalytic power of the enzymes is high. The rate enhancement factor compared to the noncatalyzed reaction is estimated to be 10^{10} to 10^{12} (Bedford et al., 1995). The SULT enzymes display generally broad substrate specificity, with overlap to varying extents, especially between enzymes of the same subfamily. Many SULT isoenzymes are supposed to be class specific, showing preference for certain structural types represented by an endogenous substrate. This is reflected in the names used for some SULT isoforms, like dopamine (catecholamine) SULT (1A3), estrogen SULT (1E1), and dehydroepiandrosterone ([DHEA] hydroxysteroid) SULT (2A1). In practice, SULT specificity promotes selectivity, favoring some structures and discriminating against others. Regioselectivity and stereoselectivity have also been observed with certain substrates and enzymes.

How the SULTs manage the two basic tasks of enzymes, catalysis and specificity, is the topic of this review. Understanding of catalysis and specificity is based on structural studies by x-ray crystallography, studies on enzyme kinetics of the bisubstrate reaction, structure–function relationships studies using mutant enzymes, and structure–activity relationships studies based on kinetics of varying substrates. Recent advances with human cytosolic SULTs in these areas are reviewed, with occasional reference to studies with animal cytosolic SULTs for comparison. The emerging picture of the catalytic and kinetic mechanism and the control of substrate specificity of the SULT enzymes are discussed. Several reviews have been published recently, partly covering the areas discussed in this chapter (Coughtrie, 2002; Duffel et al., 2001; Negishi et al., 2001; Yoshinari et al., 2001).

RESULTS AND DISCUSSION

PROTEIN STRUCTURE

Most cloned human cytosolic SULTs contain 284 to 296 amino acids, giving them calculated subunit molecular masses between 30,000 and 35,000 Da. An exception is SULT2B1, the a and b variants of which consist of 350 and 365 amino acids, respectively. The SULT enzymes are generally present as dimers in solution, as evidenced by molecular masses between 60,000 and 70,000 Da observed by experimental methods (Dajani et al., 1999b; Zhang et al., 1998). Most SULTs have a conserved consensus sequence KXXXTVXXE near the C terminus of their primary structure, designated as the KTVE motif. It has been shown by Petrotchenko et al. (2001) that mutation of a single amino acid in this sequence, Val269Glu in the case of hSULT1E1 and Val260E in the case of hSULT2A1, converts these dimeric enzymes to monomers. Mouse Sult1e1 is naturally a monomer in solution and has PE instead of TV in the corresponding sequence. Mutation of PE to TV results in dimerization of the mouse enzyme. Heterodimerization of hSULT2A1 was observed with the wild type hSULT1E1 but not with the V269E mutant. These findings have led to the conclusion that the KTVE motif is the common protein–protein interaction motif for homo- and heterodimerization of SULTs in solution *in vitro* (Petrotchenko et al., 2001).

Three-dimensional (3D) structures are known at present for human SULTs 1A3, 1E1, 2A1, 1A1, and 2B1a and b and for mouse SULT1E1. The knowledge of the tertiary structures comes from x-ray crystallographic studies with recombinant proteins expressed in and purified from bacterial cells. The first crystal structure of a mammalian cytosolic SULT was published in 1997 by the group of Pedersen and Negishi (Kakuta et al., 1997) for mouse estrogen SULT (mSult1e1). Recently, the same group solved the structure of human SULT1E1 (Pedersen et al., 2002). The structure of human dopamine SULT (SULT1A3) was solved independently by two groups (Bidwell et al., 1999; Dajani et al., 1999a). DHEA SULT, SULT2A1, is the third human cytosolic SULT form, for which experimental 3D structures are available. The two crystal structures reported represent variants of SULT2A1 (Pedersen et al., 2000; Rehse et al., 2002). Recently, the structure of human SULT1A1, in complex with substrate (4-nitrophenol) and PAP, was solved (Gamage et al., 2003). In this structure, two molecules of 4-nitrophenol can be seen in two binding sites within the active site of the enzyme, providing an explanation for the partial substrate inhibition commonly observed with this enzyme. The structures of SULT2B1a and SULT2B1b (Lee et al., 2003) show the molecular basis for the substrate specificities of these two enzymes. In SULT2B1b, the substrate is located in a different orientation in the active site to that seen with SULT2A1, and the amino terminal helix (Asp19 – Lys26) controls substrate specificity by becoming ordered on binding of pregnenolone and covering the substrate binding pocket.

The folding pattern and other essential structural features are conserved in all cytosolic human and mouse SULTs for which the structures have been solved. The overall structure is a central five-stranded parallel β-sheet surrounded by α-helices (Figure 2.2; see color photo insert following p. 210). One end of the β-sheet extends to the surface; the other end is buried inside the protein. The shape of a SULT monomer is roughly spherical as demonstrated in the surface models (Figure 2.3). Interestingly, the SULT fold is very similar to that of uridylate kinase (Kakuta et al., 1997), suggesting an evolutionary relationship between sulfation and phosphorylation.

ACTIVE SITE

The solved SULT crystal structures represent various complexes of the enzymes with one or two ligands. As a result, crystallographic studies have revealed many details of the enzyme function. The active site of SULTs is a long groove at the buried end of the β-sheet. The groove contains binding sites for both substrates and the catalytic machinery. In the SULT1E1, SULT1A1, and SULT2A1 crystal structures, the whole active site is covered by a long loop (residues 233 to 269 in 1E1) connecting helices 12 and 13. Both substrates are buried almost completely under the surface of the protein as demonstrated for mSult1E1 in a surface model (Figure 2.3). The entrance of the substrates necessitates that the loop covering the active site can change its conformation like an opening lid. Flexibility of the loop is

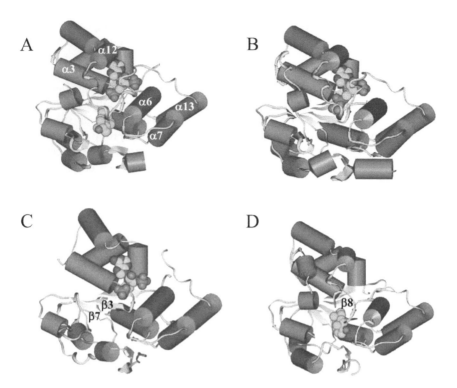

FIGURE 2.2 Folding pattern, secondary structural elements, and location of bound ligands revealed by x-ray crystal structures of cytosolic SULTs. (A) Mouse Sult1e1 complexed with PAP and estradiol (Kakuta et al., 1997). (B) Human SULT1E1 complexed with PAPS (Pedersen et al., 2002). (C) Human SULT1A3 complexed with PAP (Dajani et al., 1999a). (D) Human SULT2A1 complexed with DHEA (Rehse et al., 2002). Red cylinders represent α-helices; yellow arrows represent β-strands; ligands are shown as CPK-models. The figure was created with InsightII. Crystal coordinates for (A), (B), and (D) are from PDB entry 1aqu.ent, 1hy3.ENT, and 1j99.ENT, respectively. (See color photo insert following p. 210.)

supported by the fact that this region is disordered in SULT1A3 crystals and cannot be seen in the structures although this loop is not disordered in the other structures.

The binding site of PAPS is well characterized due to availability of several SULT crystal structures complexed with PAP (Dajani et al., 1999a; Gamage et al., 2003; Kakuta et al., 1997; Pedersen et al., 2000) and of hSULT1E1 complexed with PAPS (Pedersen et al., 2002). Structural features of PAPS binding appear to be well conserved in all SULTs. The region of nine amino acid residues (residues 45 to 53 in mSult1e1, hSULT 1A3, and SULT1A1, TYPKSGT) connecting strand 3 and helix 3 has been named the phosphate–sulfate binding loop (PSB-loop). The 5′-phosphate group of PAPS is positioned accurately in a rigid cage by hydrogen bonding of three oxygen atoms to the main chain NH groups of the PSB-loop in all crystal complexes with PAP or PAPS (Figure 2.4). The phosphate–sulfate bridging oxygen is hydrogen bonded to the side chain amino group of Lys48 in mSult1e1 and the corresponding Lys44 in hSULT2A1. Residues from strand 8 (Arg130) and helix 6 (Ser138) bind

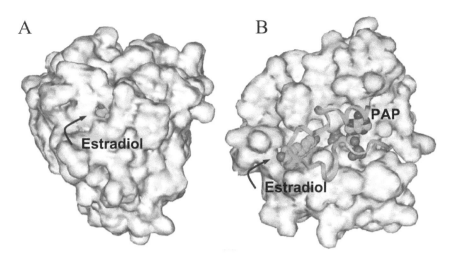

FIGURE 2.3 (A) Surface model of the crystal structure of mSult1e1 complexed with estradiol and PAP. (B) Surface removed from the loop connecting helices α-12 and α-13.

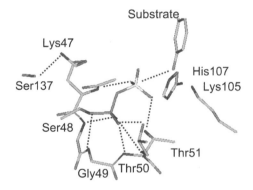

FIGURE 2.4 The reaction center and binding of sulfate and 5′-phosphate of PAPS in the PSB-loop of hSULT1E1. The position of the phenol ring of the substrate was modeled by superimposition on the mSult1e1 structure complexed with estradiol. Crystal coordinates from PDB entries 1aqu.ENT and 1hy3.ENT (Kakuta et al., 1997; Pedersen et al., 2002).

the 3′-phosphate group in all structures with PAP or PAPS. The adenine of PAPS is sandwiched between the heterocyclic ring of Trp53 from helix 3 and Phe229 from the binding site covering loop. Several residues of this loop are found at hydrogen-bonding distance from oxygens of the 3′-phosphate.

The substrate binding site contains variable regions and displaced loops and is generally not as well characterized as the cosubstrate binding site. The SULT1A3 structures do not contain bound substrates. However, crystal structures of mSult1e1 and hSULT2A1 have been solved with complexed substrates, estradiol and DHEA, respectively, making it possible to identify the regions involved and characterize several important features of substrate binding (Kakuta et al., 1997; Rehse et al.,

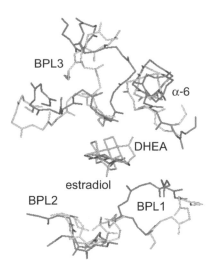

FIGURE 2.5 Peptide backbone structures of the regions lining the substrate binding pocket in mSult1e1 (cyan) with bound estradiol (red) and hSULT2A1 (magenta) with bound DHEA (green). (See color photo insert following p. 210.)

2002). The crystal structure of SULT1A1 was solved with two molecules of 4-nitro-phenol bound in the active site, giving insight into substrate inhibition as well (Gamage et al., 2003). The sulfating end of the substrate molecule is buried near the PSB-loop, between helix 6 and the end of strand 7, and is in contact with highly conserved residues. Closer to the protein surface, the binding pocket is lined by regions, which have been named (Yoshinari et al., 2001) BPL1 (binding pocket loop 1, residues 20 to 25), BPL2 (residues 80 to 89), and BPL3 (residues 236 to 250). In addition to these regions, the beginning of the loop connecting helices 6 and 7 (residues 146 to 149) is a part of the binding pocket wall (Figure 2.5; see color photo insert following p. 210). All these regions are variable and some of them, at least BPL3 covering the whole active site, are flexible and may move upon substrate binding.

MOLECULAR MECHANISMS OF SULFURYL TRANSFER

Crystal structure studies show that the transfer of the sulfate group from the PAPS to the acceptor substrate happens in a ternary complex in which both substrates are bound in the active site of the enzyme. No structures have been solved with both substrates bound in the ternary complex. However, superimposition of the structures of hSULT1E1-PAPS complex and hSULT1E1-PAP-estradiol complex shows that the reacting sulfur and oxygen atoms are positioned in a geometry (S-O distance about 3 Å, O-S-O angle about 162°) that makes direct SN2 displacement feasible (Pedersen et al., 2002). In the transition state of the SN2 reaction, the nucleophilic atom of the substrate and the leaving oxygen of PAP are partially bound to the sulfur atom, which is then surrounded by five oxygens. The transition state for such an electronic arrangement is expected to have a trigonal bipyramidal structure in which the attacking oxygen atom, the sulfur atom, and the leaving oxygen are in a straight line and the three other oxygens of sulfate are in a perpendicular plane. A complex of the enzyme with vanadate ion can be used to probe the transition state interactions because vanadate can adopt the trigonal bipyramidal geometry. The crystal structure of the mSult1e1-PAP-vanadate complex indicated that the vanadate ion adopted the expected geometry with its equatorial oxygens coordinated to the side chain nitrogens of Lys48, Lys106, and His108, evidently participating in stabilization of the transition state (Kakuta et al., 1998).

His108 is also hydrogen bonded to the reacting hydroxyl of estradiol in the mSult1e1-PAP-estradiol complex. His108 is conserved in all SULTs and mutating it abolishes sulfation activity completely (Kakuta et al., 1998; Liu et al., 2000). Due to its position and crucial importance, it is supposed to act as a general base catalyst, activating the reacting hydroxyl by proton abstraction. Mutation of Lys48Met led to the loss of sulfation activity, while a Lys48Arg mutant displayed lowered k_{cat} value (Kakuta et al., 1998). The mSult1e1 structures complexed with PAP and estradiol or with PAP and vanadate suggest that the side chain amino group of Lys48 is hydrogen bonded to the bridging oxygen between the 5'-phosphate and sulfate moiety of PAPS. This interaction implies that Lys48 may act as a general acid catalyst facilitating the dissociation of PAP. The side chain of Lys48 adopts a different conformation in hSULT1A3-PAP and hSULT1E1-PAPS complexes and is hydrogen bonded to the Ser138 hydroxyl which, in turn, is hydrogen bonded to the 3'-phosphate group of PAPS. It has been hypothesized that this interaction, in the absence of substrate, slows down the premature hydrolysis of PAPS (Pedersen et al., 2002). Following the binding of substrate, the catalytic base His108 abstracts the proton from the reacting hydroxyl, thereby increasing its nucleophilicity. Subsequently, the oxygen anion attacks the sulfur atom. Accumulation of negative charge in the activated complex then switches Lys48 hydrogen bonding from Ser138 to protonate the P-S bridging oxygen of PAPS, leading to the completion of the sulfuryl transfer (Figure 2.6).

KINETICS OF SULFATION

The crystallographic studies establish that the SULT enzyme reaction follows a ternary-complex mechanism but leave open the details of the kinetic mechanism, which can be approached by enzyme kinetic experiments. The most intriguing questions have been the order of binding of PAPS and the acceptor substrate and the strong substrate inhibition observed in certain cases.

Early kinetic studies with mammalian SULTs isolated from biological sources have been interpreted to support a rapid-equilibrium random-order (Duffel and Jakoby, 1981) and a steady-state compulsory-order mechanism (Wittemore et al., 1986). Recent studies with recombinant SULTs have not resolved this controversy.

The crystallographic evidence on the mechanism of the sulfuryl transfer seems to support a compulsory-order mechanism with PAPS as the first binding substrate (Negishi et al., 2001; Pedersen et al., 2002). A careful kinetic study published by Zhang et al. (1998) on sulfation of estradiol by hSULT1E1 supports the random-order mechanism, including substrate inhibition through binding to an allosteric site. Substrate inhibition was found to be significant even at concentrations below 100

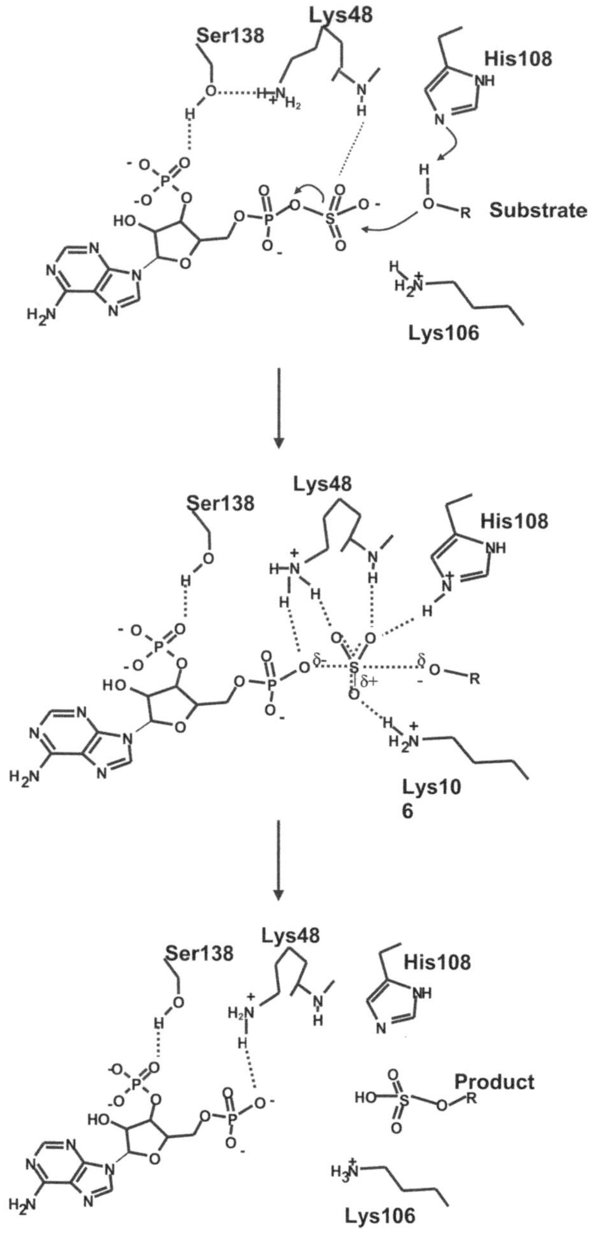

FIGURE 2.6 Sulfuryl transfer mechanism.

nM. The two-substrate initial rate experiment gave linear double reciprocal plots for an estradiol concentration range of 2 to 16 nM. In these conditions, the exceptionally high affinity (K_m = 5 nM) and low turnover (k_{cat} = 1.3 min^{-1}) for estradiol sulfation could be established. The initial rate results rule out the rapid-equilibrium ordered mechanism but cannot be used to differentiate the steady-state compulsory-order from the rapid-equilibrium random-order mechanism. However, equilibrium binding studies showed that the enzyme forms binary complexes independently with both substrates. Two estradiol molecules were found to be bound per enzyme subunit. Moreover, PAPS and PAP were found to form ternary complexes with the enzyme–estradiol–sulfate complex and the enzyme–estradiol complex, respectively.

Further insight into the mechanism of substrate inhibition was provided by the crystallographic and kinetic results of Gamage et al. (2003) on hSULT1A1. The presence of two 4-nitrophenol molecules in the active site was revealed by the crystal structure. One molecule is bound in a catalytically competent manner. The result suggests that two catalytically competent species may be formed, with one and two bound 4-nitrophenol molecules, respectively, and these may form sulfated product with different efficiencies. A kinetic equation giving a good fit to the experimental data was derived, supposing that product can be released from the two-substrate species only after dissociation of the second substrate.

The experiments of Zhang et al. (1998) were carried out in conditions in which the cysteine residues should be reduced. Jakoby and co-workers have recently provided evidence that the kinetic mechanism and substrate inhibition may depend on the oxidation state of the enzyme. Substrate inhibition was observed even at the lowest substrate concentrations in the sulfation of 4-nitrophenol by the reduced form of rat arylsulfotransferase ASTIV, known as rSULT1A1 (Duffel et al., 2001; Marshall et al., 2000). Initial rates in two-substrate experiments were found to be consistent with the kinetics derived for steady-state compulsory-order mechanism with substrate inhibition. This type of kinetics is expected for transferase reactions in which the acceptor substrate can bind in the enzyme form containing the product of first-bound donor substrate, in this case PAP (Cornish-Bowden, 1995). It was found that 4-nitrophenol binds tightly in the enzyme-PAP complex. Radiolabeled PAP was immediately displaced from the binary enzyme–PAP complex by cold PAP or PAPS but not from the ternary complex of enzyme–PAP-substrate.

Oxidation was found to change profoundly the catalytic and kinetic behavior of ASTIV/rSULT1A1. Partial oxidation by formation of Cys66-Cys232 disulfide or mixed disulfide of Cys66 and glutathione resulted in substantially higher sulfation activity for 4-nitrophenol. The reaction followed Michaelis–Menten kinetics and showed no substrate inhibition. Binding experiments showed that labeled PAP is freely exchanged for cold PAP or PAPS from the ternary enzyme—PAP-substrate complex.

The effect of the phenol ionization on the catalytic rate was also found to depend on the oxidation state. The sign of the slope of the Hammet plot for 4-substituted phenols changed from negative to positive on oxidation. This means that increasing ionization of the phenol increases the catalytic rate with the oxidized ASTIV/rSULT1A1. The opposite is true for the reduced enzyme. The result was interpreted to support mechanisms in which the SN2 sulfuryl transfer is the rate limiting step. In the case of the reduced enzyme, the rate limiting step was suggested to be recovery of free enzyme from the enzyme–PAP complex or breakdown of the ternary dead-end complex.

The effect of oxidation on the kinetics of human SULTs has not been studied. However, the cysteine residue corresponding to Cys66 of ASTIV/rSULT1A1 is conserved in most human cytosolic SULTs (Zheng et al., 1994).

SULT enzymes do not require a divalent cation for activity, unlike many enzymes using nucleotide cosubstrates. However, the activity of SULTs may be stimulated by the presence of divalent cations. Mn^{2+} has been shown to stimulate the sulfation of dopamine, dopa, and tyrosine catalyzed by SULT1A3 (Sakakibara et al., 1997; Suiko et al., 1996). Stimulation of dopa and tyrosine sulfation was remarkably stereospecific for the D-form. Studies with point-mutated SULT1A3 led to the suggestion that Mn^{2+} exerts its stimulatory effect by binding in the Glu86 residue at the substrate binding site (Pai et al., 2002). Divalent cations may also affect the activity of SULT1E1. Initial rate estradiol sulfation as a function of $MgCl_2$ concentration was found to follow a bell-shaped curve with a maximum at 7 mM $MgCl_2$ (Zhang et al., 1998). The structural basis of this effect has not been studied. It should be noted that the residue Lys85 in hSULT1E1 corresponds to Glu86.

STRUCTURAL BASIS FOR SUBSTRATE SPECIFICITY

Enzymes of the SULT1 and SULT2 families can transfer the sulfuryl group to a hydroxyl group, but they catalyze preferentially the sulfation of compounds from different chemical classes. Basically, the SULT1 enzymes catalyze sulfuryl transfer to a phenolic hydroxyl in an aromatic ring, and SULT2 enzymes to a secondary alcoholic hydroxyl in a saturated ring. The crystal structures with bound substrates, mSult1e1 complexed with estradiol, hSULT2A1 with DHEA, and hSULT1A1 with 4-nitrophenol, represent these two cases, making possible the direct comparison of the binding interactions to identify the determinants of the class specificity.

The phenolic 3-hydroxyl of estradiol binds close to the sulfate of PAPS and is hydrogen bonded to His108 and Lys106 in the mSult1e1–estradiol-PAP complex. The steroid ring system is bound between hydrophobic walls of the binding pocket. On one side, the hydrophobic parts of Lys106, Tyr81, Cys84, and Phe24 are packed tightly, positioning the edge of the aromatic ring of Tyr81 toward the aromatic plane of the substrate A ring (Figure 2.7). On the opposite side, the hydrophobic wall toward the steroid plane is formed by helix 6, especially Phe142 and Ile146. The edge of the Phe 142 aromatic ring is positioned against the aromatic A ring of estradiol. Superimposition of DHEA on estradiol shows that binding of the non-planar saturated A ring between Tyr81 and Phe142 is sterically hindered, especially due to the C-19 methyl group of DHEA, which clashes on Tyr81. Therefore Tyr81 and Phe142 seem to form a gate that allows only flat structures to enter deep enough in the active site to interact effectively with His108 and form a close transition state complex. It has been shown by Petrotchenko et al. (1999) that placing a smaller amino acid at position 81 increases the DHEA sulfation activity of mSult1e1, apparently by widening the Tyr81/Phe142 gate.

Phe142 is conserved in all human SULTs. Tyr81 is substituted by Phe in those SULT forms (1A1, 1A2, 1A3, and 1E1) that are known to prefer phenolic substrates. It has been shown that the mutation Tyr81Phe has no significant effect on the kinetic constants for estradiol sulfation by m1E1 (Petrotchenko et al., 1999). Phe24 and

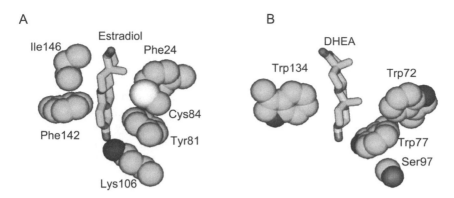

FIGURE 2.7 The amino acid side chains located in opposite sides of the substrate plain in m1E1 (A) and SULT2A1 (B). Hydrogens are omitted for clarity. The figure was created with InsightII. Crystal coordinates for (A) and (B) are from PDB entry 1aqu.ENT and 1j99.ENT, respectively.

Lys106 are also conserved in all these SULTs, but Cys84 is substituted by a hydrophobic residue in SULTs 1A1, 1A2, and 1A3. All these residues may be important for supporting the gate for flat structures.

The 3α-hydroxyl group of DHEA and the catalytic histidine of hSULT2A1 (His99) have nearly identical positions as the corresponding groups in the m1E1 structure. The steroid plane is rotated about 60°, which can be seen in Figure 2.5. The protein structure surrounding the substrate has a number of substitutions and displacements. Most importantly, the gate for flat structures does not exist. Tyr81 is substituted by Trp77, and it has moved from the side of the steroid A ring towards the 3-OH of the substrate. This is probably linked to the substitution of Lys109 of SULT1 enzymes by Ser97 in 2A1. The binding site wall formed by helix 6 on the opposite side of the steroid plane is also changed. Trp134 instead of the conserved Phe133 (corresponding to Phe142 in hSULT1E1) is positioned against the A ring of the steroid. These changes result in a wider channel, allowing thicker substrate structures close to the catalytic base. The replacement of Tyr81 by tryptophan and of Lys106 by serine residue is common to all three SULT2 enzymes.

The four SULT isoforms (1A1, 1A2, 1A3, and 1E1) that seem to have conserved the gate arrangement for selecting flat structures prefer phenolic substrates from chemically different classes. Two isoforms, SULT1A1 and 1A2, display a broad substrate specificity for neutral phenolic compounds. SULT1A3 preferentially sulfates dopamine and other catecholamine derivatives, while SULT1E1 prefers estrogens and their analogs.

The first case for which the basis of substrate specificity was characterized at the structural level was the distinct specificities of hSULT1A1 and hSULT1A3 toward neutral phenols and catecholamines, respectively. Despite high structural similarity (>93 % primary structure identity), these enzymes display opposite specificities with respect to 4-nitrophenol and dopamine. The specificity constant of SULT1A3 for dopamine is reported to be 100 to 200 times higher than the specificity constant for 4-nitrophenol (Brix et al., 1999a; Dajani et al., 1999a; Liu et al., 2000).

It was shown by the group of Coughtrie that mutating Glu146 of SULT1A3 to Ala, the corresponding residue in SULT1A1, changed the substrate specificity of SULT1A3 to resemble that of SULT1A1 (Dajani et al., 1998). This single mutation decreased the specificity constant for dopamine 100 times and increased the specificity constant 50 times for 4-nitrophenol. The essence of this finding has been confirmed by two other groups (Brix et al., 1999a; Liu et al., 2000). Modeling of dopamine in the crystal structure of SULT1A3 shows that Glu146 is located in a position where its carboxylate group can interact favorably with the protonated amino group of dopamine (Dajani et al., 1999a). However, this is not the only factor regulating dopamine–4-nitrophenol specificity of SULT1A1 and 1A3. It was shown by Sakakibara et al. (1998) by constructing a series of chimeric enzymes that the two highly variable regions, spanning amino acid residues 84 to 89 and 143 to 148, contain the structural determinants for the distinct substrate specificities of SULT1A1 and SULT1A3 for 4-nitrophenol and dopamine. Changing both the variable regions to those of the other enzyme reversed completely the substrate specificities. Changing either region abolished the high specificity of 1A3 towards dopamine but could cause high specificity towards 4-nitrophenol. Similarly, changing either region abolished the specificity of 1A1 toward 4-nitrophenol but did not produce high specificity towards dopamine. Site-directed mutagenesis studies have shown that the specificity of SULT1A1 cannot be converted to resemble that of SULT1A3 by the mutation Ala146Glu or any other single amino acid mutations (Brix et al., 1999a, 1999b; Liu et al., 2000).

Although SULT1A1 prefers neutral and SULT1A3 positively charged phenols, both enzymes can accept a wide assortment of molecules for sulfation, including rigid polycyclic compounds such as estrogens, apomorphine, and dihydrexidine (Taskinen et al., 2003). The crystal structure and modeling studies of SULT1A1 could not explain how extended fused ring systems, such as estradiol, could interact with the enzyme (Gamage et al., 2003). Gamage et al. therefore proposed that the substrate binding site of SULT1A1 is plastic, allowing it to adapt to the specific shape of the substrate during the process of binding. They proposed that the apo-form of SULT1A1 has a disordered binding site and that during catalysis the enzyme undergoes a disorder–order transition. The hypothesis is based on comparison of SULT1A1 and SULT1A3 crystal structures. Both SULT1A3 structures solved in the absence of substrate (Bidwell et al., 1999; Dajani et al., 1999a) have a disordered substrate binding site, whereas the SULT1A1-PAP–nitrophenol crystal structure complex (Gamage et al., 2003) has an ordered binding site.

Sulfation of dopa by SULT1A3 is reported to be stereoselective, with D-dopa sulfated at higher rates than L-dopa (Sakakibara et al., 1997). It has been suggested that the primary reason for this stereoselectivity is the presence of Glu146 in the active site (Pai et al., 2002). Point mutations of selected residues in the region 84 to 92 (BPL2) did not significantly affect the stereoselectivity, but a residues 84 to 90 deletional mutant showed no stereoselectivity. This suggests that the loop is necessary for steric reasons rather than for specific interactions. Stereoselectivity may be mediated by specific electrostatic interactions with the carboxylate group of Glu146 in cooperation with a steric effect of the BPL2.

The three enzymes of the SULT2 family (2A1, 2B1a, and 2B1b) sulfate alcoholic hydroxyls but prefer different steroid structures. Neither 2B1 variant efficiently sulfates DHEA, which is the main substrate of SULT2A1. Both of them sulfate pregnenolone, but SULT2B1b preferentially sulfates cholesterol, which is not sulfated at all by SULT2B1a (Fuda et al., 2002; Javitt et al., 2001). Structural determinants of the SULT2B1 specificity at the level of the primary structure have been elucidated by the group of Strott using mutational analysis (Fuda et al., 2002) and x-ray crystallography (Lee et al., 2003). SULT2B1 enzymes have extended amino- and carboxyterminal ends compared to SULT2A1. The 53 extra amino acids of the carboxy terminal end are common to the SULT2B1 forms. Removal of this extension does not change the activity or specificity of the enzymes. The amino terminal extensions are unique for each 2B1 enzyme form. Removal of the eight amino acids along the amino terminal extension of 2B1a does not alter its specificity for pregnenolone. However, removal of the 23 amino acids forming the unique aminoterminal extension of 2B1b results in loss of SULT activity toward cholesterol. It was shown by progressive truncation and alanine scanning mutagenesis that Ile20 and Ile23 are critical for cholesterol specificity. Ile20Ala and Ile23Ala mutations result in almost complete loss of cholesterol SULT activity, but have no effect on the ability of this enzyme to sulfate pregnenolone. The crystal structure revealed the nature of these differences, with a conformational change occurring on binding of pregnenolone.

CONCLUSIONS

The crystal structure and mutagenesis studies reviewed here have revealed some factors that probably are involved in the regulation of the substrate specificity of SULTs. Nevertheless, the matter is still poorly understood at the level of 3D molecular structure. The differences in substrate specificity within a class of similar compounds are due to subtle differences in the complementarity of shape and surface interaction properties of substrates and the binding pocket. The problem is further complicated by the possible plasticity of the substrate binding site. First attempts to tackle the problem with 3D QSAR modeling have been published recently (Sharma and Duffel, 2002; Sipilä et al., 2003). However, structural, kinetic, and computational studies with many different substrates are still needed to advance the understanding of substrate specificity.

REFERENCES

Bedford, C.T., Kirby, A.J., Logan, C.J., and Drummond, J.N., (1995) Structure-activity studies of sulfate transfer: the hydrolysis and aminolysis of 3'-phosphoadenosine 5'-phosphosulfate (PAPS). *Bioorg Med Chem* 3:167–172.

Bidwell, L.M., McManus, M.E., Gaedigk, A., Kakuta, Y., Negishi, M., Pedersen, L., and Martin, J.L., (1999) Crystal structure of human catecholamine sulfotransferase. *J Mol Biol* 293:521–530.

Brix, L.A., Barnett, A.C., Duggleby, R.G., Leggett, B., and McManus, M.E., (1999a) Analysis of the substrate specificity of human sulfotransferases SULT1A1 and SULT1A3: site-directed mutagenesis and kinetic studies. *Biochemistry* 38:10474–10479.

Brix, L.A., Duggleby, R.G., Gaedigk, A., and McManus, M.E., (1999b) Structural character-ization of human aryl sulphotransferases. *Biochem J* 337:337–343.

Cornish-Bowden, A., *Fundamentals of Enzyme Kinetics*. Portland Press, London, 1995.

Coughtrie, M.W.H., (2002) Sulfation through the looking glass — recent advances in sul-fotransferase research for the curious. *Pharmacogenomics* J 2:277–283.

Dajani, R., Cleasby, A., Neu, M., Wonacott, A.J., Jhoti, H., Hood, A.M., Modi, S., Hersey, A., Taskinen, J., Cooke, R.M., Manchee, G.R., and Coughtrie, M.W., (1999a) X-ray crystal structure of human dopamine sulfotransferase, SULT1A3. Molecular model-ing and quantitative structure-activity relationship analysis demonstrate a molecular basis for sulfotransferase substrate specificity. *J Biol Chem* 274:37862–37868.

Dajani, R., Hood, A.M., and Coughtrie, M.W., (1998) A single amino acid, glu146, governs the substrate specificity of a human dopamine sulfotransferase, SULT1A3. *Mol Phar-macol* 54:942–948.

Dajani, R., Sharp, S., Graham, S., Bethell, S.S., Cooke, R.M., Jamieson, D.J., and Coughtrie, M.W., (1999b) Kinetic properties of human dopamine sulfotransferase (SULT1A3) expressed in prokaryotic and eukaryotic systems: comparison with the recombinant enzyme purified from *Escherichia coli*. *Protein Express Purif* 16:11–18.

Duffel, M.W. and Jakoby, W.B., (1981) On the mechansim of aryl sulfotransferase. *J Biol Chem* 256:11123–11127.

Duffel, M.W., Marshall, A.D., McPhie, P., Sharma, V., and Jakoby, W.B., (2001) Enzymatic aspects of the phenol (aryl) sulfotransferases. *Drug Metab Rev* 33:369–395.

Fuda, H., Lee, Y.C., Shimizu, C., Javitt, N.B., and Strott, C.A., (2002) Mutational analysis of human hydroxysteroid sulfotransferase SULT2B1 isoforms reveals that exon 1B of the SULT2B1 gene produces cholesterol sulfotransferase, whereas exon 1A yields pregnenolone sulfotransferase. *J Biol Chem* 277:36161–36166.

Gamage, N.U., Duggleby, R.G., Barnett, A.C., Tresillian, M., Latham, C.F., Liyou, N.E., McManus, M.E., and Martin, J.L., (2003) Structure of a human carcinogen-converting enzyme, SULT1A1: structural and kinetic implications of substrate inhibition. *J Biol Chem* 278:7655–7662.

Javitt, N.B., Lee, Y.C., Shimizu, C., Fuda, H., and Strott, C.A., (2001) Cholesterol and hydroxycholesterol sulfotransferases: identification, distinction from dehydroepi-androsterone sulfotransferase, and differential tissue expression. *Endocrinology* 142:2978–2984.

Kakuta, Y., Pedersen, L.G., Carter, C.W., Negishi, M., and Pedersen, L.C., (1997) Crystal structure of estrogen sulphotransferase. *Nature Struct Biol* 4:904–908.

Kakuta, Y., Petrotchenko, E.V., Pedersen, L.C., and Negishi, M., (1998) The sulfuryl transfer mechanism: crystal structure of a vanadate complex of estrogen sulfotransferase and mutational analysis. *J Biol Chem* 273:27325–27330.

Lee, K.A., Fuda, H., Lee, Y.C., Negishi, M., Strott, C.A., and Pedersen, L.C., (2003) Crystal structure of human cholesterol sulfotransferase (SULT2B1b) in the presence of preg-nenolone and 3′-phosphoadenosine 5′-phosphosulfate: rationale for specificity differ-ences between prototypical SULT2A1 and the SULT2B1 isoforms. *J Biol Chem* 278:44593–44599.

Liu, M.C., Suiko, M., and Sakakibara, Y., (2000) Mutational analysis of the substrate bind-ing/catalytic domains of human M form and P form phenol sulfotransferases. *J Biol Chem* 275:13460–13464.

Marshall, A.D., McPhie, P., and Jakoby, W.B., (2000) Redox control of aryl sulfotransferase specificity. *Arch Biochem Biophys* 382:95–104.

Negishi, M., Pedersen, L.G., Petrotchenko, E.V., Shevtsov, S., Gorokhov, A., Kakuta, Y., and Pedersen, L.C., (2001) Structure and function of sulfotransferases. *Arch Biochem Biophys* 390:149–157.

Pai, T.G., Okhimoto, K., Sakakibara, Y., Suiko, M., Sugahara, T., and Liu, M.C., (2002) Manganese stimulation and stereospecificity of the Dopa/tyrosine-sulfating activity of human monoamine-form phenol sulfotransferase: kinetic studies of the mechanism using wild type and mutant enzymes. *J Biol Chem* 277:43813–43820.

Pedersen, L.C., Petrotchenko, E., Shevtsov, S., and Negishi, M., (2002) Crystal structure of the human estrogen sulfotrtansferase-PAPS complex. *J Biol Chem* 277:17928–17932.

Pedersen, L.C., Petrotchenko, E.V., and Negishi, M., (2000) Crystal structure of SULT2A3, human hydroxysteroid sulfotransferase. *FEBS Lett* 475:61–64.

Petrotchenko, E.V., Doerflein, M.E., Kakuta, Y., Pedersen, L.C., and Negishi, M., (1999) Substrate gating confers steroid specificity to estrogen sulfotransferase. *J Biol Chem* 274:30019–30022.

Petrotchenko, E.V., Pedersen, L.C., Borchers, C.H., Tomer, K.B., and Negishi, M., (2001) The dimerization motif of cytosolic sulfotransferases. *FEBS Lett* 490:39–43.

Rehse, P.H., Zhou, M., and Lin, S.-X., (2002) Crystal structure of human dehydroepiandrosterone sulphotransferase in complex with substrate. *Biochem J* 364:165–171.

Sakakibara, Y., Katafuchi, J., Takami, Y., Nakayama, T., Suiko, M., Nakajima, H., and Liu, M.-C., (1997) Manganese-dependent dopa/tyrosine sulfation in HepG2 human hepatoma cells: novel dopa/tyrosine sulfotransferase activities associated with the human monoamine-form phenol sulfotransferase. *Biochim Biophys Acta* 1355: 102–106.

Sakakibara, Y., Takami, Y., Nakayama, T., Suiko, M., and Liu, M.-C., (1998) Localization and functional analysis of the substrate specificity/catalytic domains of human M-form and P-form phenol sulfotransferases. *J Biol Chem* 273:6242–6247.

Sharma, V. and Duffel, M.W., (2002) Comparative molecular field analysis of substrates for an aryl sulfotransferase based on catalytic mechanism and protein homology modeling. *J Med Chem* 45:5514–5522.

Sipilä, J., Hood, A.M., Coughtrie, M.W.H., and Taskinen, J., (2003) CoMFA modeling of enzyme kinetics: K_m values for sulfation of diverse phenolic substrates by human catecholamine sulfotransferase SULT1A3. *J Chem Inf Comput Sci* 43:1563–1569.

Suiko, M., Sakakibara, Y., Nakajima, H., and Liu, M.C., (1996) Enzymic sulphation of dopa and tyrosine isomers by HepG2 human hepatoma cells: stereoselectivity and stimulation by Mn^{2+}. *Biochem J* 314:151–158.

Taskinen, J., Ethell, B.T., Hood, A.M., Pihlavisto, P., Burchell, B., and Coughtrie, M.W.H., (2003) Conjugation of catechols by recombinant human sulfotransferases, UDP-glucuronosyltranserases and soluble catechol O-methyltransferase: structure-conjugation relationships and predictive models. *Drug Metab Dispos* 30:1187–1197.

Wittemore, R.M., Pearce, L.B., and Roth, J.A., (1986) Purification and kinetic characterization of a phenol-sulfating form of phenolsulfotransferase from human brain. *Arch Biochem Biophys* 249:464–471.

Yoshinari, K., Petrotchenko, E.V., Pedersen, L.C., and Negishi, M., (2001) Crystal structure-based studies of cytosolic sulfotransferase. *J Biochem Mol Toxicol* 15:67–75.

Zhang, H., Varmalova, O., Vargas, F.M., Falany, C.N., and Leyh, T.S., (1998) Sulfuryl transfer: the catalytic mechanism of human estrogen sulfotransferase. *J Biol Chem* 272: 10888–10892.

Zheng, Y., Bergold, A., and Duffel, M.W., (1994) Affinity labeling of aryl sulfotransferase IV: identification of a peptide sequence at the binding site for 3'-phosphoadenosine-5'-phosphate. *J Biol Chem* 269:30313–30319.

3 PAPS and Sulfoconjugation

Nancy B. Schwartz

CONTENTS

INTRODUCTION

Sulfation has long been recognized as a critical modification of macromolecules integral to many different physiological processes, e.g., elimination of end products of catabolism, inactivation of hormones, and bioactivation of xenobiotics (Klaassen and Boles, 1997). More recently, there has been increased interest in sulfation as a post-translational modification of many secreted and membrane-bound proteins due to studies implicating sulfate as a determinant in protein–protein interactions mediating leucocyte adhesion, hemostasis, and chemokine signaling (Kehoe and Bertozzi, 2000). Concomitantly, a significant amount of information has accumulated suggesting that carbohydrate sulfation is a common modification of biomolecules used to address the challenges of multicellular communication (Bowman and Bertozzi, 1999). The mechanism of biotransformation of sulfate was elucidated by discovery of the donor substrate of sulfation, 3'-phosphoadenosine 5'-phosphosulfate (PAPS; Lipmann, 1958). From *Escherichia coli* to man, the sulfate activation system responsible for synthesis of PAPS consists of two activities, ATP-sulfurylase (EC 2.7.74), which catalyzes production of adenosine 5'-phosphosulfate (APS) from ATP, and SO^{4-}_2 and APS-kinase (EC 2.7.1.25), which phosphorylates APS using another molecule of ATP to produce donor PAPS, as shown:

$$ATP + SO^{-2} \longrightarrow APS + PPi$$

ATP-sulfurylase

$$APS + ATP \longrightarrow ADP + PAPS$$

APS-kinase

The level of sulfation achieved by all sulforeactions is limited by the bioavailability of PAPS, which in turn depends on the rates of its synthesis, degradation, and overall utilization. The integrated pathway for SO_4^{-2} uptake, activation, and utilization, shown in Figure 3.1, encompasses multiple components and intracellular compartments (Schwartz et al., 1998), only some of which have been identified and fully characterized. Although there are tissue- and species-specific differences, PAPS tissue concentrations are low compared to the level of PAPS required to sustain high rates of sulfation, and the steady-state PAPS concentration cannot be increased by increasing tissue sulfate concentration (Huxtable, 1986). Therefore, high rates of PAPS biosynthesis are required to effect sufficient sulfation. This review focuses on how expression of PAPS synthetase is regulated, as well as how this complex bifunctional enzyme, the keystone of the integrated sulfation pathway, functions in a multicomponent, multicompartment system to achieve the biological efficiency necessary to maintain adequate levels of PAPS.

SULFATE TRANSPORTERS

In all instances examined, the cellular mechanism of sulfate uptake is a protein-facilitated active transport process. Two distinct classes of vertebrate sulfate transporters have been identified and characterized. One superfamily of anion exchangers comprises Na^+-independent sulfate transporters sensitive to the anion exchange inhibitor, 4,4'-diisothio-cyanostilbene-2-2'-disulfonic acid (DIDS). Members of this family (Bissig et al., 1994) include the rat hepatocyte Sat-1, human diastrophic dysplasia sulfate transporter (DTDST; Hastbacka et al., 1994), and mouse down regulated in adenomas (DRA; Silberg et al., 1995) transporters, which share significant homology at the amino acid level and contain 12 membrane-spanning domains. Mutations in the human DTD gene result in undersulfated proteoglycans, leading to chondrodysplasias that range in severity from mild to lethal (Superti-Furga et al., 1996). A second superfamily of Na^+-coupled transporters function as DIDS-resistant, Na^+-dependent sulfate transporters in kidney and intestine (Markovich et al., 1993). The genes for all of these transporters have been cloned, expression patterns for

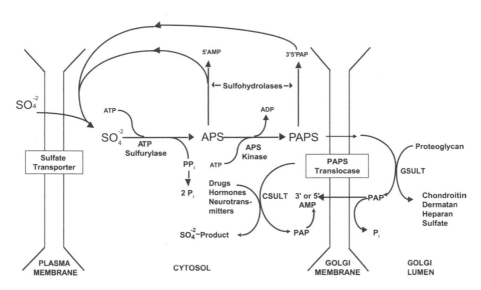

FIGURE 3.1 Integrated pathway for sulfate uptake, activation, and delivery to the golgi.

these genes have been determined, and the roles of the various transporters in the regulation of the overall sulfate uptake process are being elucidated. For instance, the high level of DTDST mRNA in cartilage, as well as the identification of three human congenital chondrodysplasias, i.e., diastrophic dysplasia, atelosteogenesis type II, and achondrogenesis type 1B, which are caused by mutations in the single DTDST gene, support the notion that cartilage depends predominantly on the DTDST system for its sulfate uptake (Satoh et al., 1998). By extrapolation, high levels of expression of specific transporters in only certain cells/tissues may indicate a unique ion dependence for sulfate transport by those transporters, and thus non-redundant tissue-specific roles.

ENZYMOLOGY OF PAPS SYNTHETASE

Once inside the cell, inorganic sulfate is transformed into the active sulfate donor PAPS (Figure 3.1). In unicellular organisms and in plants, the two activities involved in sulfate activation are borne by relatively small, separate proteins. In contrast, extensive evidence from mammals suggested that ATP-sulfurylase and APS-kinase constitute a single bifunctional polypeptide; i.e., they fractionated together through extensive purification and exhibited common properties and behavior (Geller et al., 1987). Most importantly, the brachymorphic (bm) mouse, which has a sulfation deficit, exhibited reduced activity of both PAPS synthesizing enzymes (Sugahara and Schwartz, 1979, 1982a, 1982b).

REACTION MECHANISM

Mechanistic studies of the enzymes involved in synthesis and utilization of APS and PAPS have long been hampered by the lability of APS and PAPS, the vulnerability of these compounds to enzymatic degradation, the unfavorable equilibrium constant for the initial sulfate activation reaction, and the lack of suitable inhibitors. Therefore, we generated the stable substrate analogs, β-methylene APS and β-methylene PAPS, in which the bridging oxygen of the phosphoanhydride linkage was replaced by a methylene group (Callahan et al., 1989; Ng et al., 1991). These analogs have facilitated kinetic investigations of ATP-sulfurylase and APS-kinase and have allowed affinity purification and characterization of other PAPS-utilizing proteins, e.g., sulfotransferases and PAPS translocases (Ozeran et al., 1996a, 1996b). To explore the mechanistic aspects of the PAPS synthetase, kinetic analyses of the activities individually and in coupled reactions were performed, providing the first indication for a functional interaction between these two activities (Lyle et al., 1994a, 1994b; Schwartz et al., 1998).The kinetic mechanism of rat chondrosarcoma ATP-sulfurylase was investigated by steady-state methods in both the physiologically forward reaction direction and the reverse direction. In the forward direction, K_m^{ATP} was 200 μM and $K_m SO_4^{2-}$ was 97 μM. Chlorate, a competitive inhibitor with respect to sulfate, showed noncompetitive inhibition with respect to ATP with an apparent K_i of 1.97 μM. In the physiologically reverse direction, K_m^{APS} was 39 μM and $K_m^{pyrophosphate}$ was 18 μM. The results of steady-state experiments using magnesium indicated that the true substrate is the magnesium–pyrophosphate complex. The APS analog, 5′ β-methylene APS, was an inhibitor competitive with APS and noncompetitive with respect to magnesium–pyrophosphate. The simplest formal mechanism in accord with all the data is an ordered, steady-state single displacement with MgATP as the leading substrate in the forward direction and APS as the leading substrate in the reverse direction (Lyle et al., 1994a).

A complete kinetic characterization of APS-kinase was also carried out. Initial velocity patterns for rat chondrosarcoma APS-kinase suggested a single-displacement formal mechanism with K_m^{APS} = 75 nM and K_m^{ATP} = 24 μM (Lyle et al., 1994b). Recent studies from our laboratory and others (Venkatachalam et al., 1998) have reported kinetic analyses on expressed recombinant murine and human PAPS synthetases, with findings of similar kinetic constants. Inhibition studies with analogs of substrates and products were performed in order to determine the complete formal mechanism of APS-kinase. The PAPS analog, 5′-β-methylene PAPS, exhibited inhibition competitive with APS and noncompetitive with ATP. The APS analog was also competitive with APS and noncompetitive with ATP. Imido-ATP showed inhibition competitive with ATP and mixed-type inhibition with respect to APS. These results suggest a steady-state, ordered mechanism with APS as the leading substrate and PAPS as the final product released (Lyle et al., 1994b). Our kinetic investigations also provided the first indication that a functional interaction between these two activities occurs. Since APS is the last product to be released from the sulfurylase reaction and APS is the first substrate to bind to the kinase activity, the APS intermediate may be transferred between the active sites, or APS may remain bound to the same site for action by the phosphorylation step. The specific order of substrate addition and product release results in more efficient generation of PAPS and helps to overcome the many obstacles to the synthesis of PAPS through this pathway (Schwartz et al., 1998). These results have prompted subsequent studies designed to elucidate the physical and functional relationships between ATP-sulfurylase and APS-kinase.

CHANNELING

The relationship of these two sequential activities was examined by several approaches. First, the accumulation of both APS and PAPS in the presence of both activities yielded a PAPS to APS ratio corresponding to a channeling efficiency of 96%. Second, the velocity of the APS-kinase reaction measured in the overall system, utilizing endogenously synthesized APS, was eight-fold greater than that of the isolated kinase reaction using exogenous APS. Third, addition of a 10,000-fold higher concentration of exogenous unlabeled APS to a system initiated with ATP and labeled SO_4^{-2} produced only a 26% reduction in production of labeled PAPS, indicating that the APS intermediate was not released into the bulk medium but remained bound in the overall system. These results suggest that APS is channeled between the active sites of ATP-sulfurylase and APS-kinase during the production of PAPS (Lyle et al., 1994d, 1995). Similar approaches were used to demonstrate that the genetic defect in the bm mouse affects the functioning of the unique coupling mechanism between the two activities and causes a decrease in the ability of the enzyme to channel APS and so produce PAPS efficiently (Lyle et al., 1995). Taking the channeling data together with the kinetic analyses described above, interaction between the ATP-sulfurylase and APS-kinase activities is seen to be crucial to the function of the overall sulfate activation pathway (Schwartz et al., 1998).

STRUCTURE AND FUNCTION RELATIONSHIP OF PAPS SYNTHETASE

Early studies showed that both the sulfurylase and kinase activities from rat chondrosarcoma co-purified through gel filtration, ion exchange, and ATP affinity chromatography, which first suggested that the activities are inseparable (Geller et al., 1987). Subsequently, both activities co-purified through substrate affinity chromatography and reversed-phase chromatography, yielding a single band on denaturing polyacrylamide gels. The molecular mass of the native active unit containing both activities in the purified preparation corresponded to approximately 56 kDa by analytical gel filtration. Coincident binding and elution of ATP-sulfurylase and APS-kinase by immunoaffinity chromatography, using an antiserum generated against the 56-kDa protein, also demonstrated that the enzyme contains both activities. Finally, a single N-terminal amino acid sequence was obtained from the 56-kDa band isolated by gel electrophoresis. These biochemical results all suggest that ATP-sulfurylase and APS-kinase constitute a single bifunctional protein (Lyle et al., 1994c).

The existence of a fused mammalian sulfurylase/kinase was confirmed by cDNA cloning, sequencing, and expression of bifunctional enzymes from mouse and human (Li et al., 1995). Comparison against sequence databases revealed that the predicted amino acid sequences of these enzymes showed extensive homology to known separate sequences of both ATP-sulfurylase and APS-kinase from several sources. The first 199 amino acids corresponded to known APS-kinase sequences, followed by a tract of 37 residues not recognizable as homologous to either enzyme, and in turn by a 388 amino acid sequence highly homologous to known ATP-sulfurylase sequences. Recombinant enzyme expressed in COS-1 cells and *E. coli* exhibited

SK2 Gene Structure

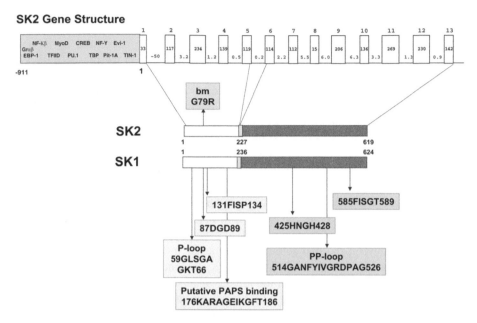

FIGURE 3.2 Schematic of SK2 Genomic Structure with Transcriptions Sites in 5'UTR and SK1 Protein with Conserved Functional Motifs.

both ATP-sulfurylase and APS-kinase activities, verifying our biochemical studies (Geller et al., 1987; Lyle et al., 1994c) that mouse ATP-sulfurylase and APS-kinase constitute a bifunctional enzyme.

Our cloning of the first PAPS synthetase (Li et al., 1995) was subsequently repeated for several organisms including Drosophila (Julien et al., 1997), thermophilic worm (Rosenthal and Leustek, 1995), and humans (Venkatachalam et al., 1998; Xu et al., 2000) and verified our early suggestion that PAPS synthetase is a single bifunctional polypeptide in higher animals (Lyle et al., 1994c). Curiously, the structural order of the fused functional domains (kinase–sulfurylase) is the reverse of the gene order in *E. coli* (Leyh et al., 1992) and *Rhizobium melioti* (Schwedock et al., 1994) cys operons, with the APS-kinase portion N terminus to the ATP-sulfurylase portion (Figure 3.2), suggesting an origin other than the simple fusion of adjacent genes. Moreover, fusion of the two activities into a single bifunctional polypeptide results in kinetic mechanisms distinct from those of the monofunctional sulfurylases or kinases. The three-dimensional structure of monofunctional APS-kinase from *Penicillium chrysogenum* (MacRae et al., 2000) has been reported and a plausible model proposed for the reaction mechanism in which the order of substrate addition and product release is entirely different from that established for the mammalian bifunctional PAPS synthetase APS-kinase activity (Lyle, 1994a, 1994b). Additionally, the structures of ATP-sulfurylase from *P. chrysogenum* (MacRae et al., 2001) and *Saccharomyces cerevisiae* (Ullrich et al., 2001) were recently published, providing insights into the molecular mechanisms of these monofunctional enzymes. This work provides a starting point for structural and functional studies on the more complex bifunctional mammalian PAPS synthetase, which performs PAPS synthesis in a concerted reaction system that synthesizes APS and channels it to the APS-kinase domain for final phosphorylation (Lyle, 1994d). To support and extend our previous biochemical and molecular studies, the mechanistic studies were continued and structural investigations initiated to elucidate the architecture of PAPS synthetase with the aims of identifying the active site residues and motifs involved in ATP sulfation, as well as of understanding the kinetic features of the obligatory ordered double reaction sequence and APS channeling.

ACTIVITY AND STABILITY OF REARRANGED AND MONOFUNCTIONAL DOMAINS

A comprehensive analysis of PAPS synthetase was conducted by synthesizing and expressing over 50 altered recombinant constructs with a strategy based on two general principles: to target putative functional groups and residues conserved over evolution. These were characterized with respect to ATP-sulfurylase and APS-kinase assays as well as overall PAPS synthesis from SO_4^{-2} and ATP. Possible advantages of bifunctionality were probed by separating the domains on the cDNA level and expressing them as monofunctional proteins. Expressed recombinant ATP-sulfurylase domain was active in both the forward and reverse sulfurylase assays. APS-kinase-only recombinants exhibited no kinase activity, while extension of the kinase domain at the C terminus by inclusion of the 37-residue linker region restored kinase activity. The importance of the domain order and organization was demonstrated by generating a series of rearranged recombinants in which the order of the two active domains was reversed or altered relative to the linker region. The critical role of the linker region was established by generating constructs that had the linker deleted or rearranged relative to the two active domains. The intrinsic stability of the various recombinants was also investigated by measuring enzyme deactivation as a function of time of incubation at 25°C or 37°C. Fusion of the two active domains was found not only to enhance the intrinsic stability of the sulfurylase, but also to promote the catalytic activity of the kinase and therefore the overall efficiency of PAPS synthesis (Deyrup, 1999a). Monofunctional derivatives of the human PAPS synthetase were also analyzed with somewhat different results that were not readily understood (Venkatachalam et al., 1998); possibly the important role of the linker region, i.e., as crucial to kinase activity and sulfurylase stability (Deyrup et al., 1999a), was not recognized.

STRUCTURE AND FUNCTION STUDIES OF THE APS-KINASE DOMAIN

The human and mouse PAPS synthetases contain a P-loop, a common sequence motif found in ATP- and GTP-binding proteins, at residues 59–66 in the APS-kinase portion of the bifunctional protein (Deyrup, 1998). Indeed, all known APS-kinases and PAPS synthetases possess this motif (Figure 3.2). Assays of the multiple functions of the enzyme were used to determine the effect of deleting or altering specific residues constituting this motif (Figure 3.2). In addition to the full-length cDNA construct (1MSK), two deletion mutants that progressively truncated the N terminus by 34 amino acids (2MSK) and 70 amino acids (3MSK) were designed to examine the effects of translation initiation before (2MSK) and after (3MSK) the P-loop codons. The 2MSK protein possessed sulfurylase and kinase activities equivalent to those of the full-length construct, but 3MSK protein exhibited no kinase activity and had reduced sulfurylase activity. In light of the evident importance of this motif, a number of site-directed mutants were designed to investigate the contribution of key residues. Mutation of a highly conserved P-loop lysine to alanine or arginine (K65A or K65R) or of the following threonine to alanine (T66A) ablated APS-kinase activity while leaving ATP-sulfurylase activity intact. Three mutations (G59A, G62A, and G64A) addressed the roles of the conserved P-loop glycines: G64A showed diminished APS-kinase activity only, G62A had no effect on either activity, and G59A caused a significant decrease in ATP-sulfurylase activity without effect on APS-kinase activity. A series of highly conserved flanking cysteines (C53A, C77A, and C83A) were mutated, but these changes did not result in decreased activity (Deyrup et al., 1998).

In systematically mutating every conserved P-loop residue, this study revealed which of the eight positions can tolerate mutations and the chemical nature of the amino acid side chain allowed. It also demonstrated an advantage to the fused PAPS synthetase of placing the kinase at the N terminus. This arrangement most likely allows the same flexible orientation of the P-loop motif seen in other ATP- and GTP-binding proteins in which this sequence is located near the N terminus where it forms a loop between a β-strand and α-helix (Hasemann et al., 1996; Stehle and Schulz, 1992). Had the sulfurylase and kinase been fused in the order found in the arrangement of the genomes of bacteria, the P-loop would be buried in the sequence and therefore more conformationally constrained.

FUNCTIONAL MOTIFS IN THE ATP-SULFURYLASE DOMAIN

The sulfurylase domains of the mouse and human bifunctional PAPS synthetases contain HxxH and PP-loop motifs that feature a number of highly conserved arginines and histidines (Figure 3.2). Of these conserved residues, H425, H428, or R421 are present within or near the HxxH motif, though the H506, R510, and R522 residues are present in and around the PP-loop. To elucidate the functional importance of these motifs, chemical modification and site-directed mutagenesis studies were performed. Modification of either arginines or histidines, with phenylglyoxal or diethyl pyrocarbonate, respectively, rendered the enzyme inactive in sulfurylase, kinase, and overall assays. Site-directed mutagenesis was used to replace individual conserved arginines and histidines with alanine or, in the case of R522, lysine. In the HxxH motif, none of the H425A, H428A, and R421A mutant proteins had sulfurylase or overall activity though they all exhibited normal kinase activity, suggesting that these residues are critical only for sulfurylase activity. Of mutations designed to probe the PP-loop requirements, R510A exhibited normal sulfurylase and kinase activity, R522A and R522K showed no sulfurylase activity, and H506A had normal sulfurylase activity but produced a severe loss of kinase activity. Mutation of the single aspartate that is part of the highly conserved GRD sequence of the PP-loop (D523A) affected both sulfurylase and kinase activities. This mutational analysis indicated that the HxxH motif plays a role only in the sulfurylase activity, whereas the PP-loop is involved in both sulfurylase and kinase activities, and identified sulfurylase domain residues specific for sulfurylase activity distinct from others involved in kinase activity (Deyrup et al., 1999b). Independently, a mutagenesis study of the HXGH motif in human PAPS synthetase, assessed for overall activity only, also implicated this motif in the sulfurylase reaction of PAPS synthetase (Venkatachalam et al., 1999). In sum, our series of studies (Deyrup et al., 1998, 1999a, 1999b) represents the most comprehensive analyses of the PAPS synthetase or monofunctional sulfurylases or kinases to date and supports the case made by the previous kinetic data for an interaction between the functional domains (Lyle et al., 1994c, 1994d).

MOLECULAR BIOLOGY AND REGULATION

PAPS Synthetase Gene Family

Our earlier cloning of the bifunctional murine sulfurylase-kinase (SK1; Li et al., 1995), which did not allow the identification the underlying etiology of the murine bm phenotype, together with the differential mapping of SK1 to mouse chromosome 3 and the bm mutation to chromosome 19 (Kurima et al., 1998; Lane and Dickie, 1968; Rusiniak et al., 1996) led us to consider the possibility that there might be more than one SK gene. A second SK isoform was detected following sequence similarity searching using murine SK1 as the query sequence, which yielded a group of expressed sequence flags (ESTs) with lower nucleotide sequence similarity but high predicted amino acid sequence identity, potentially representing novel SK isoforms. One of these ESTs was used to isolate the complete cDNA of a new isoform, referred to as SK2, from mouse liver total RNA by 5'-inverse PCR. SK2 contains an uninterrupted coding sequence of 1863 bps predicting a protein with 76% amino acid identity to SK1 and containing ATP-sulfurylase and APS-kinase domain functional motifs (Li et al., 1995). Northern-blot analysis revealed a single SK2 transcript of 3.7 kb in liver, cartilage, skin, and brain slightly larger than the 3.3 kb SK1 transcript. SK1 co-segregates with *D3Xrf10*, a locus on the distal end of chromosome 3, while SK2 was mapped to chromosome 19 using an intragenic single strand conformational polymorphism on an interspecific backcross mapping panel. Radiation hybrid mapping of the human homologues of SK1 and SK2 placed them on chromosomes 4q and 10q, respectively, consistent with the locations of other human genes whose mouse homologues are syntenic with the SK loci (Kurima et al., 1998). Human PAPS synthetase 1 and 2 (Figure 3.2) were subsequently cloned and characterized by others (Girard and Amalric, 1998; Venkatachalam et al., 1998; Xu et al., 2000).

Expression of PAPS Synthetase Isoforms

The physiological significance of two isoforms of PAPS synthetase remains unclear although they do appear to be expressed in a tissue and developmental-specific manner. Immunodetection of SK1 and SK2 was carried out on limb sections from normal C57 mice at postnatal day 10, and protein expression of both SK isoforms was present in the epiphyseal growth plates with some staining in muscle. Antibody reaction was observed in the resting, proliferative, and hypertrophic regions, with stronger staining by anti-SK2 than anti-SK1 in sections of all zones. In particular there was very little SK1 but significant SK2 staining in the hypertrophic zone, suggesting cell-subtype-specific expression. To determine whether the observed reduction in SK1 enzyme protein is due to reduced mRNA levels, expression was assessed by complementary *in situ* hybridization performed on consecutive limb sections hybridized with SK1 and SK2 probes at postnatal day 6 (P6). SK1 message was moderately and uniformly expressed throughout the growth plate. In contrast, SK2 was more strongly expressed in growth plate and predominantly localized to the hypertrophic chondrocytes and late proliferative chondrocytes (also called pre-hypertrophic chondrocytes) of the wild-type mouse, suggesting that the SK2 isoform is selectively expressed in these regions. The increased levels of SK2 may be due to a higher requirement for sulfation of extracellular matrix components (proteogly-cans) in prehypertrophic and hypertrophic chondrocytes. The expression patterns of SK1 and SK2 correlate well with the tissue-specific pattern of the bm mutation detected more than 20 years ago (Sugahara and Schwartz, 1979, 1982a, 1982b).

Another study localized PAPS synthetase 1 to the nucleus of mammalian and yeast cells, while PAPS synthetase 2 localized to the cytoplasm when ectopically expressed (Besset et al., 2000). However, when PAPS synthetase 2 was co-expressed with PAPS synthetase 1, it was relocated to the nucleus. Thus, the authors suggested that subcellular localization of PAPS synthetases may provide an additional level of regulation of the sulfation pathway. Comparative mRNA expression of human PAPS synthetase isoforms by reverse transcription-PCR was recently reported, showing wide-spread expression of PAPS synthetase 1 compared to PAPS synthetase 2. Interestingly, in that study a higher catalytic efficiency was observed for PAPS synthetase 2 (Fuda et al., 2002). Thus, differential mRNA, protein, and activity patterns of the two isoforms may contribute to modulation of PAPS production during development.

SK GENOMIC STRUCTURE AND REGULATORY REGION

As mentioned previously, in contrast to the mammalian PAPS synthetase, in plants and simpler organisms the ATP-sulfurylase and APS-kinase reactions are catalyzed by separate enzymes encoded by two or even (in bacteria) three genes, suggesting that a fusion of separate genes during the course of evolution might have generated the bifunctional enzyme. At some point in animal evolution, the PAPS synthetase gene apparently was duplicated, giving rise to the SK1 and SK2 isoforms seen in mouse and humans. We have characterized the genomic structure of the PAPS synthetase SK2 isoform genes from mouse (MSK2) and human (HSK2) and ana-lyzed the possible fusion region and the 5'-flanking regions (Kurima et al., 1999). The MSK2 and HSK2 genes exhibit a common structure of 13 exons, with intron/exon boundaries, sizes of exons, and the phase of splicing conserved between the two mammalian species (Figure 3.2).

We also identified two splicing variants in mouse and human PAPS synthetase 2 cDNAs; the alternatively spliced 15nt was in exon 8 in both genomes, leading to the deletion of the peptide sequences GVVPR and GMACP from the ATP-sulfurylase domains of mouse and human PAPS synthesases, respectively (Kurima et al, 1999). A similar 15nt/5 amino acid insertion is found in the *Drosophila melanogaster* and an even larger insertion in the *Caenorhabditis elegans* PAPS synthetases at the same location. Consistent occurrence of a splice variant at this position among the different species implies some functional significance; however, we found no differences in the abilities of the splice variants to catalyze ATP-sulfurylase, APS-kinase, or PAPS synthetase overall reactions. Therefore, the insertions appear not to influence enzyme function directly. However PCR analysis showed that the deletion splice variant is the major form in cartilage, while the originally described PAPS synthetase 2 is the major form in liver of both mouse and human. The potential importance of this splice variant was recently reinvestigated by Xu et al. (2002) and Fuda et al. (2002), who provided some additional information on subtype distribution among other tissues. Thus, the physiological significance of the existence of PAPS synthetase subtypes still remains an enigma.

In additional studies, enzyme activities of several bacterially expressed exon assemblages showed exons 1 through 6 encode APS-kinase, while exons 6 through 13 encode ATP-sulfurylase. Interestingly, the MSK2 construct designed to exclude the exon 6-encoded peptide showed no kinase or sulfurylase activity, demonstrating that exon 6 encodes sequences required for both activities. Exon 1 and its 5'-flanking sequence were highly divergent between the two species, and intron 1 of the HSK2 gene contains a region similar to the MSK2 promoter sequence, suggesting that it may be the remnant of a now-superceded regulatory region. The HSK2 promoter contained a GC-rich region not present in the mouse promoter and had few transcription factor binding sites in common with MSK2. Analysis of 1 kb of the mouse PAPS synthetase 2 gene immediately flanking the translation start site indicated potential binding motifs for a variety of transcription factors, including response elements for progesterone receptor, EBP-1, NF-$_k$B, MyoD, PU.1, CREB, Pit-la, CCAAT-binding protein, NF-Y, Evi-1, TIN-1, and Grαβ (Figure 3.2). With regard to the human PAPS synthetase gene, the proximal promoter region (2000 bp) has a subregion with high content of G + C (73% in the 500-bp proximal region) and at least nine potential Sp-1 binding sites, present in many constitutive gene promoters, and sites for NFE2, LyF-1, MEF-2, Pit-1a, AP-1, TFE3-S, USF, CAC-binding protein, NFAT-1, and CCAAT-binding protein. Only a few detected motifs (Pit-1a, CCAAT-binding protein) were common to both murine and human PAPS synthetase (Kurima et al., 1999). The differences between the two species' promoter regions suggest that species-specific mechanisms regulate expression of the SK2 isoform (Kurima et al., 1999) and highlight the necessity for elucidating how the expression of PAPS synthetase isoforms is regulated at the transcriptional level in order to understand how PAPS concentrations are so efficiently and exquisitely controlled. More recently, additional descriptions of the human PAPS synthetase 1 (Shimizu et al., 2001) and 2 (Shimizu et al., 2002) gene promoter regions have appeared.

PAPS SYNTHETASE DISORDERS

The dependence of cartilage on a single PAPS synthetase isoform for production of PAPS has recently been shown in an animal model, the bm mouse, and in a human disorder, spondyloepimetaphyseal dysplasia. bm mice are characterized by dome-shaped skulls, short thick tails, and shortened, but not widened, limbs (Lane and Dickie, 1968). They breed and have a lifespan comparable to that of normal mice. Homozygous mutant and normal neonates are indistinguishable at birth, but the size differential becomes more apparent over the first few weeks of growth and results in a 25% reduction in axial skeleton size and 50% reduction in limb bone length (Schwartz and Domowicz, 1998, 2002). Histological and ultrastructural studies on homozygous bm mice suggest that a defect in cartilage matrix is associated with their shortened limbs (Orkin et al., 1977). The growth plate of homozygous bm neonates is organized normally and maintains the proper alignment of chondrocytes and appropriate demarcation of zones as growth proceeds, while the difference in sizes of the proliferative and hypertrophic zones in normal and bm neonates becomes more pronounced over the first month of life and is reflected in an overall reduction in skeletal dimensions (Schwartz and Domowicz, 1998). The mutant cartilage matrix contains normal cartilage fibrils, but the proteoglycan aggregate granules are smaller than normal and are present in reduced numbers, particularly in the columnar hypertrophic zones of the growth plate. Initial biochemical analysis demonstrated that bm cartilage contained normal levels of glycosaminoglycans (Orkin et al., 1976); however, the incorporation of sulfate was lower in mutant than in normal cartilage and considerable amounts of undersulfated glycosaminoglycan disaccharides were obtained only from mutant cartilage (Schwartz et al., 1978; Sugahara and Schwartz, 1979). The availability of PAPS was limiting and the specific activities of the ATP-sulfurylase and APS-kinase functions were reduced 50% and 85%, respectively, in the bm mouse (Sugahara and Schwartz, 1982b).

Since chromosome mapping showed that SK2 is tightly linked to *D19Mit13* (Kurima et al., 1998), as is the bm locus (Rusiniak et al., 1996), SK2 was tested as a candidate brachymorphism gene. Sequence analysis of wild-type and bm SK2 cDNA and genomic DNA revealed a G to A transition in the bm allele that changes a glycine residue to an arginine residue (G79A) in the APS-kinase domain of the bifunctional enzyme. This glycine is conserved in all previously characterized bifunctional SK proteins and in all monofunctional APS-kinases (Li et al., 1995). The identification of this mutation has allowed the characterization of a novel bm motif in all PAPS synthetases that function in endogenous APS transfer (Singh and Schwartz, 2002). Further support for this mutation being the cause of the bm phenotype derives from *in vitro* assays of wild-type SK2, which catalyzed both ATP-sulfurylase and APS-kinase reactions comparably to SK1. In contrast, recombinantly expressed SK2 protein containing the G79A mutation has no APS-kinase activity, suggesting that the naturally occurring mutation is responsible for murine brachy-morphism (Kurima et al., 1998). A mutation in SK2 gene was subsequently found to be associated with human spondyloepimetaphyseal dysplasia (Pakistani type), characterized by short and bowed lower limbs, enlarged knee joints, and early onset of degenerative joint disease in the hands and knees (Ahmad et al., 1998).

Since PAPS is the universal sulfate donor for all naturally occurring sulfated compounds, it was somewhat surprising that the defect in PAPS synthesis in the bm mouse did not result in a more severe phenotype. Of those tissues where a significant demand for PAPS might be predicted — i.e., liver, which is rich in heparan sulfate and uses sulfoconjugation for detoxification; skin, which is rich in dermatan sulfate; or kidney and brain, which have high concentrations of sulfatides — only liver exhibited a reduction in PAPS synthesis (Sugahara and Schwartz, 1982b) similar to cartilage. Tissue-specific isoform expression analyses on either the protein or mRNA level showed tissue distribution that mirrored the bm defect, e.g., brain and skin exhibited only SK1, while SK2 was higher in cartilage and liver. The high demand for PAPS in cartilage thus results in the predominant manifestation of the PAPS-synthesis defect as a skeletal disorder.

It is intriguing that the SK2 mutation specifically affects postnatal development of skeletal and other selected tissues. Newborn bm mice appear normal, suggesting that the SK2 gene product's principal role during skeletogenesis is during postnatal growth and that other sulfurylase/kinases may substitute for SK2 *in utero*. It has been shown that hepatic detoxification in bm mice is principally via glucuronidation, while it is via sulfation in wild-type mice (Schwartz, 1983; Sugahara and Schwartz, 1982a). Bleeding times are also abnormal in bm mice (Rusiniak et al., 1996). However, other tissues that contain important sulfated biomolecules, such as the kidney and brain, appear normal in bm mice, suggesting tissue-specific PAPS segregation mechanisms probably due to the predominant expression of SK1. As mentioned previously, the resulting progressive reduction in sulfated proteoglycans in the bm extracellular matrix (ECM) occurs concomitantly with a reduction in the size of this region of the growth plate and leads to the overall bm phenotype (i.e., growth reduction). Since SK1 appears not to be produced at appreciable levels by hypertrophic chondrocytes, SK1 is not able to rescue or compensate for loss of function by mutant SK2 in bm mice, and cartilage is predominantly dependent on the SK2 isoform for its sulfate utilization.

Ultimately, the contribution of each SK isoform during growth and development must await the identification of all family members and careful quantitation of their mRNA and protein levels. However, it is now clear that the PAPS synthetases constitute a gene family, with known members lacking complete functional redundancy at the tissue level and with perhaps more members yet to be discovered. Furthermore, the PAPS synthetase genes appear to make different contributions over various periods of development and growth and are expressed predominantly in tissue-specific patterns, as appears to be the case for sulfate transporters (Satoh et al., 1998). While the key steps governing biomolecular sulfation have been assumed to be at the level of the myriad of sulfotransferases, recent findings on the PAPS synthetase family demonstrate that regulation at the level of forming the sulfate donor is also important, suggesting the possibility of regulatory mechanisms at other steps along the overall sulfation pathway as well (Schwartz et al., 1998; Schwartz, 2002).

PAPS TRANSLOCASE

A major segment of the sulfate uptake, activation, and utilization pathway occurs in the Golgi, which requires transport of the product of the activation process, PAPS, from the site of synthesis in the cytosol to the site of utilization in the Golgi lumen (Figure 3.1). Therefore, we have sought to identify the PAPS Golgi-membrane receptor (translocase) that transports PAPS through the membrane, thus making it available for the multitude of intralumenal sulfotransferases.

In early studies, we studied PAPS transport in a partially purified system, a reconstituted system, and in an intact-cell system in which we could access the cytosol in both rat liver and rat chondrosarcoma (Kearns et al., 1993; Vertel et al., 1993). To further these studies, the stable β-methylene analog of PAPS was used in a number of ways, i.e., as an inhibitor in kinetic characterization studies, as a ligand for affinity chromatography, and, following labeling with ^{32}P in the 3' position, as an affinity label for identification. ^{35}S-PAPS is transported across intact Golgi vesicle membranes in a time- and concentration-dependent manner, which is saturable, proportional to protein concentration, and sensitive to inhibition by other nucleotides and by inhibitors of nucleotide transport typical of protein-mediated transport systems.

We identified the putative Golgi membrane PAPS translocase from rat liver by chromatographic methods and affinity labeling. The translocase is a ~230-kDa integral membrane protein that can be labeled with 3'-[^{32}P] β-methylene PAPS by photoaffinity crosslinkers in intact membranes. In solubilized membranes, additional proteins, ~97 kDa, corresponding to the N-heparan sulfate sulfotransferase, and ~30 kDa and 50 kDa, corresponding to some of the arylsulfotransferases and steroid sulfotransferases, are also labeled. A major advance in understanding the transport of PAPS between sites of its synthesis and utilization was made by solubilizing and then reconstituting the PAPS translocase activity, since transport is a phenomenon that can be measured only in a system made up of sealed vesicles with incorporated protein (Ozeran et al., 1996b). When PAPS translocase activity was characterized in the reconstituted system and compared to the original Golgi-enriched membrane vesicles, the apparent K_m and V_{max} were comparable, 0.8 mM and 10.7 pmol/min/mg protein, respectively. The competitive nature of inhibition by several purine nucleotides and the relative affinities of the PAPS translocase for each of the inhibitors suggest that PAPS binding is selective for adenine nucleotides containing 3'-phosphate groups although specificity is limited with regard to 5' substitutions on the substrate. Our more rigorous criteria for the identification and characterization of PAPS translocase from Golgi membrane support the conclusion that PAPS translocase is a ~230 kDa Golgi membrane protein (Ozeran et al., 1996a).

Identifying the major components of the sulfation process will allow a more comprehensive understanding of the integrated pathway and will suggest candidate genes and gene products that may produce human genetic disorders of sulfate metabolism. To distinguish between the several possible molecular components in which defects may result in nearly identical phenotypes, all of the components of the sulfate metabolic pathway must be elucidated. The human and animal models with mutations already identified clearly highlight the importance of biosulfation and the multiplicity of genes that might affect the sulfate uptake, activation, and utilization pathway (Schwartz and Domowicz, 2002).

ACKNOWLEDGMENTS

Supported by NIH grants HD–17332 and HD–09402. Expert secretarial assistance was provided by Glenn Burrell. Appreciation is expressed to all the members of the Schwartz Lab who contributed to these studies over the past 20 years.

REFERENCES

Ahmad, M., Haque, M.F., Ahmad, W., Abbas, H., Haque, S., Krakow, D., Rimoin, D.L., Lachman, R.S., and Cohn, D.H., 1998, Distinct, autosomal recessive form of spondyloepimetaphyseal dysplasia segregating in an inbred Pakistani kindred, *Am. J. Med. Genet.*, 78, 468–473.

Besset, S., Vincourt, J.B., Amalric, F., and Girard, J.P., 2000, Nuclear localization of PAPS synthetase 1: a sulfate activation pathway in the nucleus of eukaryotic cells, *Faseb J.*, 14, 345–354.

Bissig, M., Hagenbuch, B., Stieger, B., Koller, T., and Meier, P.J., 1994, Functional expression cloning of the canalicular sulfate transport system of rat hepatocytes, *J. Biol. Chem.*, 28, 3017–3021.

Bowman, K.G. and Bertozzi, C.R., 1999, Carbohydrate sulfotransferases: mediators of extracellular communication, *Chem. Biol.*, 6, R9–R22.

Callahan, L., Ng, K., Geller, D.H., Agarwal, K., and Schwartz, N.B., 1989, Synthesis and properties of a nonhydrolyzable adenosine phosphosulfate analog, *Anal. Biochem.*, 177, 67–71.

Deyrup, A.T., Krishnan, S., Cockburn, B.N., and Schwartz, N.B., 1998, Deletion and site-directed mutagenesis of the ATP-binding motif (P-loop) in the bifunctional murine ATP-sulfurylase/adenosine 5′-phosphosulfate kinase enzyme, *J. Biol. Chem.*, 273, 9450–9456.

Deyrup, A.T., Krishnan, S., Singh, B., and Schwartz, N.B., 1999a, Activity and stability of recombinant bifunctional rearranged and monofunctional domains of ATP-sulfurylase and adenosine 5′-phosphosulfate kinase, *J. Biol. Chem.*, 274, 10751–10757.

Deyrup, A.T., Singh, B., Krishnan, S., Lyle, S., and Schwartz, N.B., 1999b, Chemical modification and site-directed mutagenesis of conserved HXXH and PP-loop motif arginines and histidines in the murine bifunctional ATP-sulfurylase/adenosine 5′-phosphosulfate kinase, *J. Biol. Chem.*, 274, 28929–28936.

Fuda, H., Shimizu, C., Lee, Y.C., Akita, H., and Strott, C.A., 2002, Characterization and expression of human bifunctional 3′-phosphoadenosine 5′-phosphosulphate synthase isoforms, *Biochem. J.*, 365, 497–504.

Geller, D., Henry, J.G., Belch, J., and Schwartz, N.B., 1987, Co-purification and characterization of ATP-sulfurylase and adenosine-5′-phosphosulfate kinase from rat chondrosarcoma, *J. Biol. Chem.*, 262, 7374–7382.

Girard, J.P. and Amalric, F., 1998, Biosynthesis of sulfated L-selectin ligands in human high endothelial venules (HEV), *Adv. Exp. Med. Biol.*, 435, 55–62.

Hasemann, C.A., Istvan, E.S., Uyeda, K., and Deisenhofer, J., 1996, The crystal structure of the bifunctional enzyme 6-phosphofructo-2-kinase/fructose-2,6-bisphosphatase reveals distinct domain homologies, *Structure*, 4, 1017–1029.

Hastbacka, J., Chapelle, A.D.L., Mahtani, M.M., Clines, G., Reeve-Daly, M.P., Hamilton, B.A., Kusumi, K., Trivedi, B., Weaver, A., Coloma, A., Lovett, M., Buckler, A., Kaitila, I., and Lander, E.S., 1994, The diastrophic dysplasia gene encodes a novel sulfate transporter: positional cloning by fine-structure linkage disequilibrium mapping, *Cell*, 78, 1073–1078.

Huxtable, R.J., 1986, *Biochemistry of Sulfur*, Plenum Publishing, New York.

Julien, D., Crozatier, M., and Kas, E., 1997, cDNA sequence and expression pattern of the *Drosophila melanogaster* PAPS synthetase gene: a new salivary gland marker, *Mech. Dev.*, 68, 179–186.

Kearns, A.E., Vertel, B.M., and Schwartz, N.B., 1993, Topography of glycosylation and UDP-xylose production, *J. Biol. Chem.*, 268, 11097–11104.

Kehoe, J.W. and Bertozzi, C.R., 2000, Tyrosine sulfation: a modulator of extracellular protein-protein interactions, *Chem. Biol.*, 7, R57–R61.

Klaassen, C.D. and Boles, J.W., 1997, Sulfation and sulfotransferases 5: the importance of 3'- phosphoadenosine 5'-phosphosulfate (PAPS) in the regulation of sulfation, *Faseb J.*, 11, 404–418.

Kurima, K., Warman, M.L., Krishnan, S., Domowicz, M., Krueger, R.C., Jr., Deyrup, A., and Schwartz, N.B., 1998, A member of a family of sulfate-activating enzymes causes murine brachymorphism, *Proc. Natl. Acad. Sci. U.S.A.*, 95, 8681–8685.

Kurima, K., Singh, B., and Schwartz, N.B., 1999, Genomic organization of the mouse and human genes encoding the ATP-sulfurylase/adenosine 5'-phosphosulfate kinase isoform SK2, *J. Biol. Chem.*, 274, 33306–33312.

Lane, P.W. and Dickie, M.M., 1968, Three recessive mutations producing disproportionate dwarfing in mice, *J. Hered.*, 65, 297–300.

Leyh, T.S., Vogt, T.F., and Suo, Y., 1992, The DNA sequence of the sulfate activation locus from *E. coli* K-12, *J. Biol. Chem.*, 267, 10405–10410.

Li, H., Deyrup, A., Mensch, J., Domowicz, M., Konstantinidis, A., and Schwartz, N.B., 1995, The isolation and characterization of cDNA encoding the mouse bifunctional ATP-sulfurylase -adenosine 5'-phosphosulfate kinase, *J. Biol. Chem.*, 270, 29453–29459.

Lipmann, F., 1958, Biological sulfate activation and transfer, *Science*, 128, 575–580.

Lyle, S., Geller, D.H., Ng, K., Westley, J., and Schwartz, N.B., 1994a, Kinetic mechanism of ATP-sulfurylase from rat chondrosarcoma, *Biochem. J.*, 301, 349–354.

Lyle, S., Geller, D.H., Ng, K., Stanzak, J., Westley, J., and Schwartz, N.B., 1994b, Kinetic mechanism of adenosine 5'-phosphosulfate kinase from rat chondrosarcoma, *Biochem. J.*, 301, 355–359.

Lyle, S., Stanzack, J., Ng, K., and Schwartz, N.B., 1994c, Rat chondrosarcoma ATP-sulfurylase and adenosine 5'-phosphosulfate kinase reside on a single bifunctional protein, *Biochemistry*, 33, 5920–5925.

Lyle, S., Ozeran, J.D., Stanzak, J., Westley, J., and Schwartz, N.B., 1994d, Intermediate channeling between ATP-sulfurylase and adenosine 5'-phosphosulfate kinase from rat chondrosarcoma, *Biochemistry*, 33, 6822–6827.

Lyle, S., Stanzak, J., Westley, J., and Schwartz, N.B., 1995, Sulfate activating enzymes in normal and brachymorphic mice: evidence for a channeling defect, *Biochemistry*, 34, 940–945.

MacRae, I.J., Segel, I.H., and Fisher, A.J., 2000, Crystal structure of adenosine 5'-phosphosulfate kinase from Penicillium chrysogenum, *Biochemistry*, 39, 1613–1621.

MacRae, I.J., Segel, I.H., and Fisher, A.J., 2001, Crystal structure of ATP-sulfurylase from Penicillium chrysogenum: insights into the allosteric regulation of sulfate assimilation, *Biochemistry*, 40, 6795–6804.

Markovich, D., Forgo, J., Stange, G., Biber, J., and Murer, H., 1993, Expression cloning of rat renal Na$^+$/SO$_4$$^{(2-)}$ cotransport, *Proc. Natl. Acad. Sci. U.S.A.*, 90, 8073–8077.

Ng, K., D'souza, M., Callahan, L., Geller, D.H., Kearns, A.E., Lyle, S., and Schwartz, N.B., 1991, Synthesis and utilization of a nonhydrolyzable phosphoadenosine phosphosulfate analog, *Anal. Biochem.*, 198, 60–67.

Orkin, R.W., Pratt, R.M., and Martin, G.R., 1976, Undersulfated chondroitin sulfate in the cartilage matrix of brachymorphic mice, *Dev. Biol.*, 50, 82–94.

Orkin, R.W., Williams, B.R., Cranley, R.E., Poppke, D.C., and Brown, K.S., 1977, Defects in the cartilaginous growth plates of brachymorphic mice, *J. Cell Biol.*, 73, 287–299.

Ozeran, J.D., Westley, J., and Schwartz, N., 1996a, Kinetics of PAPS translocase: evidence for an antiport mechanism, *Biochemistry*, 35, 3685–3694.

Ozeran, J.D., Westley, J., and Schwartz, N.B., 1996b, Identification and partial purification of PAPS translocase, *Biochemistry*, 35, 3695–3703.

Rosenthal, E. and Leustek, T., 1995, A multifunctional *Urechis caupo* protein, PAPS synthetase, has both ATP-sulfurylase and APS-kinase activities, *Gene*, 165, 243–248.

Rusiniak, M.E., O'Brien, E.P., Novak, E.K., Barone, S.M., McGarry, M.P., Reddington, M., and Swank, R.T., 1996, Molecular markers near the mouse brachymorphic (bm) gene, which affects connective tissues and bleeding time, *Mamm. Genome*, 7, 98–102.

Satoh, H., Susaki, M., Shukunami, C., Iyama, K., Negoro, T., and Hiraki, Y., 1998, Functional analysis of diastrophic dysplasia sulfate transporter: its involvement in growth regulation of chondrocytes mediated by sulfated proteoglycans, *J. Biol. Chem.*, 273, 12307–12315.

Schwartz, N.B., Ostrowski, V., Brown, K.S., and Pratt, R., 1978, Defective PAPS synthesis on epiphyseal cartilage from brachymorphic mice, *Biochem. Biophys. Res. Commun.*, 82, 173–178.

Schwartz, N.B., 1983, Defect in proteoglycan synthesis in brachymorphic mice, in *Limb Development and Regeneration*, Fallon, J.F. and Caplan, A.I., Eds., A.R. Liss, New York, pp. 97–103.

Schwartz, N.B., Lyle, S., Ozeran, J.D., Li, H., Deyrup, A., Ng, K., and Westley, J., 1998, Sulfate activation and transport in mammals: system components and mechanisms, *Chemico-Biol. Interactions*, 109, 143–151.

Schwartz, N.B. and Domowicz, M., 1998, Proteoglycan gene mutations and impaired skeletal development, in *Skeletal Growth and Development*, Buckwalter, J.A., Ehrlich, M.G., Sandell, L.J., and Trippel, S.B., Eds., American Association of Orthopedic Surgeon Publications, Rosemont, IL, pp. 413–433.

Schwartz, N.B. and Domowicz, M., 2002, Chondrodysplasias due to proteoglycan defects, *Glycobiology*, 12, 57R–68R.

Schwartz, N.B., 2002, PAPS synthetase, in *Encyclopedia of Molecular Medicine*, J. Wiley and Sons, New York, pp. 284–287.

Schwedock, J.S., Liu, C., Leyh, T.S., and Long, S.R., 1994, *Rhizobium meliloti* NodP and NodQ form a multifunctional sulfate-activating complex requiring GTP for activity, *J. Bacteriol.*, 176, 7055–7064.

Shimizu, C., Fuda, H., Lee, Y.C., and Strott, C.A., 2001, Transcriptional regulation of human 3'-phosphoadenosine 5'- phosphosulfate synthase 1, *Biochem. Biophys. Res. Commun.*, 284, 763–770.

Shimizu, C., Fuda, H., Lee, Y.C., and Strott, C.A., 2002, Transcriptional regulation of human 3'-phosphoadenosine 5'- phosphosulphate synthase 2, *Biochem. J.*, 363, 263–271.

Silberg, D.G., Wang, W., Moseley, R.H., and Traber, P.G., 1995, The Down regulated in Adenoma (dra) gene encodes an intestine-specific membrane sulfate transport protein, *J. Biol. Chem.*, 270, 11897–11902.

Singh, B. and Schwartz, N.B., 2002, Identification and functional characterization of the novel BM motif in the murine PAPS synthetase, *J. Biol. Chem.*, 31, 71–75.

Stehle, T. and Schulz, G.E., 1992, Refined structure of the complex between guanylate kinase and its substrate GMP at 2.0 A resolution, *J. Mol. Biol.*, 224, 1127–1141.

Sugahara, K. and Schwartz, N.B., 1979, Defect in 3'-phosphoadenosine 5'-phosphosulfate formation in brachymorphic mice, *Proc. Natl. Acad. Sci. U.S.A.*, 76, 6615–6618.

Sugahara, K. and Schwartz, N.B., 1982a, Tissue distribution of defective PAPS synthesis and decreased sulfoconjugation of a phenolic compound in brachymorphic mice, in *Proceedings of the 6th International Symposium on Glycoconjugate*, Yamakawa, T., Oswa, T., and Hauda, S., Eds., Sci. Soc. Press, Tokyo, pp. 493–495.

Sugahara, K. and Schwartz, N.B., 1982b, Defect in 3'-phosphoadenosine 5'-phosphosulfate synthesis in brachymorphic mice. I. Characterization of the defect, *Arch. Biochem. Biophys.*, 214, 589–601.

Superti-Furga, A., Hastbacka, J., Wilcox, W.R., Cohn, D.H., van der Harten, H.J., Rossi, A., Blau, N., Rimoin, D.L., Steinmann, B., Lander, E.S., and Gitzelmann, R., 1996, Achondrogenesis type IB is caused by mutations in the diastrophic dysplasia sulfate transporter gene, *Nat. Genet.*, 12, 100–102.

Ullrich, T.C., Blaesse, M., and Huber, R., 2001, Crystal structure of ATP-sulfurylase from *Saccharomyces cerevisiae*, a key enzyme in sulfate activation, *Embo. J.*, 20, 316–329.

Venkatachalam, K.V., Akita, H., and Strott, C.A., 1998, Molecular cloning, expression, and characterization of human bifunctional 3'-phosphoadenosine 5'-phosphosulfate synthase and its functional domains, *J. Biol. Chem.*, 273, 19311–19320.

Venkatachalam, K.V., Fuda, H., Koonin, E.V., and Strott, C.A., 1999, Site-selected mutagenesis of a conserved nucleotide binding HXGH motif located in the ATP-sulfurylase domain of human bifunctional 3'- phosphoadenosine 5'-phosphosulfate synthase, *J. Biol. Chem.*, 274, 2601–2604.

Vertel, B.M., Walters, L.M., Flay, N., Kearns, A.E., and Schwartz, N.B., 1993, Xylosylation is an endoplasmic reticulum to Golgi event, *J. Biol. Chem.*, 268, 11105–11112.

Xu, Z.H., Otterness, D.M., Freimuth, R.R., Carlini, E.J., Wood, T.C., Mitchell, S., Moon, E., Kim, U.J., Xu, J.P., Siciliano, M.J., and Weinshilboum, R.M., 2000, Human 3'-phosphoadenosine 5'-phosphosulfate synthetase 1 (PAPSS1) and PAPSS2: gene cloning, characterization and chromosomal localization, *Biochem. Biophys. Res. Commun.*, 268, 437–444.

Xu, Z.H., Freimuth, R.R., Eckloff, B., Wieben, E., and Weinshilboum, R.M., 2002, Human 3'-phosphoadenosine 5'-phosphosulfate synthetase 2 (PAPSS2) pharmacogenetics: gene resequencing, genetic polymorphisms and functional characterization of variant allozymes, *Pharmacogenetics*, 12, 11–21.

4 Sulfate Conjugation: Pharmacogenetics and Pharmacogenomics

Richard M. Weinshilboum, M.D. and
Araba A. Adjei, Ph.D.

CONTENTS

INTRODUCTION

Pharmacogenetics is the study of the role of inheritance in individual variation in drug response — variation in either efficacy or toxicity (Weinshilboum, 2003). Pharmacogenomics, as used here, is the convergence of pharmacogenetics with the rapid advances that have occurred in human genomics — as illustrated in this chapter for sulfate conjugation. Many examples of functionally and clinically significant pharmacogenetic effects result from inherited variation in drug metabolism (Weinshilboum, 2003). Sulfation, or more accurately, sulfonation, is an important pathway in the biotransformation of a large number of drugs (Falany, 1997; Weinshilboum and Otterness, 1994). The sulfotransferase (SULT) enzymes that catalyze the sulfate conjugation of drugs also catalyze the sulfation of other xenobiotics and of endogenous compounds such as steroid

61

hormones and neurotransmitters (Falany, 1997; Weinshilboum and Otterness, 1994). Therefore, inherited variation in sulfate conjugation could potentially influence not just drug metabolism but also the biotransformation of endogenous compounds such as steroid hormones and, as a result, might contribute to risk for disease.

Pharmacogenetic studies of sulfate conjugation date back at least two decades. That research began with biochemical genetic experiments designed to determine the possible contribution of inheritance to individual differences in level of enzyme activity or other enzyme properties (Price et al., 1988, 1989; Reveley et al., 1982/1983; Van Loon and Weinshilboum, 1984). That pharmacogenetic research strategy then moved beyond the demonstration of genetically determined variation in phenotype to include attempts to understand the molecular basis for that variation at the level of the genome. As our knowledge of the structures and DNA sequences of genes encoding proteins involved in sulfate conjugation grew, this "phenotype-to-genotype" research strategy was complemented by the application of a "genotype-to-phenotype" strategy. That approach began with knowledge of gene structure and sequence, followed by a determination of common variation in gene sequence — most often by "resequencing" the gene of interest using DNA from a large number of subjects. Often different ethnic groups were studied because allele types and frequencies can differ greatly on the basis of ethnicity. This genotype-to-phenotype strategy then progressively moved back from genotype to phenotype — either cellular phenotype or phenotype at the level of the whole organism. We now know the sequences and structures of all, or virtually all, human cytosolic SULTs. In addition, the genes that encode the enzymes that catalyze synthesis of the sulfate donor for these reactions, 3′-phosphoadenosine 5′-phosphosulfate (PAPS), have also been cloned and characterized (Kurima et al., 1999; Xu et al., 2000). As a result, we now know of common genetic polymorphisms that can influence the function of PAPS synthetase 1 and 2 (PAPSS1 and PAPSS2), the two PAPS synthesizing enzymes in humans as well as many of the human SULT genes.

The dendrogram shown in Figure 4.1 demonstrates that, since the cloning and characterization of the first human SULT cDNA just over a decade ago (Otterness et al., 1992), there has been a dramatic increase in our understanding of the genes encoding these cDNAs. Many of the human SULT genes are organized in clusters (Figure 4.1). For example, there is a SULT1A cluster on the short arm of chromosome 16 that includes SULT1A1, 1A2, and 1A3, a SULT1C cluster on the long arm of chromosome 2, a cluster on the long arm of chromosome 4 that includes SULT1B1 and SULT1E1, and a SULT2 cluster on the long arm of chromosome 19 (Freimuth et al., 2004). A recent human genomic database mining exercise resulted in the identification of SULT6B1 as well as a series of human SULT pseudogenes — both processed pseudogenes and pseudogenes with an exon–intron structure (Freimuth et al., 2004). Data with respect to human SULT gene structures, plus the character-ization of PAPSS1 on the long arm of human chromosome 4 and PAPSS2 on the long arm of human chromosome 10 (Xu et al., 2000), have made it possible to rapidly characterize common sequence variation in these genes — followed by a determination of the potential functional implications of that variation. In fact, detailed gene resequencing data for all of the SULT genes listed in Figure 4.1 can

Human SULT Isoforms

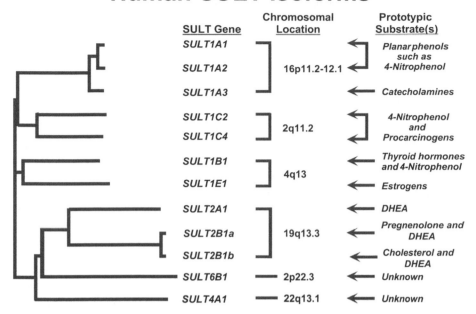

SULT Gene	Chromosomal Location	Prototypic Substrate(s)
SULT1A1		Planar phenols such as
SULT1A2	16p11.2-12.1	4-Nitrophenol
SULT1A3		Catecholamines
SULT1C2		4-Nitrophenol and
SULT1C4	2q11.2	Procarcinogens
SULT1B1		Thyroid hormones and 4-Nitrophenol
SULT1E1	4q13	Estrogens
SULT2A1		DHEA
SULT2B1a	19q13.3	Pregnenolone and DHEA
SULT2B1b		Cholesterol and DHEA
SULT6B1	2p22.3	Unknown
SULT4A1	22q13.1	Unknown

FIGURE 4.1 Human SULT dendrogram. The dendrogram depicts relationships among the amino acid sequences of human cytosolic SULTs that are encoded by the genes listed. Chromosomal locations of genes as well as prototypic substrates for each isoform are also listed.

be obtained at the United States National Institutes of Health-funded public website www.PharmGKB.org.

This chapter will briefly review our current understanding of the nature, extent, and functional significance of variation in the genes encoding human SULTs as well as PAPSS1 and PAPSS2. The focus will be on human data although a similar degree of genetic variation can be anticipated in experimental, domestic, and companion animals, and veterinary pharmacogenetic data of that type have recently begun to appear (Salavaggione et al., 2002, 2004). In the course of this brief review, sulfation pharmacogenetic studies that began with phenotype and moved to genotype as well as more recent experiments that have used a genotype-to-phenotype research strategy will be described. In addition, underlying mechanisms responsible for the functional effects of common polymorphisms in genes that influence sulfate conjugation will be described when that information is available. The material included in this chapter will move progressively through the SULTs, beginning with SULT1A1 (see Figure 4.1). Common genetic variation in PAPSS1 and PAPSS2 will also be described. Although polymorphisms that alter the encoded amino acid sequences of SULT and PAPSS isoforms will be emphasized, that is only one of many ways in which genetic polymorphisms might alter function. Finally, pharmacogenetic information with regard to sulfation is being translated rapidly into increased understanding of the possible contribution of inheritance to individual variation in drug metabolism, drug effect, and disease pathophysiology. Of equal importance, basic mechanisms by which genetic polymorphisms influence function — e.g., decreased levels of enzyme protein in response to inherited alteration of only a single amino acid — have also begun to be understood as a result of this research (Weinshilboum and Wang, 2004).

SULT PHARMACOGENETICS

SULT1A1

SULT1A1 catalyzes the sulfate conjugation of many planar phenolic compounds, including a large number of drugs (Falany, 1997). It is also capable of catalyzing the sulfation of steroids, including estrone (E1), 17β-estradiol (E2), and cate-cholestrogens (Adjei and Weinshilboum, 2002). SULT1A1 was one of the first SULT isoforms studied for the possibility of inherited variation in level of activity or other properties. Those studies began two decades ago with the measurement of what was then referred to as "phenol" (P) or "thermostable" (TS) phenol sulfotransferase (PST) activity in an easily accessible human tissue, the human blood platelet. Levels of SULT1A1 activity in the human platelet measured with micromolar concentrations of 4-nitrophenol as a substrate varied over 50-fold (Figure 4.2A; Price et al., 1989; Raftogianis et al., 1997; Van Loon and Weinshilboum, 1984). Furthermore, that variation was — as described subsequently — primarily due to the effects of inheritance, i.e., it displayed high heritability (Price et al., 1989; Reveley et al., 1982/1983). Those original biochemical genetic studies included the determination of another SULT1A1 property, thermal stability, measured as a heated/control (H/C) ratio (Figure 4.2A). Thermal stability was a phenotypic marker that was thought to reflect alterations in amino acid sequence (Van Loon and Weinshilboum, 1984; Weinshilboum, 1981), a hypothesis that was subsequently shown to be correct for SULT1A1 (Raftogianis et al., 1997). Population and family data indicated that the trait of thermolabile SULT1A1 activity, like level of activity, was inherited. The allele responsible for this trait had a frequency of approximately 30% (Price et al., 1989; Van Loon and Weinshilboum, 1984).

Studies of monozygotic and dizygotic twins showed that individual variation in level of SULT1A1 activity in the platelet had a heritability of approximately 0.8 (i.e., approximately 80% of the variance was due to the effects of inheritance; Reveley et al., 1982/1983). Segregation analysis of family data for 237 individuals in 50 nuclear families confirmed the high heritability of levels of both SULT1A1 activity and H/C ratio and indicated that a three allele model might be required to account for the effects of inheritance on both basal level of activity and enzyme thermal stability (Price et al., 1989). A molecular explanation for those early biochemical genetic observations became possible only after the cloning and characterization of the human SULT1A1 cDNA and gene (Raftogianis et al., 1996; Wilborn et al., 1993). Once that had occurred, *SULT1A1* was resequenced using DNA samples from a large number of Caucasian subjects (Raftogianis et al., 1997). Those gene resequencing experiments resulted in the identification of a series of *SULT1A1* single nucleotide polymorphisms (SNPs). Of particular interest and importance was a

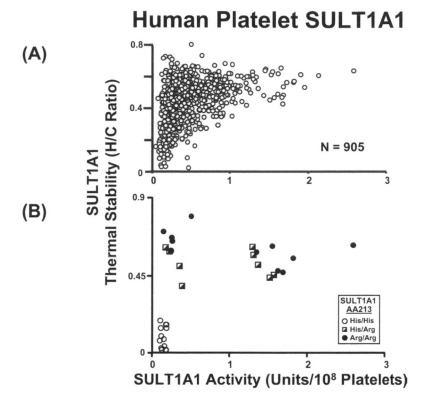

FIGURE 4.2 Human platelet SULT1A1 activity and thermal stability measured as an H/C ratio. (A) Data for 905 individual human platelet samples. (B) Data for 33 platelet samples selected from among the 905 samples in panel (A). These samples were selected to represent phenotypes with high activity–high H/C ratios, low activity–high H/C ratios, or low activity–low H/C ratios. DNA from these samples was then used to genotype each sample for the *SULT1A1*2* allele that alters the amino acid at codon 213. Genotypes at codon 213 are also shown. (From Raftogianis et al., 1997, *Biochemical Biophysical Research Communications*, 239, 298–304. With the permission of Academic Press.)

nonsynonymous coding SNP (cSNP) that changed the amino acid at codon 213 from Arg to His (Raftogianis et al., 1997).

The allele encoding His213 had a frequency in Caucasians of approximately 31% in 150 DNA samples obtained from randomly selected blood donors — very similar to the estimated frequency of the allele responsible for the trait of thermolabile SULT1A1. Two other *SULT1A1* nonsynonymous cSNPs with lower allele frequencies were also identified (Raftogianis et al., 1997). When platelet samples selected for level of SULT1A1 activity and thermal stability were resequenced for the codon 213 alleles, DNA from subjects with very low levels of activity and thermolabile enzyme was uniformly homozygous for the His 213 variant allele (Figure 4.2B) (Raftogianis et al., 1997). Therefore, these genotype–phenotype correlation studies showed that the His213 variant was responsible for the inherited trait of thermolabile SULT1A1. However, samples with higher thermal stability had a wide range of activities that could not be explained entirely by the codon 213 genetic polymorphism (Figure 4.2B). More recent studies have resulted in the identification of two polymorphisms within the *SULT1A1* promoter that appear to be responsible for at least some of this additional variation in level of activity (Prondzinski et al., 2003). These results, when taken together, indicate that large, genetically determined individual variation in SULT1A1 activity and thermal stability in the human platelet, variation that is correlated with those same phenotypes in the liver, intestine, and brain (Campbell and Weinshilboum, 1986; Sundaram et al., 1989; Young et al., 1985), is due to SNPs within both the *SULT1A1* ORF and promoter. These results also serve to emphasize that the functional implications of intragene haplotype — the set of polymorphisms present within an allele — will have to be understood if we are ultimately to correlate genotype with phenotype. SULT1A1 has also been the isoform studied most intensively for the possible clinical implications of genetic variation in sulfation, and the *SULT1A1*2* allele encoding the His213 variant allozyme has been reported to be a risk factor for breast cancer (Zheng et al., 2001) as well as a factor associated with survival in breast cancer patients treated with the antiestrogen drug tamoxifen (Nowell et al., 2002).

SULT1A2

SULT1A2, as described in other chapters in this volume, encodes a protein that is 96 to 97% identical to that encoded by *SULT1A1* (Her et al., 1996). When *SULT1A2* was resequenced using DNA from 61 liver biopsy samples, a series of genetic polymorphisms and haplotypes were also found for this gene (Raftogianis et al., 1999). A total of six nonsynonymous cSNPs were identified, but only three had allele frequencies of greater than 1%. Although these polymorphisms were shown to influence apparent K_m values for the two cosubstrates for the reaction as well as thermal stability for recombinant allozymes, no significant correlation was observed between these polymorphisms and either level of SULT activity or enzyme thermal stability in human liver biopsy samples (Raftogianis et al., 1999). However, it should be emphasized that *SULT1A2* remains somewhat of an "orphan" SULT with regard to its functional importance because of uncertainty with regard to tissue expression patterns or level of expression. However, the functional importance of the third member of the human SULT1A gene subfamily, *SULT1A3*, is not subject to question.

SULT1A3

SULT1A3 catalyzes the sulfate conjugation of catecholamines such as dopamine, as well as drugs that are structurally related to catecholamines (Reiter et al., 1983). Biochemical genetic studies of twins performed by measuring platelet SULT1A3 activity, originally referred to as the monoamine (M) or thermolabile (TL) form of PST (Rein et al., 1982; Reiter et al., 1983), showed that variation in this enzyme activity, like that for SULT1A1, had very high heritability — greater than 70% (Reveley et al., 1982/1983). Family studies of SULT1A3 activity in platelet samples from 232 subjects in 49 nuclear families confirmed the high heritability of level of activity (77%), with evidence for a common genetic polymorphism (Price et al., 1988). However, unlike SULT1A1 for which genetically determined variation in platelet activity was significantly correlated with levels of this enzyme activity in other tissues (r_s = 0.7 to 0.8), platelet SULT1A3 activity showed correlation coefficients of only approximately 0.4 with SULT1A3 activity in the small intestine, liver, and cerebral cortex (Campbell and Weinshilboum, 1986; Sundaram et al., 1989; Young et al., 1985).

After the human SULT1A3 cDNA and gene had been cloned and characterized (Aksoy and Weinshilboum, 1995; Wood et al., 1994), the gene was resequenced using DNA samples from both African-American and Caucasian-American subjects (Thomae et al., 2003). DNA samples from 60 Caucasian-American subjects (120 alleles) had no nonsynonymous cSNPs although a total of 5 SULT1A3 SNPs were observed. Eight SNPs were present in DNA from African-American subjects, and one of those polymorphisms, with an allele frequency of 4.2%, was nonsynonymous and resulted in a Lys234 → Asn change in the encoded amino acid (Thomae et al., 2003). The allozyme encoded by this variant allele displayed significant decreases in levels of both enzyme activity and immunoreactive protein after expression in COS-1 cells, corrected for transfection efficiency (Figure 4.3), but without significant changes in substrate kinetics. Furthermore, those decreases appeared to be due to accelerated degradation of the variant allozyme — as shown by experiments performed with a rabbit reticulocyte lysate (Figure 4.4; Thomae et al., 2003). As will be emphasized in subsequent paragraphs, decreased level of enzyme protein is a common mechanism responsible for the functional effects of nonsynonymous cSNPs (Weinshilboum and Wang, 2004). The underlying reason for the decrease, when it has been determined, is most often accelerated degradation of the variant allozyme (Siegel et al., 2001; Tai et al., 1999). In addition, there is evidence that molecular chaperones such as the heat shock proteins are involved in targeting variant allozymes for degradation (Wang et al., 2003; Weinshilboum and Wang, 2004). The *in vivo* functional significance of these SULT1A3 pharmacogenetic data remain to be determined. However, they raise the possibility of inherited variation in ability to catalyze the sulfate conjugation of catecholamines and drugs that are structurally related to catecholamines as a result of the common Lys234Asn polymorphism present in African-American subjects. This example helps to emphasize the importance of the inclusion of samples from multiple ethnic groups in studies of this type.

SULT1C2

SULT1C2 (originally referred to as SULT1C1) was identified on the basis of sequence homology of its cDNA to those of other SULTs (Her et al., 1997). In adults, this enzyme is most highly expressed in the stomach, thyroid, and kidney, and it is also expressed in the fetal liver (Her et al., 1997). Cloning and characterization of the human SULT1C2 gene (Freimuth et al., 2000) made it possible to perform a genotype-to-phenotype resequencing study of this gene. That experiment utilized DNA extracted from blood samples obtained from 89 randomly selected Caucasian blood donors (Freimuth et al., 2001) and resulted in the identification of 19 polymorphisms, including 4 nonsynonymous cSNPs and 5 insertions/deletions

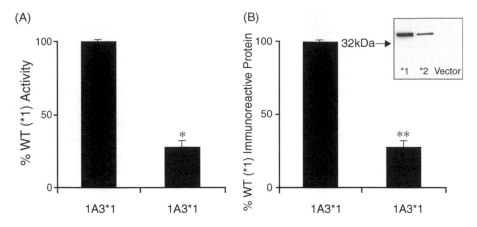

FIGURE 4.3 SULT1A3 COS-1 cell expression. Results for the expression of *SULT1A3*1* (Lys234) and *2 (Asn234) in COS-1 cells. (A) Average levels of SULT1A3 enzyme activity for the *1 and *2 allozymes with dopamine as the sulfate acceptor substrate. Each bar represents the average of nine independent transfections that have been corrected for transfection efficiency (mean ± SEM). * = p < 0.01 when compared to the *1 construct. (B) Average levels of SULT1A3 immunoreactive protein for the *1 and *2 allozymes using a SULT1A3-specific antibody to perform western-blot analyses. Each bar represents the average of nine independent transfections (mean ± SEM). ** = p < 0.001. The gels were loaded on the basis of cotransfected β-galactosidase activity to correct for transfection efficiency. The insert is a representative western blot used to obtain these data. (From Thomae et al., 2003, *Journal of Neurochemistry,* 87, 809–819. With the permission of Blackwell Publishing.)

(indels). When the four nonsynonymous cSNPs were expressed in COS-1 cells, three of the four were associated with striking decreases in levels of enzyme activity, with parallel decreases in levels of immunoreactive protein — a situation similar to that observed for the nonsynonymous *SULT1A3* cSNP present in African-American subjects. This same theme will recur for many of the other genes described in this chapter. No endogenous substrate has been identified for SULT1C2, but it can catalyze the sulfation of high concentrations of 4-nitrophenol, and it can metabolically activate procarcinogens such as N-hydroxyl-2-acetylaminofluorene (Freimuth et al., 2000; Sakakibara et al., 1998). The next SULT isoform to be discussed, SULT1E1, has important endogenous substrates.

SULT1E1

SULT1E1 was originally named estrogen sulfotransferase (EST), and this enzyme has the lowest apparent K_m values for E1, E2, and the catecholestrogens of any human cytosolic SULT isoform (Adjei and Weinshilboum, 2002). The *SULT1E1* exons, splice junctions, and 5′-flanking region were resequenced using 60 DNA samples from African-American and 60 samples from Caucasian-American subjects (Adjei et al., 2003). The arrows in Figure 4.5 show the locations of the SNPs observed

SULT1A3 RRL Degradation

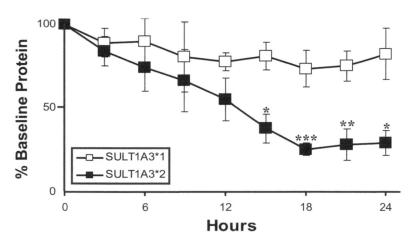

FIGURE 4.4 SULT1A3 rabbit reticulocyte lysate (RRL) protein degradation study. ^{35}S-Methionine radioactively labeled SULT1A3*1 and *2 were incubated in an RRL for 24 hr, and loss of the protein was determined by SDS-PAGE. Each point represents the average of four independent experiments (mean ± SEM). * = $p < 0.02$, ** = $p < 0.05$, *** = $p < 0.005$ when compared with the *1 allozyme at the same incubation time. (From Thomae et al., 2003, *Journal of Neurochemistry*, 87, 809–819. With the permission of Blackwell Publishing.)

Human *SULT1E1* Polymorphisms

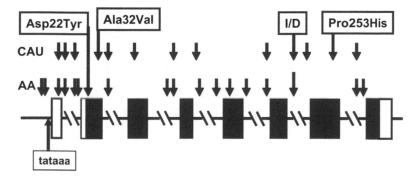

FIGURE 4.5 Human *SULT1E1* genetic polymorphisms. A schematic representation of the human SULT1E1 gene with the locations of polymorphisms indicated by arrows. Black rectangles represent the open reading frame, and white rectangles represent portions of exons that encode an untranslated region (UTR) sequence. Arrows indicate the locations of SNPs and a single indel (I/D). AA represents data obtained with DNA from African-American subjects, and CAU represents data obtained using DNA from Caucasian-American subjects. Changes in encoded amino acids resulting from the presence of nonsynonymous cSNPs are also indicated. (From Adjei et al., 2003, *British Journal of Pharmacology*, 139, 1373–1382. With the permission of Nature Publishing Group.)

in these samples. Three of the *SULT1E1* cSNPs were nonsynonymous and resulted in the following alterations in encoded amino acids: Asp22Tyr, Ala32Val, and Pro253His (Adjei et al., 2003). Those three alleles, two of which were observed in DNA from Caucasian-American subjects and one, that encoding Tyr22, in DNA from an African-American subject, were transiently expressed in COS-1 cells. The Tyr22 allozyme showed a striking decrease in level of enzyme activity and immunoreactive protein; the Val32 variant displayed a decrease of approximately 50% in both; and the His253 allozyme had no changes in level of either enzyme activity or immunoreactive protein as compared to the wild type (WT) allozyme (Figure 4.6; Adjei et al., 2003). Although apparent K_m values for both cosubstrates showed slight differences among these variant allozymes, the most striking alteration — other than

Recombinant Human SULT1E1 Allozymes

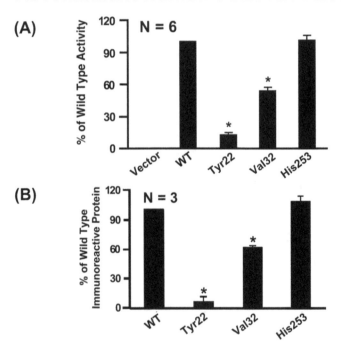

FIGURE 4.6 Human SULT1E1 pharmacogenetics. (A) Recombinant human SULT1E1 allozyme enzyme activity. Average levels of enzyme activity for each of the recombinant SULT1E1 allozymes assayed with E2 as the sulfate acceptor substrate. All values have been corrected for transfection efficiency. Each bar represents the average of six independent transfections (mean ± SEM). * = $p < 0.001$ when compared to the WT allozyme. (B) Recombinant human SULT1E1 allozyme western-blot analysis. Average levels of immunoreactive SULT1E1 protein for each of the recombinant allozymes, expressed as a percentage of WT protein. Each bar represents the average of three independent transfections (mean ± SEM). * = $p < 0.0006$ when compared to the WT level of immunoreactive activity. (From Adjei et al., 2003, *British Journal of Pharmacology*, 139, 1373–1382. With the permission of Nature Publishing Group.)

the differences in level of activity — was a significant decrease in thermal stability for the Tyr22 variant, the allozyme that displayed striking decreases in levels of both enzyme activity and immunoreactive protein (Figure 4.6). Preliminary, unpublished studies showed that the SULT1E1 Tyr22 variant, like the SULT1A3 Asn234 variant allozyme, is more rapidly degraded in a rabbit reticulocyte lysate than is the WT allozyme (Liewei Wang and Yvette Martin, 2003, unpublished observations).

SULT2A1

SULT2A1 was originally referred to as dehydroepiandrosterone sulfotransferase (DHEA ST; Otterness et al., 1992). This SULT isoform catalyzes the sulfate conjugation of a variety of steroid compounds, but DHEA has been the prototypic substrate used in most studies (Falany, 1997). SULT2A1 was the first human SULT for which a cDNA was cloned and characterized (Otterness et al., 1992), and the gene encoding this isoform maps to the long arm of chromosome 19 (Figure 4.1; Otterness et al., 1995). As described in other chapters in this volume, SULT2A1 is highly expressed in the human adrenal cortex, liver, and intestine (Otterness et al., 1992; Otterness and Weinshilboum, 1994). When *SULT2A1* was resequenced using DNA samples from 60 African-American and 60 Caucasian-American subjects, 3 nonsynonymous cSNPs were observed — but all 3 of those polymorphisms were present only in DNA from African-American subjects (Thomae et al., 2002). These SNPs resulted in variant allozymes with the following alterations in encoded amino acid: Ala63Pro, Lys227Glu, and Ala261Thr; with allele frequencies of approximately 5%, 1%, and 13%, respectively. Furthermore, the codon 63 and 261 cSNPs were in tight linkage disequilibrium. When expression constructs for all three of these polymorphisms, including the double variant at codons 63 and 261, were transiently expressed in COS-1 cells, the Pro63 allozyme showed a decrease in enzyme activity of approximately 50%, while the Glu227 allozyme showed an 85% decrease in enzyme activity — with parallel decreases in the quantity of immunoreactive protein for both variants (Figure 4.7; Thomae et al., 2002). The more common Thr261 variant did not show significant changes in levels of either enzyme activity or immunoreactive protein, but the common double variant displayed a decrease of approximately 60% in levels of both activity and protein. Even though the common Thr261 variant allozyme did not alter level of enzyme activity after transient expression in COS-1 cells, it disrupted the SULT2A1 dimerization interface, resulting in inability to form a homodimer (Thomae et al., 2002). The functional implications of these *SULT2A1* polymorphisms remain to be explored in either human biopsy tissue samples or in clinical studies. These observations could be of clinical importance in light of epidemiologic studies that indicate that circulating levels of DHEA and DHEA sulfate might be related to a variety of pathophysiologic changes that are associated with aging (Lane et al., 1997; Orentreich et al., 1992; Villareal et al., 2000). Testing the hypothesis that genetic variation in SULT2A1 or other SULT isoforms might have clinical implications will be a major focus for future pharmacogenetic research. However, sulfation does not involve just the SULTs, but it also requires the synthesis of the sulfate donor molecule, PAPS. There are also common sequence variations in the two human PAPS synthetase genes, which will be discussed subsequently.

Recombinant Human SULT2A1

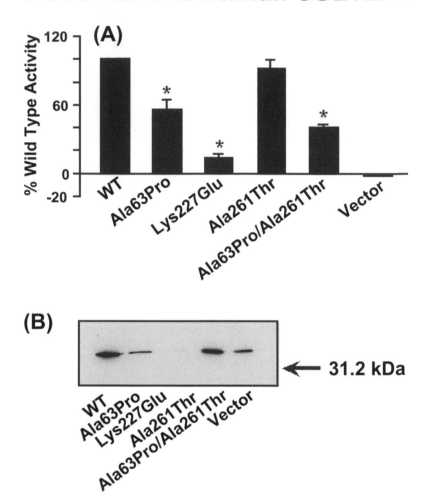

FIGURE 4.7 Recombinant human SULT2A1 allozyme activity and western-blot analysis. (A) Average levels of enzyme activity for each of the SULT2A1 recombinant allozymes with DHEA as the sulfate acceptor substrate. All values have been corrected for transfection efficiency. Each bar represents the average of six independent transfections (mean ± SEM). * = $p < 0.01$ when compared with the WT construct. (B) SULT2A1 recombinant allozyme western-blot analysis. The gel was loaded on the basis of β-galactosidase activity to correct for variations in transfection efficiency. (From Thomae et al., 2002, *Pharmacogenomics Journal*, 2, 48–56. With the permission of Nature Publishing Group.)

PAPSS PHARMACOGENETICS

PAPSS1

The genes encoding human PAPSS1 and PAPSS2 were cloned and characterized —
in part — as a step toward pharmacogenetic studies (Xu et al., 2000). Because PAPS
is essential for the reactions catalyzed by all SULT isoforms, functional genetic
polymorphisms for PAPSS1, PAPSS2, or both might have significant functional
implications. That hypothesis has already been shown to be true since, as described
subsequently, inactivating mutations in *PAPSS2* are associated with an inherited
musculoskeletal disorder — spondyloepimetaphaseal dysplasia (ul Haque et al.,
1998). When the human PAPSS1 gene was resequenced using 60 DNA samples
from African-American subjects and 58 samples from Caucasian-American subjects,
21 polymorphisms were observed, including one indel and 20 SNPs (Xu et al., 2003).
Two of the SNPs were nonsynonymous and resulted in Arg333Cys and Glu531Gln
alterations in encoded amino acids. The polymorphism encoding the Cys333 allozyme had a frequency of 2.5% but was observed only in Caucasian-American
subjects, while that encoding the Gln531 was rare and was seen in only a single
African-American subject. When these two variant allozymes were expressed in
COS-1 cells, there were no changes in levels of basal enzyme activity measured
under optimal conditions for either allozyme, but the Gln531 polymorphism displayed altered substrate kinetics, with a 5-fold higher apparent K_m value for SO_4^{2-}
than that observed for the WT allozyme (Xu et al., 2003). The functional implications
of this change in cosubstrate affinity remain to be determined.

PAPSS2

PAPSS2 is known to be of functional importance — at least for the musculoskeletal
system — since inactivating mutations for this gene result in severe joint disease,
with the early occurrence of degenerative arthritis as a result of genetically determined spondyloepimetaphaseal dysplasia (ul Haque et al., 1998). When *PAPSS2*
was resequenced using 90 ethnically diverse but anonymized Polymorphism Discovery Resource DNA samples, 22 SNPs were observed, including 4 nonsynonymous cSNPs that altered the following amino acids: Glu10Lys, Met281Leu,
Val291Met, and Arg432Lys (Xu et al., 2002). There were also 4 indels, and one
sample was homozygous for an 81 bp deletion in the 5′-flanking region of the gene
that was located 286 bp upstream from the site of transcription initiation. The
nonsynonymous cSNPs that resulted in Lys10 and Met291 variant allozymes both
displayed significant decreases, 27% and 63%, respectively, in level of enzyme
activity after transient expression in COS-1 cells. There was also a decrease in the
quantity of immunoreactive protein for the Lys10 variant allozyme. The Met291
variant, in contrast, displayed a significant decrease in affinity for both ATP and
SO_4^{2-}, the two cosubstrates for the reaction, when compared with the WT allozyme,
but without an alteration in the level of immunoreactive protein (Xu et al., 2002).

CONCLUSIONS

The human cytosolic SULT gene superfamily displays significant genetic variation, as do the genes that encode PAPSS1 and PAPSS2. Two decades of research were required to discover and functionally characterize genetic polymorphisms for *SULT1A1* and *SULT1A3*. However, very rapid progress in SULT genomics during the past decade, made possible in part by the Human Genome Project, has resulted in the recent identification and characterization of a series of human cytosolic SULT genes and the genes encoding PAPSS1 and PAPSS2. Those advances, in turn, made it possible to use a genotype-to-phenotype pharmacogenetic research strategy to study these genes. Application of that strategy has resulted in the identification of a series of nonsynonymous cSNPs and other polymorphisms that are of functional significance. Pharmacogenetic research on sulfate conjugation has also helped to provide insight into mechanisms by which nonsynonymous cSNPs influence enzyme function. For example, studies of the SULTs and PAPSSs that are described in this chapter, as well as similar studies of other enzyme systems — especially the methyltransferases — have demonstrated that the most common mechanism by which inherited variation in amino acid sequence influences function involves a decrease in the quantity of enzyme protein, usually as a result of accelerated protein degradation (Weinshilboum and Wang, 2004).

The implications of SULT and PAPSS genetic polymorphisms for clinical medicine still remain to be determined. However, epidemiologic studies of the *SULT1A1*2* genetic polymorphism have already provided evidence that this polymorphism may represent a risk factor for hormone-related disease such as breast cancer (Zheng et al., 2001). It may also have pharmacogenetic implications for the treatment of patients with breast cancer (Nowell et al., 2002). The possible clinical importance of common genetic polymorphisms for the other human SULT genes described in this chapter await systematic study. However, the speed with which common, functionally significant polymorphisms in genes encoding enzymes that play a role in sulfation have been discovered and characterized will make it possible to rapidly apply this information to test clinically relevant hypotheses with regard to both disease pathophysiology and variation in response to drugs that undergo sulfate conjugation. We are only beginning to understand the role of inheritance in individual variation in sulfate conjugation, but we are now in a position to determine the extent to which genetic variation might contribute to individual differences in biotransformation catalyzed by this important group of phase II drug, neurotransmitter, and hormone-metabolizing enzymes.

ACKNOWLEDGMENT

We thank Luanne Wussow for her assistance with the preparation of this manuscript.

REFERENCES

Adjei, A.A. and Weinshilboum, R.M., 2002, Catecholestrogen sulfation: possible role in carcinogenesis. *Biochemical Biophysical Research Communications*, 292, 402–408.

Adjei, A.A., Thomae, B.A., Prondzinski, J.L., Eckloff, B.W., Wieben, E.D., and Weinshilboum, R.M., 2003, Human estrogen sulfotransferase (SULT1E1) pharmacogenetics: gene resequencing and functional genomics. *British Journal of Pharmacology*, 139, 1373–1382.

Aksoy, I.A. and Weinshilboum, R.M., 1995, Human thermolabile phenol sulfotransferase gene (STM): molecular cloning and structural characterization. *Biochemical Biophysical Research Communications*, 208, 786–795.

Campbell, N.R.C. and Weinshilboum, R., 1986, Human phenol sulfotransferase (PST): correlation of liver and platelet activities. *Canadian Society for Clinical Investigation*, 9 (Suppl.), A14.

Falany, C.N., 1997, Enzymology of human cytosolic sulfotransferases. *FASEB Journal*, 11, 206–216.

Freimuth, R.R., Raftogianis, R.B., Wood, T.C., Moon, E., Kim, U.-J., Xu, J., Siciliano, M.J., and Weinshilboum, R.M., 2000, Human sulfotransferases SULT1C1 and SULT1C2: cDNA characterization, gene cloning, and chromosomal localization. *Genomics*, 65, 157–165.

Freimuth, R.R., Eckloff, B., Wieben, E.D., and Weinshilboum, R.M., 2001, Human sulfotransferase SULT1C1 pharmacogenetics: gene resequencing and functional genomic studies. *Pharmacogenetics*, 11, 747–756.

Freimuth, R.R., Wiepert, M., Chute, C.G., Wieben, E.D., and Weinshilboum, R.M., 2004, Human cytosolic sulfotransferase database mining: identification of seven novel genes and pseudogenes. *Pharmacogenomics Journal*, 4, 54–65.

Her, C., Raftogianis, R., and Weinshilboum, R.M., 1996, Human phenol sulfotransferase STP2 gene: molecular cloning, structural characterization and chromosomal localization. *Genomics*, 33, 409–420.

Her, C., Kaur, G.P., Athwal, R.S., and Weinshilboum, R.M., 1997, Human sulfotransferase SULT1C1: cDNA cloning, tissue-specific expression and chromosomal localization. *Genomics*, 41, 467–470.

Kurima, K., Singh, B., and Schartz, N.B., 1999, Genomic organization of the mouse and human genes encoding the ATP sulfurylase/adenosine 5'-phophosulfate kinase isoform SK2. *Journal of Biological Chemistry*, 274, 33306–33312.

Lane, M.A., Ingram, D.K., Ball, S.S., and Roth, G.S., 1997, Dehydroepiandrosterone sulfate: a biomarker of primate aging slowed by calorie restriction. *Journal of Clinical Endocrinology and Metabolism*, 82, 2093–2096.

Nowell, S., Sweeney, C., Winters, M., Stone, A., Lang, N.P., Hutchins, L.F., Kadlubar, F.F., and Ambrosone, C.B., 2002, Association between sulfotransferase 1A1 genotype and survival of breast cancer patients receiving tamoxifen therapy. *Journal of the National Cancer Institute*, 94, 1635–1640.

Orentreich, N., Brind, J.L., Vogelman, J.H., Andres, R., and Baldwin, H., 1992, Long-term longitudinal measurements of plasma dehydroepiandrosterone sulfate in man. *Journal of Clinical Endocrinology and Metabolism*, 75, 1002–1004.

Otterness, D.M., Wieben, E.D., Wood, T.C., Watson, R.W.G., Madden, B.J., McCormick, D.J., and Weinshilboum, R.M., 1992, Human liver dehydroepiandrosterone sulfotransferase: molecular cloning and expression of cDNA. *Molecular Pharmacology*, 41, 865–872.

Otterness, D.M. and Weinshilboum, R., 1994, Human dehydroepiandrosterone sulfotransferase: molecular cloning of cDNA and genomic DNA. *Chemico-Biological Interactions*, 92, 145–159.

Otterness, D.M., Mohrenweiser, H.W., Brandriff, B.F., and Weinshilboum, R.M., 1995, Dehydroepiandrosterone sulfotransferase gene (STD): localization to human chromosome 19q13.3. *Cytogenetics and Cell Genetics*, 70, 45–47.

Price, R.A., Cox, N.J., Spielman, R.S., Van Loon, J., Maidak, B.L., and Weinshilboum, R.M., 1988, Inheritance of human platelet thermolabile phenol sulfotransferase (TL PST) activity. *Genetic Epidemiology*, 5, 1–15.

Price, R.A., Spielman, R.S., Lucena, A.L., Van Loon, J.A., Maidak, B.L., and Weinshilboum, R.M., 1989, Genetic polymorphism for human platelet thermostable phenol sulfotransferase (TS PST) activity. *Genetics*, 122, 905–914.

Prondzinski, J., Thomae, B., Wang, L., Eckloff, B., Wieben, E., and Weinshilboum, R., 2003, Sulfotransferase (SULT) 1A1 pharmacogenetics: functional 5'-flanking region (5'-FR) polymorphisms. *Clinical Pharmacology Therapeutics*, 73, P77.

Raftogianis, R., Her, C., and Weinshilboum, R.M., 1996, Human phenol sulfotransferase pharmacogenetics: STP1 gene cloning and structural characterization. *Pharmacogenetics*, 6, 473–487.

Raftogianis, R.B., Wood, T.C., Otterness, D.M., Van Loon, J.A., and Weinshilboum, R.M., 1997, Phenol sulfotransferase pharmacogenetics in humans: association of common SULT1A1 alleles with TS PST phenotype. *Biochemical Biophysical Research Communications*, 239, 298–304.

Raftogianis, R.B., Wood, T.C., and Weinshilboum, R.M., 1999, Human phenol sulfotransferases SULT1A2 and SULT1A1: genetic polymorphisms, allozyme properties and human liver genotype-phenotype correlations. *Biochemical Pharmacology*, 58, 605–610.

Rein, G., Glover, V., and Sandler, M., 1982, Multiple forms of phenolsulfotransferase in human tissues: selective inhibition by dichloronitrophenol. *Biochemical Pharmacology*, 31, 1893–1897.

Reiter, C., Mwaluko, G., Dunnette, J., Van Loon, J., and Weinshilboum, R., 1983, Thermolabile and thermostable human platelet phenol sulfotransferase: substrate specificity and physical separation. *Naunyn-Schmiedeberg's Archives of Pharmacology*, 324, 140–147.

Reveley, A.M., Carter, S.M.B., Reveley, M.A., and Sandler, M., 1982/1983, A genetic study of platelet phenolsulfotransferase activity in normal and schizophrenic twins. *Journal of Psychiatric Research*, 17, 303–307.

Salavaggione, O.E., Kidd, L., Prondzinski, J.L., Szumlanski, C.L., Pankratz, V.S., Wang, L., Trepanier, L., and Weinshilboum, R.M., 2002, Canine red blood cell thiopurine S-methyltransferase: companion animal pharmacogenetics. *Pharmacogenetics*, 12, 713–724.

Salavaggione, O.E., Yang, C., Kidd, L.B., Thomae, B.A., Pankratz, V.S., Trepanier, L.A., and Weinshilboum, R.M., 2004, Cat red blood cell thiopurine S-methyltransferase: companion animal pharmacogenetics. *Journal of Pharmacology and Experimental Therapeutics*, 308, 617–626.

Sakakibara, Y., Yanagisawa, K., Katafuchi, J., Ringer, D.P., Takami, Y., Nakayama, T., Suiko, M., and Liu, M.-C., 1998, Molecular cloning, expression, and characterization of novel human SULT1C sulfotransferases that catalyze the sulfonation of N-hydroxy-2-acetylaminofluorene. *Journal of Biological Chemistry*, 273, 33929–33935.

Siegel, D., Anwar, A., Winski, S.L., Kepa, J.K., Zolman, K.L., and Ross, D., 2001, Rapid polyubiquitination and proteosomal degradation of a mutant form of NAD(P)H: quinone oxioreductase 1. *Molecular Pharmacology*, 59, 263–268.

Sundaram, R.S., Van Loon, J.A., Tucker, R., and Weinshilboum, R.M., 1989, Sulfation pharmacogenetics: correlation of human platelet and small intestinal phenol sulfotransferase. *Clinical Pharmacology and Therapeutics*, 46, 501–509.

Tai, H.-L., Fessing, M.Y., Bonten, E.J., Yanishevsky, Y., d'Azzo, A., Krynetski, E.Y., and Evans, W.E., 1999, Enhanced proteasomal degradation of mutant human thiopurine S-methyltransferase (TPMT) in mammalian cells: mechanism for TPMT protein deficiency inherited by TPMT*2, TPMT*3A, TPMT*3B or TPMT*3C. *Pharmacogenetics*, 9, 641–650.

Thomae, B.A., Eckloff, B.W., Freimuth, R.R., Wieben, E.D., and Weinshilboum, R.M., 2002, Human sulfotransferase SULT2A1 pharmacogenetics: genotype-to-phenotype studies. *Pharmacogenomics Journal*, 2, 48–56.

Thomae, B.A., Rifki, O.F., Theobald, M.A., Eckloff, B.W., Wieben, E.D., and Weinshilboum, R.M., 2003, Human catecholamine sulfotransferase (SULT1A3) pharmacogenetics: common functional genetic polymorphism. *Journal of Neurochemistry*, 87, 809–819.

ul Haque, M.F., King, L.M., Krakow, D., Cantor, R.M., Rusiniak, M.E., Swank, R.T., Superti-Furga, A., Haque, S., Abbas, H., Ahmad, W., Ahmad, M., and Cohn, D.H., 1998, Mutations in orthologous genes in human spondyloepimetaphyseal dysplasia and the brachymorphic mouse. *Nature Genetics*, 20, 157–162.

Van Loon, J.A. and Weinshilboum, R.M., 1984, Human platelet phenol sulfotransferase: familial variations in the thermal stability of the TS form. *Biochemical Genetics*, 22, 997–1014.

Villareal, D.T., Holloszy, J.O., and Kohrt, W.M., 2000, Effects of DHEA replacement on bone mineral density and body composition in elderly women and men. *Clinical Endocrinology*, 53, 561–568.

Wang, L., Sullivan, W., Toft, D., and Weinshilboum, R., 2003, Thiopurine S-methyltransferase pharmacogenetics: chaperone protein association and allozyme degradation. *Pharmacogenetics*, 13, 555–564.

Weinshilboum, R.M., 1981, Enzyme thermal stability and population genetic studies: application to erythrocyte catechol-O-methyltransferase and plasma dopamine ß-hydroxylase, in *Genetic Strategies in Psychobiology and Psychiatry*, Gershon, E.S., Matthysee, S., Breakfield, X.O., and Ciaranello, R.D., Eds., Boxwood Press, Pacific Grove, CA, pp. 79–94.

Weinshilboum, R., 2003, Inheritance and drug response. *New England Journal of Medicine*, 348, 529–537.

Weinshilboum, R. and Otterness, D., 1994, Sulfotransferase enzymes, in *Conjugation-Deconjugation Reactions in Drug Metabolism and Toxicity*, Kauffman, F.C., Ed., Springer-Verlag, Berlin, chap. 2.

Weinshilboum, R. and Wang, L., 2004, Pharmacogenetics: inherited variation in amino acid sequence and altered protein quantity. *Clinical Pharmacology and Therapeutics*, 75, 253–258.

Wilborn, T.W., Comer, K.A., Dooley, T.P., Reardon, I.M., Heinrikson, R.L., and Falany, C.N., 1993, Sequence analysis and expression of the cDNA for the phenol sulfating form of human liver phenol sulfotransferase. *Molecular Pharmacology*, 43, 70–77.

Wood, T.C., Aksoy, I.A., Aksoy, S., and Weinshilboum, R.M., 1994, Human liver thermolabile phenol sulfotransferase: cDNA cloning, expression and characterization. *Biochemical Biophysical Research Communications*, 198, 1119–1127.

Xu, Z.-H., Otterness, D.M., Freimuth, R.R., Carlini, E.J., Wood, T.C., Mitchell, S., Moon, E., Kim, U.-J., Xu, J.-P., Siciliano, M.J., and Weinshilboum, R.M., 2000, Human 3'-phosphoadenosine 5'-phosphosulfate synthetase 1 (PAPSS1) and PAPSS2: gene cloning, characterization and chromosomal localization. *Biochemical Biophysical Research Communications*, 268, 437–444.

Xu, Z.-H., Freimuth, R.R., Eckloff, B., Wieben, E., and Weinshilboum, R.M., 2002, Human 3'-phosphoadenosine 5'-phosphosulfate synthetase 2 (PAPSS2) pharmacogenetics: gene resequencing, genetic polymorphisms and functional characterization of variant allozymes. *Pharmacogenetics*, 12, 11–21.

Xu, Z.-H., Thomae, B.A., Eckloff, B.W., Wieben, E.D., and Weinshilboum, R.M., 2003, Pharmacogenetics of human 3'-phosphoadenosine 5'-phosphosulfate synthetase 1 (PAPSS1): gene resequencing, sequence variation and functional genomics. *Biochemical Pharmacology*, 65, 1787–1796.

Young, W.F., Jr., Laws, E.R., Jr., Sharbrough, F.W., and Weinshilboum, R.M., 1985, Human phenol sulfotransferase: correlation of brain and platelet activities. *Journal of Neurochemistry*, 44, 1131–1137.

Zheng, W., Xie, D., Cerhan, J.R., Sellers, T.A., Wen, W., and Folsom, A.R., 2001, Sulfotransferase 1A1 polymorphism, endogenous estrogen exposure, well-done meat intake, and breast cancer risk. *Cancer Epidemiology Biomarkers & Prevention*, 10, 89–94.

5 Molecular Cloning of the Human Cytosolic Sulfotransferases

Charles N. Falany, Connie A. Meloche,
Dongning He, and Josie L. Falany

CONTENTS

INTRODUCTION

For over 150 years, the conjugation of xenobiotic compounds with sulfate has been recognized as an important conjugation reaction in human metabolism. Since the identification of phenyl sulfate in patients treated with phenol by Baumann in 1876 (Baumann, 1876), it has been demonstrated that a broad spectrum of drug, xenobiotic, and endogenous compounds are substrates for conjugation with sulfate. The variety of chemical structures that are sulfated has strongly indicated that a family of distinct enzymes is involved in the formation of sulfate conjugates. Identification and characterization of the individual sulfotransferase (SULT) isoenzymes involved in sulfate conjugation has been one of the major aspects in the investigation of sulfation in humans. Understanding the properties and genetic variability of the individual SULT isoforms will greatly add to our knowledge of the physiological properties and functions of sulfation.

Sulfate conjugation involves the transfer of the sulfonate group from 3'-phosphoadenosine 5'-phosphosulfate (PAPS) to an acceptor group, generally a hydroxyl or an amine group. Although the reaction involves the transfer of a sulfonate (SO_3) group, historically the reaction has been termed sulfation due to the identification of sulfate esters that are the major products of the reaction. Conjugation of a hydroxyl group results in the formation of a sulfate ester, whereas conjugation of an amine forms a sulfamate. Both reactions form charged conjugates that, in most instances, are highly water soluble and readily excreted from the body. In general, sulfates are not biologically active although sulfation can generate reactive electrophilic products following the sulfation of hydroxymethyl polyaromatic hydrocarbons or N-hydroxy heterocyclic aromatic amines (Glatt et al., 1994; Glatt, 1997). It has also been reported that sulfation increases the biological activity of minoxidil in stimulating hair growth (Meisheri et al., 1988; Waldon et al., 1989). The broad range of small compounds capable of undergoing sulfation is a reflection of the ability of alcohols, phenols, and amines to be sulfated as well as of the multiplicity and overlapping substrate reactivity of the cytosolic SULTs catalyzing their sulfation.

The SULT enzymes capable of catalyzing the sulfonation reaction include both membrane-associated forms found in the Golgi apparatus and isoforms found in the cytosolic fraction of cell fractionation preparations. The membrane-associated SULTs are primarily involved with glycosaminoglycan and glycoprotein sulfation in the Golgi apparatus and are not generally considered to be important in the conjugation of drugs and xenobiotics (Fukuda et al., 2001; Habuchi, 2000). Therefore, this chapter will focus on the molecular cloning and properties of the human cytosolic SULTs associated with drug, xenobiotic, and small endogenous compound sulfate conjugation.

MULTIPLICITY OF HUMAN SULT ACTIVITIES AND PURIFICATION STUDIES

Initial studies of the heterogeneity of the human SULTs focused on two types of SULT activity, the high levels of hydroxysteroid sulfation in the human adrenal gland and multiple forms of phenol SULT activity in human platelets. The human adrenal is responsible for the high levels of dehydroepiandrosterone (DHEA) sulfate present in human fetal plasma. Micromolar concentrations of DHEA-sulfate in adult plasma are due to the synthesis and sulfation of DHEA in the reticular layer of the adrenal cortex (Parker, 1999). The synthesis of such large amounts of DHEA-sulfate in the adrenal gland is limited to humans and some higher primates (Parker, 1999). In humans, large amounts of DHEA-sulfate are synthesized and released by the fetal adrenal; however, the fetal adrenal layer disappears shortly after birth and DHEA synthesis almost ceases. High levels of DHEA-sulfate synthesis begin again at adrenarche with the highest production in young adults. Adrenarche refers to the increase in synthesis of adrenal hormones that occurs shortly before puberty in humans. Plasma DHEA-sulfate levels are highest at 20 to 25 years of age and slowly decrease with age until levels are about 10% of the maximal levels in the seventh to eighth decades of life. DHEA in the circulation is utilized as a precursor for the synthesis of androgens and estrogens in different tissues (Arlt et al., 1999; Mortola and Yen, 1990). The high levels of estrogens in pregnant women are due to the conversion of DHEA-sulfate synthesized in the fetal adrenal cortex to estrogens in the placenta (Baxter and Tyrell, 1987; Casey et al., 1987). Characterization of human adrenal DHEA sulfation activity indicated that this enzymatic activity was different from the phenol SULT activities identified in human platelet cytosol. The observation that human blood platelets contained readily measurable levels of phenol SULT activity opened the way for characterization of these enzymes in human tissues (Hart et al., 1979). Numerous studies identified two distinct isoenzymes of phenol SULT activity based on a number of characteristics, including substrate reactivity, thermal stability, and chromatographic separation (Weinshilboum, 1986; Falany, 1997). The conclusion that multiple isoenzymes of the SULTs exist in human tissues was confirmed by the purification and characterization of these three distinct SULT isoenzymes (Falany et al., 1989, 1990; Heroux and Roth, 1988).

The two distinct isoenzymes of phenol SULT activity in blood platelets were initially termed monoamine-sulfating phenol SULT (MPST) and phenol-sulfating phenol SULT (PPST), based on the selective activity of the two enzymes with dopamine and 4-nitrophenol (PNP), respectively (Weinshilboum, 1986). MPST, which is more abundant than PPST in blood platelets, was purified to homogeneity from platelet cytosol by Heroux and Roth (1988) using a mixture of DEAE anion-exchange chromatography and PAP-agarose affinity chromatography. PPST was subsequently purified from human liver cytosol using similar procedures, demonstrating that the phenol SULTs were unique enzymes. Polyclonal antibodies raised in rabbits to purified MPST cross-react strongly with both enzymes (Falany et al., 1990; Ganguly et al., 1995; Heroux et al., 1989). Immunoblot analysis of both human liver and platelet cytosols demonstrated that MPST migrated with a mass of 34,000 Da, whereas PPST displayed a molecular mass of 32,000 Da during SDS-polyacrylamide gel electrophoresis (Falany et al., 1990; Ganguly et al., 1995). This differential migration during SDS-polyacrylamide gel electrophoresis permits discrimination of the two proteins in human tissue samples.

Characterization of the properties of DHEA sulfation activity in human tissues indicated that it was catalyzed by a SULT isoform distinct from the phenol SULTs. This was demonstrated by the separation of DHEA SULT activity and phenol SULT activity in human liver cytosol and the purification of liver DHEA SULT (Falany et al., 1989). DHEA SULT is one of the major SULTs expressed in human liver cytosol and has an important role in the sulfation of bile acids (Radominska et al., 1991). Polyclonal rabbit antihuman liver DHEA SULT antibody reacted with identical proteins in human adrenal and liver cytosols and did not react with the human platelet or liver phenol SULT isoforms. During SDS-polyacrylamide gel electrophoresis, DHEA SULT was resolved from the phenol SULTS due to its slightly larger molecular mass aiding in identification of the proteins (Comer et al., 1993). These studies indicated that the three human SULTs were distinct enzymes with different tissue expression and regulation.

Overall, the characterization and purification studies of human platelet and liver SULTs confirmed that multiple distinct isoforms of the cytosolic SULTs were present in these tissues. Extensive studies of the pharmacogenetic properties of the platelet phenol SULTs by Weinshilboum's group at the Mayo Clinic established that these isoforms were independently regulated unique gene products (Weinshilboum, 1988, 1990). Characterization of the physical properties of the platelet phenol SULTs, such as their thermostability, indicated that functional genetic variants of the phenol SULTs were expressed in humans, which set the stage for pharmacogenetic studies. The purification studies had previously established that the human SULTs were functionally and genetically distinct (Falany et al., 1990; Heroux and Roth, 1988). However, these procedures were slow, laborious, and not very adaptable to the identification of novel isoforms of human SULT. These studies were also limited by the availability of human tissues and the appearance of HIV in the blood supply. Therefore, the investigation of the heterogeneity of the human cytosolic SULTs became associated with the molecular analysis of the human genome.

MOLECULAR CLONING OF THE NONHUMAN SULTS

The investigation of human cytosolic SULTs was dependent on the prior isolation of SULT cDNAs from both rat and cow. The isolation and sequencing of a number of cDNAs for rat and bovine cytosolic SULTs provided valuable information as well as molecular probes for the subsequent cloning and identification of the human SULTs. The first report of a SULT cDNA was that of the rat senescence marker protein-2 (rSMP-2; Chatterjee et al., 1987) although it was not recognized as a SULT at that time. The SMP-2 cDNA was initially described as an androgen-regulated protein whose expression was influenced by aging. Within a year, Nash et al. (1988) reported the cloning of a bovine estrogen SULT. These authors used degenerate oligonucleotides synthesized using the sequence of peptides generated by tryptic digestion of the purified estrogen SULT protein to screen a λgt11 bovine pancreas cDNA library (Moore et al., 1988). The cloning of bovine estrogen SULT was followed shortly by a report describing the cloning of a rat liver hydroxysteroid SULT (Sta; Ogura et al., 1989). Sequence comparison studies detected the similarity between Sta and SMP-2, indicating that these cDNAs were members of the hydroxysteroid SULT gene family (Ogura et al., 1990). Subsequently, Ozawa et al. (1990) reported the cloning of a rat liver phenol SULT (PST-1) and Demyan et al. (1992) described the cloning and expression of a rat liver estrogen SULT. The initial isolation and characterization of multiple isoforms of rat SULT cDNA confirmed the heterogeneity of the SULTs and provided cDNAs to use as molecular probes for the isolation of human SULT cDNAs as well as for their subsequent identification and comparison. Comparison of the amino acid and nucleotide sequences of the initially cloned SULTs identified conserved sequences and motifs that were useful in the identification of the human SULTs.

HUMAN CYTOSOLIC SULTS

The human cytosolic SULTs comprise a family of 11 related proteins encoded by 10 distinct genes. The human SULTs that have been identified to date and the names used to refer to the proteins are shown in Table 5.1. Further characterization and

TABLE 5.1
Human Cytosolic SULTS

Suggested Nomenclature	Previously Used Names
SULT1A1	Phenol-sulfating phenol sulfotransferase (P-PST, P-PST-1; Wilborn et al., 1993; Jones et al., 1995)
	Thermostable phenol sulfotransferase (TS-PST, TS-PST-1; Raftogianis et al., 1997)
	Human aryl sulfotransferase 1 (HAST1; Zhu et al., 1993)
SULT1A2	Phenol-sulfating phenol sulfotransferase-2 (P-PST-2,STP2; Dooley and Huang, 1996)
	Thermostable phenol sulfotransferase-2 (TS-PST-2; Her et al., 1996a)
	HAST4 (Zhu et al., 1996)
	ST1A2 (Ozawa et al., 1995)
SULT1A3	Monoamine-sulfating phenol sulfotransferase (M-PST; Dooley et al., 1994; Ganguly et al., 1995; Jones et al., 1995)
	Thermolabile phenol sulfotransferase (TL-PST, STM; Aksoy et al., 1994a)
	HAST3 (Zhu et al., 1993)
	ST1A5 (Yamazoe et al., 1994)
	Placental estrogen sulfotransferase (Bernier et al., 1994)
SULT1B1	Thyroid hormone sulfotransferase (ST1B2; Fujita et al., 1997; Wang et al., 1998)
SULT1C1	SULT 1C sulfotransferase 1 (Sakakibara et al., 1998a)
	SULT1C2 (Yoshinari et al., 1998)
SULT1C2	SULT 1C sulfotransferase 2 (Sakakibara et al., 1998a)
	SULT1C3 (Yoshinari et al., 1998)
SULT1E1	Estrogen sulfotransferase (EST; Aksoy et al., 1994b; Falany et al., 1995)
SULT2A1	Dehydroepiandrosterone sulfotransferase (DHEA-ST; Comer et al., 1993; Otterness et al., 1992)
	Hydroxysteroid sulfotransferase (HST; Forbes et al., 1995)
	Alcohol/hydroxysteroid sulfotransferase (hST_a; Kong et al., 1992)
SULT2B1a	Hydroxysteroid sulfotransferase (Her et al., 1998)
SULT2B1b	Hydroxysteroid sulfotransferase (Her et al., 1998)
SULT4A1	Brain sulfotransferase-like cDNA (BR-STL; Falany et al., 2000)

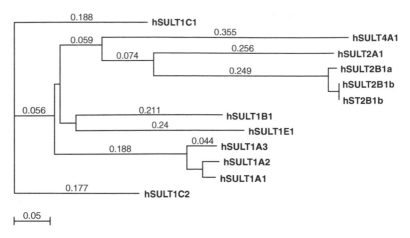

FIGURE 5.1 Phylogenetic tree of the human SULT amino acid sequence relationships. The relationships between the SULT isoforms were generated using the MacVector Sequence Analysis programs. The SULT sequences and accession numbers used in this analysis were: 1A1, L19999; 1A2, U28169; 1A3, U08032; 1B1, U95726; 1C1, U66036; 1C2, AF055584; 1E1, S77383; 2A1, L20000; 2b1a, U92314; 2B1b, U92315; 4A1, AF188698. The tree is unrooted and generated by neighbor joining.

analysis of the human genome may identify novel SULT-related genes or sequences. Figure 5.1 is a phylogenetic tree of the known human SULT sequences. Table 5.2 shows the amino acid identities of the SULT isoform sequences used to construct the phylogenetic tree. Currently, the SULT sequences represent ten genes located at five gene loci. However, even if new SULT genes were identified, it is anticipated that the number would be small because of the extensive investigation of the human genome that has already been carried out. Novel SULT cDNAs have been identified in nonhuman mammals that are not expressed in humans. A canine SULT1D1 cDNA was isolated and expressed that represented a novel member of the SULT1A family (Tsoi et al., 2001). Meinl and Glatt (2001) subsequently identified a SULT1D1 pseudogene in the gene cluster containing the human SULT1E1 and SULT1B1 genes on chromosome 4.

HUMAN SULT2A1: DHEA SULT

The initial identification of a rat hydroxysteroid SULT provided sequences for the generation of PCR primers and degenerate oligonucleotide probes to screen human cDNA libraries for related SULT sequences. Using degenerate oligonucleotide primers designed to conserved amino acid sequences in the rat and bovine SULT cDNA sequences, Otterness et al. (1992) reported the first cloning and expression of a human SULT, DHEA SULT. The translation of the DHEA SULT cDNA yielded a protein with a sequence identical to that of peptides obtained from limited proteolysis of the pure protein. Expression of the cDNA in COS-1 cells resulted in high levels of DHEA sulfation activity, confirming the identity of the cDNA. Within a year, two separate laboratories reported the isolation of the DHEA SULT cDNA by screening

TABLE 5.2
Sequence Identities of the SULT Isoforms

	1A1	1A2	1A3	1B1	1C1	1C2	1E1	2A1	2B1a	2B1b	4A1
1A1											
1A2	95.6										
1A3	92.9	89.5									
1B1	53.4	54.6	52.4								
1C1	52.2	51.9	51.2	52.7							
1C2	53.2	53.6	53.2	53.4	62.5						
1E1	50.1	49.0	49.0	55.6	47.6	44.2					
2A1	34.6	35.7	35.0	36.4	36.4	36.4	36.0				
2B1a	36.3	35.6	37.3	38.8	34.8	33.4	34.7	47.9			
2B1b	36.9	36.6	38.0	37.7	35.8	35.1	35.7	47.9	98.6		
4A1	34.2	33.4	35.9	33.1	32.7	36.3	33.8	29.6	33.5	33.1	

Note: The sequences of the SULTs were determined using the ClustalW program. The percent identity for each individual pair is presented.

human liver cDNA libraries using rat hydroxysteroid SULT cDNAs as probes under low stringency hybridization conditions. Kong et al. (1992) isolated two cDNAs with identical open reading frames but different length 3′-nontranslated regions. Comer et al. (1993) expressed the cloned DHEA SULT protein in COS-7 cells and demonstrated by immunoblot analysis that the expressed protein was similar in mass to the liver and adrenal enzymes (Comer & Falany, 1992; Comer et al., 1993). DHEA SULT is so named due to the high level of DHEA sulfation in humans (Hornsby, 1995). In the proposed SULT nomenclature, this isoform has been named SULT2A1. The enzyme is highly expressed in the reticular layer of the adult human adrenal (Falany et al., 1995) as well as in the fetal adrenal (Forbes et al., 1995; Parker et al., 1993). DHEA SULT is relatively abundant in the liver where it is apparently involved in bile acid sulfation (Radominska et al., 1991) and is present in the intestinal tract (Her et al., 1996b), including the lining of the stomach (Tashiro et al., 2000). DHEA SULT exists as a dimer with a subunit molecular mass of 33,765 Da (Comer et al., 1993) and migrates with a slightly larger mass than the human platelet phenol SULTs during SDS-polyacrylamide gel electrophoresis (Falany et al., 1990; Comer & Falany, 1992). The enzyme is capable of sulfating 3α, 3ß, and 3-phenolic hydroxysteroids as well as many aliphatic alcohols (Falany et al., 1989; Meloche et al., 2002; Radominska, 1991). The similar enzymatic and biochemical properties of DHEA SULT expressed and purified from different systems indicate that the native form of the enzyme is probably not modified posttranslationally. Analysis of purified SULT2A1 using MALDI TOF mass spectroscopy has not detected any mass changes associated with posttranslational modifications. Recent studies have indicated that hydroxysteroid SULTs responsible for bile acid sulfation in the liver of rodents and humans are regulated by bile acids via the farnesoid X receptor (FXR); (Kitada et al., 2003; Song et al., 2001). This would be consistent with the protective effects of sulfation on bile acid toxicity in cholestasis.

SULT2A1 is encoded by a single gene localized to human chromosome 19q13.3 (Otterness et al., 1995b). The gene consists of 6 exons and 5 introns and spans approximately 17 kilobases. The location of splice junctions is conserved among a number of the SULT genes, including several rodent SULT genes (Otterness et al., 1995a). The gene encoding the two isoforms of SULT2B1 is also located on chromosome 19q13.3 (Her et al., 1998).

HUMAN SULT2B1

The hydroxysteroid SULT subfamily also includes the two isoforms of SULT2B1. These enzymes represent two transcriptional products from the SULT2B1 gene that is localized at the same gene locus as the SULT2A1 gene (Her et al., 1998; Otterness et al., 1995a). The SULT2B1 gene was initially identified by screening a human placental expressed sequence tag database with an oligonucleotide encoding a highly conserved amino acid sequence motif "RKGxxGDWKNxFT" present in the SULTs (Her et al., 1998). This procedure identified two related cDNA sequences that were derived from the same gene but utilized different transcriptional start sites to incorporate different first exons. The cDNAs were termed SULT2B1a and SULT2B1b. SULT2B1b is 365 amino acids in length, whereas SULT2B1a is 350 amino acids in length. The final 344 amino acids of both sequences are identical since they are derived from the same exons. Expression of the SULT2B1 cDNAs in COS-1 cells utilizing the pCR3.1 vector generated both active proteins. Subsequent expression studies of the histidine (His)-tagged and native forms of the SULT2B1 cDNAs in *Escherichia coli* demonstrated difficulty in the expression of the native form of SULT2B1b (Meloche and Falany, 2001). However, SULT2B1b was readily expressed with an amino terminal His tag; the active native enzyme could then be obtained by cleavage of the His tag. The expressed enzymes demonstrated selectivity for the sulfation of the 3ß-OH position of hydroxysteroids such as DHEA and pregnenolone. No activity was observed using 3α-OH steroids or 3-phenolic steroids as substrates (Meloche and Falany, 2001), but both expressed enzymes were capable of sulfating dihydrotestosterone (Geese and Raftogianis, 2001).

SULT2B1a and SULT2B1b message expression occurs in a number of human tissues including prostate, placenta, trachea, and skin (Her et al., 1998; Geese and Raftogianis, 2001; Meloche and Falany, 2001). The levels of SULT2B1b specific message are generally several-fold greater than those for SULT2B1a. Immunoblot analysis of the expressed native forms of SULT2B1a and SULT2B1b has shown that these proteins can be easily distinguished by their different molecular masses (Meloche and Falany, 2001). In contrast to message levels, immunoblot analysis of several human tissues detects the presence of only SULT2B1b protein (Figure 5.2).

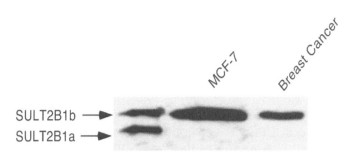

FIGURE 5.2 Expression of SULT2B1b in human MCF-7 breast cancer cells and human breast cancer tissue. Cytosol was prepared from MCF-7 cells and from a specimen of human breast cancer (white female, age 56 years). The cytosols were resolved by SDS-polyacrylamide gel electrophoresis, transferred to nitrocellulose membrane, and incubated with a rabbit antihuman SULT2B1 antibody (Meloche and Falany, 2001). The secondary antibody was goat antirabbit IgG with protein visualization by chemiluminescence using a West Pico kit (Pierce).

These results suggest that SULT2B1a and SULT2B1b expression in human tissues is controlled at both the transcriptional and translational levels and that SULT2B1b is selectively expressed in human tissues.

HUMAN PHENOL SULTS

The analysis of phenol SULT activity and expression in human tissues was significantly advanced by the presence of phenol SULT activity in blood platelets. The ability to analyze platelet phenol SULT activity resulted in the extensive characterization of the biochemical and pharmacogenetic properties of these enzymes. Purification of MPST (SULT1A3) and PPST (SULT1A1) from platelet and liver, respectively, was instrumental in providing the background for the cloning and expression of the phenol SULT gene subfamily. The phenol SULT family consists of seven members encoded by separate genes that are arranged in clusters on several chromosomes (Freimuth et al., 2000; Weinshilboum et al., 1997).

SULT1A1: PHENOL-SULFATING PHENOL SULT-1

Molecular characterization of the human SULTs began with the cloning and expression of the cDNA for PPST-1 from a liver cDNA library (Wilborn et al., 1993). With the development of our understanding of the heterogeneity of the SULT family, PPST has subsequently been termed SULT1A1 (Table 5.1). It was demonstrated that the biochemical and enzymatic properties of the expressed PPST-1 protein were similar to those of the platelet PPST enzyme (Wilborn et al., 1993). Polyclonal rabbit antibody raised to either MPST (SULT1A3) or PPST-1 (SULT1A1) reacts strongly with both proteins, but differential migration of the proteins during SDS-polyacrylamide gel electrophoresis facilitates identification of the individual enzymes (Falany et al., 1990; Heroux et al., 1989). The translated molecular mass of SULT1A1 is 34,097 Da (Wilborn et al., 1993), which is slightly less than that of SULT1A3 (MPST; Wood et al., 1994); however, SULT1A1 migrates with a mass of approximately 32 kD, whereas

SULT1A3 migrates with a mass of approximately 34,000 Da (Heroux et al., 1989; Wilborn et al., 1993). Analysis of the cDNAs encoding SULT1A1 indicates the use of multiple transcriptional start sites and incorporation of different nontranslated first exons (Weinshilboum et al., 1997). A systematic evaluation of the use of different promoters for the expression of SULT1A1 or the possible involvement of these promoters in the levels or tissue selective expression of SULT1A1 has not been reported.

Expressed liver SULT1A1 is capable of sulfating a wide variety of small phenolic compounds as well as ß-estradiol (Falany et al., 1994), iodothyronines (Kester et al., 1999b), and minoxidil (Falany et al., 1990). SULT1A1 activity is generally measured with low micromolar concentrations of PNP although other phenol SULTs are also capable of catalyzing PNP sulfation. The broad substrate reactivity of SULT1A1 also leads to the sulfation of buffer components and contaminants as well as small endogenous compounds such as tyrosine in tissue preparations and small alcohols in reaction mixtures. When (^{35}S)-PAPS is used in the assays, product identification is generally required to ensure proper kinetic analysis. SULT1A1 is recognized as the major drug/xenobiotic SULT in human tissues due to its high activity in the liver and gastrointestinal tract, its broad tissue distribution, and its broad substrate reactivity. As observed with all human SULTs, SULT1A1 is capable of sulfating and activating promutagens such as 1-hydroxymethylpyrene and N-hydroxyacetylaminofluorene to reactive electrophiles (Glatt, 1997; Glatt et al., 1994, 1995).

SULT1A1 has been expressed in COS and V79 cells as well as in *E. coli*, either as native enzyme or with a His tag to facilitate purification (Glatt et al., 1995; Lewis et al., 1996; Veronese et al., 1994b; Wilborn et al., 1993). The cloned enzyme is easily expressed and stable and possesses kinetic properties similar to those of the purified liver enzyme (Falany et al., 1994; Wilborn et al., 1993). The active enzyme is a dimer and apparently does not undergo posttranslational modification. Many of the human SULTs demonstrate substrate inhibition with increasing substrate concentrations. Substrate inhibition is observed with SULT1A1, particularly with high affinity substrates such as PNP. Substrate inhibition is due in part to the formation of nonproductive SULT enzyme–PAP complexes although allosteric modulation by substrate may be involved with some of the SULTs (Zhang et al., 1998). Gamage et al. (2003) have reported that two molecules of PNP are capable of binding the active site of SULT1A1. The active site may be flexible, permitting conjugation of a wide range of compounds by SULT1A1.

Initial studies on the variability and thermostability of PPST activity in human platelets strongly indicated that humans expressed multiple allelic variants of SULT1A1 (Weinshilboum, 1990). At least 17 allelic variants of human SULT1A1 have been reported encoding 6 different amino acid changes (Glatt et al., 2001). Two of these alleles are highly represented in the human population. The wild-type allele (SULT1A1*1) has a frequency of 0.674 in Caucasians in the U.S., whereas the Arg213His allele (SULT1A1*2) has a frequency of 0.313 (Raftogianis et al., 1997, 1999). The expressed Arg213His allele shows decreased thermostability compared to the wild-type allele and is apparently responsible for the thermostability differences described in platelet PPST activities (Raftogianis et al., 1999). The pharmacogenetic basis for a significant fraction of the phenol SULT activity in human tissues led to the investigation of the possible association of SULT1A1 expression with cancer development in several tissues (Bamber et al., 2001; Wang et al., 2002; Zheng et al., 2001).

The structural gene for SULT1A1 is located on chromosome 16q12.1-11.2 and is a member of the cluster of SULT1A genes present at this locus (Aksoy et al., 1994a; Dooley et al., 1993; Dooley and Huang, 1996). This locus contains the genes for SULT1A1, SULT1A2, and SULT1A3 that are greater than 92% identical in sequence, indicating that these genes arose by duplication of a single ancestral gene. The proximity of the structural genes is also responsible for the linkage in the expression of allelic variants of the SULT1A genes (Engelke et al., 2000). Structural analysis of the SULT1A genes demonstrates considerable conservation of exonic structure between the human SULT genes and the rodent SULT genes (Weinshilboum et al., 1997). A functional role for the conservation of the exonic structure has not been described.

SULT1A2

The characterization and purification of the human platelet phenol SULT activities established that two unique but closely related phenol SULTs were present in human tissues. During the cloning of the SULT1A1 (P-PST) and SULT1A3 (M-PST) cDNAs, another novel closely related SULT1A cDNA was isolated from human liver (Ozawa et al., 1995). This cDNA demonstrated high homology with the human PSTs (SULT1A1 and SULT1A3) and has been included in the SULT1A family as SULT1A2. Two variants of SULT1A2 were subsequently isolated from a human brain λgt10 library using the SULT1A3 cDNA as a probe. These represented allelic forms of SULT1A2 (originally named HAST4 and 4v) that differed significantly in their abilities to sulfate small phenols such as PNP (Zhu et al., 1996). Compared to SULT1A1, the expressed SULT1A2 isoforms showed a lower affinity for sulfation of the small phenolic substrates and were ineffective in the conjugation of dopamine. Although the characterization of SULT1A2 has progressed from the cloning of the cDNA, there have been no reports of the isolation or characterization of the protein from human tissues. This is apparently due to low levels of activity and expression relative to those of SULT1A1, making resolution of the two proteins difficult. The kinetic properties and the low, seemingly sporadic expression of SULT1A1 in human hepatic tissues implies that this SULT may have only a minor role in xenobiotic metabolism in humans.

As observed with SULT1A1, significant genetic allelic variability has been detected in the SULT1A2 gene. The major SULT1A2 allele termed SULT1A2*1 represents over 50% of the alleles in the Caucasian population (Glatt et al., 2001). SULT1A2*2 (Ile7Thr, Asn235Thr) represents approximately 30% of the alleles and has a K_m for PNP sulfation approximately 40-fold higher than that of SULT1A2*1 (Raftogianis et al., 1999). SULT1A2*3 (Pro9Leu) represents approximately 18% of alleles in the Caucasian population and encodes an allozyme with decreased thermostability but with kinetic properties similar to those of SULT1A2*1.

The gene for SULT1A2 has been localized to the SULT1A locus at chromosome 16p12.1-11.2 (Dooley and Huang, 1996), which contains the genes for SULT1A1, SULT1A2, and SULT1A3 (71-73, 77, 78). At least 13 allelic variants of SULT1A2 have been reported encoding 6 allozymes of SULT1A2 (Glatt et al., 2001). The SULT1A2 alleles are in linkage disequilibrium with SULT1A1, while the high activity allozymes (SULT1A1*1, SULT1A2*1) are associated (Engelke et al., 2000).

SULT1A3

SULT1A3, also referred to a thermolabile PST and MPST, was characterized and purified from human platelet cytosol prior to cloning of the cDNA (Heroux and Roth, 1988). The initial report of the cloning of SULT1A3 involved isolation of the cDNA from a human brain λgt10 library. Zhu et al. (1993) used primers designed to a putative human liver phenol (aryl) SULT cDNA sequence (Zhu et al., 1993) to perform PCR with a human brain cDNA library. Two related brain cDNAS were isolated with 93% sequence identity. Subsequently, the cDNAs were expressed in COS-7 cells and were identified as HAST-1 (SULT1A1) and HAST-3 (SULT1A3) based on enzymatic activity, thermostability, and immunoblot analysis (Veronese et al., 1994a). Concurrently, Wood et al. (1994) used a peptide sequence obtained from limited proteolysis of purified human jejunal thermolabile PST (SULT1A3, TL-PST, M-PST) to design degenerate oligonucleotide primers for isolation of the SULT1A3 cDNA from a human liver cDNA library. The cloned SULT1A3 cDNA was expressed in COS-1 cells and reported to have biochemical and physical properties identical to those of human liver TL-PST (SULT1A3). SULT1A3 has also been cloned from human ZR75 breast cancer cells (Ganguly et al., 1995). An apparent SULT1A3 cDNA was isolated from a human placenta cDNA library (Bernier et al., 1994); however, the placental cDNA was termed an estrogen sulfotransferase. Whereas it was initially reported that SULT1A3 did not sulfate estrogens (Ganguly et al., 1995), subsequent reports have indicated that it possesses low levels of estrogen sulfation activity (Faucher et al., 2002). Unlike SULT1A1 and SULT1A2, all of the reported human SULT1A3 amino acid sequences are identical, indicating a low frequency of sequence variants. A recent preliminary report has indicated that several SNPs are present in the SULT1A3 genes of Caucasian and African Americans including one nonsynonymous SNP in the African American population (Thomae et al., 2003).

The SULT1A3 cDNAs encode a 295 amino acid protein with a molecular mass of 34,194 Da. An extensive study by Dajani et al. (1999) compared the physical and kinetic properties of SULT1A3 expressed in *E. coli*, *Saccharomyces cerevisiae*, COS-7, and V79 cells with SULT1A3 activity in human platelet cytosol and in the purified recombinant enzyme expressed in *E. coli*. No significant differences were observed in the properties of the SULT1A3 enzymes, indicating that posttranslational modification does not alter the properties of the enzyme.

An interesting characteristic of the human SULT1A subfamily is the distinct substrate reactivities of SULT1A1 and SULT1A3 despite the fact that they are structurally very similar with a 93% sequence identity (Figure 5.3). SULT1A1 has a high affinity for PNP sulfation, whereas SULT1A3 has a high affinity for dopamine sulfation. To date, a high activity monoamine SULT isoform, such as SULT1A3,

```
SULT1A1  MELIQDTSRPPLEYVKGVPLIKYFAEALGPLQSFQARPDDLLISTYPKSGTTWVSQILDM  60
SULT1A2        I                                                         60
SULT1A3                                            S          N           60

SULT1A1  IYQGGDLEKCHRAPIFMRVPFLEFKAPGIPSGMETLKDTPAPRLLKTHLPLALLPQTLLD 120
SULT1A2                        V              NN                          120
SULT1A3         N    YV       VND  E     L         P    I S             120

SULT1A1  QKVKVVYVARNAKDVAVSYYHFYHMAKVHPEPGTWDSFLEKFMVGEVSYGSWYQHVQEWW 180
SULT1A2                        Y H    E        A                          180
SULT1A3          P         HR E A                A                        180

SULT1A1  ELSRTHPVLYLFYEDMKENPKREIQKILEFVGRSLPEETVDFMVQHTSFKEMKKNPMTNY 240
SULT1A2                                          L  E                     240
SULT1A3                                  M                                240

SULT1A1  TTVPQEFMDHSISPFMRKGMAGDWKTTFTVAQNERFDADYAEKMAGCSLSFRSEL      295
SULT1A2  RR                                                     I        295
SULT1A3   L                                                             295
```

FIGURE 5.3 Comparison of the sequences of the three human SULT1A isoforms. Sequences were compared using the ClustalW program. The accession numbers are the same as those for Figure 5.1. Only differences in the sequences are noted as compared to the sequence of SULT1A1.

has been identified only in humans. Eisenhofer et al. (1999) have suggested that SULT1A3 is responsible for the sulfation of dietary amines as well as dopamine generated in the human gastrointestinal tract. Using site-directed mutagenesis, Dajani et al. (1998) demonstrated that a single amino acid residue, GLU146, is responsible for the difference in substrate reactivity. Conversion of GLU146 to an ALA residue in SULT1A3 conferred the kinetic properties of SULT1A1 to SULT1A3. These results are consistent with the report of Sakakibara et al. (1998b) describing the investigation of the kinetic properties of SULT1A1 and SULT1A3 chimeras.

The SULT1A3 structural gene is located at the SULT1A locus on chromosome 16 (Aksoy et al., 1994a). No allelic variants of SULT1A3 have been reported although multiple start sites of transcription have been described (Weinshilboum et al., 1997). The role of the different transcription start sites in the regulation of the SULT1A family has not been investigated. In several reported instances, the additional 5′-untranslated sequence incorporated into both SULT1A1 and 1A3 cDNAs may be attributed to cloning artifacts.

SULT1B1

The initial cloning of human SULT1B1 from a human liver λgt11 cDNA library was reported by Fugita et al. (1997). A combination of cDNA library screening with rat SULT1B1 and 5′ and 3′ RACE was used to isolate the human SULT1B1 cDNA. Subsequently, a SULT1B1 sequence was isolated and expressed from a human liver λZap cDNA library (Wang et al., 1998) using a probe based on the nucleotide sequence for human SULT1B1 reported in abstract form at the Fourth International ISSX meeting by Yamazoe et al. (1995). The isolated human SULT1B1 cDNAs encode a 296 amino acid protein. However, the two SULT1B1 sequences differ at amino acid 186. The sequence of Fugita et al. (1997) encodes a GLY at this position, whereas that of Wang et al. (1998) encodes a GLU. Subsequent sequencing of ten human SULT1B1 cDNAs indicated that all of the SULT1B1 cDNAs possessed a GLU residue at position 186. Fifteen SULT1B1 samples examined by Glatt et al. (2001) were also reported to encode a GLU residue at position 186. The SULT1B1 protein has a calculated molecular mass of 34,897 Da. The amino acid sequence is 74% identical to a rat SULT1B1 sequence and 52 to 56% identical to the other human phenol SULTs. Northern-blot analysis detected message in poly A RNA isolated from human liver, colon, and small intestine as well as peripheral leukocytes, spleen, and thymus (Wang et al., 1998). A polyclonal rabbit antibody raised against pure human SULT1B1 also cross-reacted with human SULT1E1. Following absorption against pure SULT1E1, the specific anti-SULT1B1 antibody readily detects SULT1B1 in cytosol from human small intestine and colon with lower levels apparent in liver cytosol.

The structural gene encoding SULT1B1 has been localized to human chromosome 4q13.1 (Meinl and Glatt, 2001). The gene for SULT1E1 and a pseudogene related to canine SULT1D1 are also situated in the locus on chromosome 4. The presence of both the SULT1E1 and SULT1B1 genes at the same chromosomal locus suggests that these genes arose by duplication. Comparison of the high sequence identity between the SULT1A genes on chromosome 16 with the sequence identity between SULT1E1 and SULT1B1 on chromosome 4 would suggest that duplication of the chromosome 4 genes is evolutionarily a much earlier event. The presence of the SULT1D1 pseudogene at the chromosome 4 locus is consistent with an earlier duplication event.

SULT1B1 was initially referred to as a thyroid hormone-sulfating SULT due to its ability to sulfate a number of thyroid hormones, including T3 and T4 (Fujita et al., 1997; Wang et al., 1998). The highest levels of activity were observed with 3,3′-T2, which is consistent with the role of sulfation in increasing the rate of thyroid hormone deiodination (Visser, 1994, 1996). Thyroid hormones are also substrates for several other human SULTs. SULT1E1 efficiently sulfates thyroid hormones, and SULT1A1 has thyroid hormone sulfation activity as well (Kester et al., 1999a, b; Young et al., 1988). Expressed SULT1B1 also efficiently conjugates several prototypical phenolic substrates such as 1-naphthol and PNP. Most steroids are not readily sulfated by expressed human SULT1B1 (Fujita et al., 1999; Wang et al., 1998).

SULT1C1 AND SULT1C2

SULT activity is important for the bioactivation of a number of procarcinogens to reactive electrophiles capable of binding to cellular nucleophiles including DNA. N-OH acetylaminofluorene (N-OHAAF) is the classical procarcinogen which is activated by sulfation (Miller and Miller, 1968; Miller et al., 1985). The investigation of SULTs involved in the bioactivation of N-OH-AAF led to the identification of the SULT1C family. Initial identification of the human SULT1C1 cDNA was accomplished by Her et al. (1997), using an expressed sequence database screening procedure. A highly conserved SULT signature sequence from the carboxy-region of the proteins (RKxxGDWKNxFT) was used to screen expressed sequence tag databases. This approach resulted in the identification of a novel SULT-related sequence in a human fetal liver–spleen cDNA library. The clone was obtained from the American Type Culture Collection and encoded a 296 amino acid protein that was 62% identical in sequence to a rat SULT1C1 sequence (Nagata et al., 1993). SULT1C1 displays approximately 50% amino acid sequence identity with the human SULT1 proteins, resulting in its inclusion in the SULT1 subfamily. SULT1C1 message expression was reported in human adult stomach, kidney, and thyroid as well as in fetal kidney and liver (Her et al., 1997). The structural SULT1C1 gene was localized to chromosome 2q11.1-11.2 (Her et al., 1997). Resequencing of the SULT1C1 gene of 89 Caucasians detected 4 nonsynonymous SNPs and 5 insertion/deletions, indicating that SULT1C1 is polymorphic in humans (Freimuth et al., 2001).

Cloning of human SULT1C2 also involved a search of sequence databases for novel sequences with similarity to the human SULT1C1 sequence (Sakakibara et al., 1998a; Yoshinari et al., 1998). The translation of the isolated SULT1C2 cDNA encodes a 302 amino acid protein. SULT1C2 is 63% identical in sequence to SULT1C1 and is also localized to the chromosome 2q11.1-11.2 loci. Similar to SULT1C1, SULT1C2 message is expressed in fetal kidney and fetal liver. SULT1C2 message was also detected in adult human heart, kidney, and ovarian tissues (Yoshinari et al., 1998). The purified His-tagged SULT1C1 and 1C2 proteins were both capable of sulfating PNP but with different kinetic properties. Yoshinari et al. (1998) also reported that His tagged SULT1C1 but not His tagged SULT1C2 was capable of activating NOHAAF to a mutagen. Sakakibara et al. (1998a) expressed both SULT1C1 and SUL1C2 with thrombin-cleavable glutathione s-transferase (GST) tags. Following thrombin-catalyzed removal of the GST moieties, SULT1C2 possessed higher levels of N-OH-AAF activation than SULT1C1. The differences in kinetic activity may be related to minor sequence differences associated with the His- and GST-expression tags.

SULT1E1: ESTROGEN SULFOTRANSFERASE (EST)

Studies on the sulfation of hydroxysteroids and estrogens in human breast cancer cells and endometrial tissues indicated that separate SULT activities were most likely involved. The identification and characterization of these steroid sulfation activities were confused by a lack of consistency in assay conditions and substrate concentrations. Multiple SULTs present in human tissues are capable of conjugating various hydroxysteroid and estrogenic steroids with different affinities. Our understanding of the role and mechanism of estrogen sulfation has benefited greatly from the cloning and characterization of the SULTs involved in steroid sulfation.

SULT1E1 (estrogen SULT) is the SULT isoform responsible for the sulfation of ß-estradiol at physiological concentrations (Falany et al., 1995; Zhang et al., 1998). The SULT1E1 cDNA was initially cloned from human liver by Weinshilboum's research group, using degenerate oligonucleotides to highly conserved amino acid sequences conserved in nonmammalian estrogen SULT sequences (Aksoy et al., 1994b). Using PCR and RACE amplification, these researchers isolated the full-length human SULT1E1 cDNA. The translated protein was 294 amino acids in length with a calculated molecular mass of 35,123 Da. The human SULT1E1 sequence is

81%, 73%, and 72% identical to the amino acid sequences of guinea pig, cow, and rat SULT1E1s, respectively. Expression of the SULT1E1 cDNA in COS-1 cells demonstrated that SULT1E1 catalyzed ß-estradiol, estrone, and DHEA sulfation. Human liver SULT1E1 was subsequently also isolated by screening of a human liver λZap cDNA library using a rat SULT1E1 cDNA as a probe (Falany et al., 1995). The human SULT1E1 cDNAs encode proteins with identical sequences. A polyclonal rabbit antibody raised to purified expressed SULT1E1 protein recognizes SULT1E1 in human liver and intestinal cytosol and does not cross-react with other human SULTs. A unique characteristic of SULT1E1 is its high affinity for the sulfation of ß-estradiol, estrone, and 17α-ethinylestradiol. Expressed and purified SULT1E1 displays substrate inhibition with increasing ß-estradiol and estrone concentrations. Maximal ß-estradiol sulfation is observed at ß-estradiol concentrations of 15 to 20 nM. Detailed kinetic characterization of expressed purified SULT1E1 indicates that ß-estradiol has a K_m of 5 nM, while that of PAPS is 59 nM (Zhang et al., 1998). Substrate inhibition observed with increasing ß-estradiol concentrations is the result of the combination of ternary (dead-end) complex formation as well as allosteric inhibition of SULT1E1 by ß-estradiol. The physiological function of the allosteric modulation of SULT1E1 may be a moot point since the K_i of 80 nM is not in the physiological range of ß-estradiol concentrations. Expressed SULT1E1 is also capable of sulfating DHEA, pregnenolone, diethylstilbestrol, and equilenin although micromolar substrate concentrations are required (Falany et al., 1994, 1995).

SULT1E1 is expressed in liver, intestinal tract, testis, breast, and endometrium. In human endometrium, SULT1E1 is selectively expressed during the secretory phase of the menstrual cycle (Falany et al., 1998). No SULT1E1 message or protein is observed during the proliferative phase of the cycle. In contrast, the endometrial expression of SULT1A1 that sulfates estrogen at high concentrations does not vary during the menstrual cycle. Increasing progesterone levels occurring after ovulation may regulate SULT1E1 expression. SULT1E1 expression is selectively inducible by progestins in human Ishikawa endometrial adenocarcinoma cells (Falany and Falany, 1996). The increased levels of SULT1E1 activity in secretory endometrium may have a role in the diminution of ß-estradiol activity in the endometrium after ovulation (Kotov et al., 1999).

The gene for SULT1E1 was mapped to human chromosome 4q13.1 by fluorescence *in situ* hybridization (Her et al., 1995). This locus also contains the gene for SULT1B1 as well as a pseudogene related to the canine SULT1D1 cDNA (Meinl et al., 2001). Both genes consist of eight exons with most intron-exon splice junctions identical to those of the human SULT1A genes. Resequencing of the SULT1E1 alleles in 60 African and 60 Caucasian Americans identified 23 polymorphisms including 3 nonsynonymous SNPs (Adjei et al., 2003). These results are consistent with a trend toward the identification of a substantial number of polymorphisms in the individual isoforms of the SULT gene family.

SULT4A1: BRAIN SULTS

Expression of the SULTs in brain has been of interest since the identification of high levels of dopamine sulfation in human blood platelets (Hart et al., 1979). This interest level was elevated by the investigation of the role of sulfation in neurosteroid synthesis and activity (Baulieu, 1991, 1997). Both SULT1A1 and SULT1A3 have been identified in human brain tissue (Young et al., 1984) and localized to neurons in the central nervous system (Zou et al., 1990). Recently, a novel subfamily of SULTs has been identified that display highly selective expression in brain. Falany et al. (2000) reported the cloning and expression of novel SULT-related sequences from human and rat brain that have been termed SULT4A. Sequences related to the human enzyme, SULT4A1, were originally identified in a human pancreatic beta cell cDNA library. The pancreatic sequences contained several apparently unspliced introns. The SULT-related pancreatic sequences were then used to screen human tissues for related expressed RNA messages. These studies led to the identification of an abundant message in brain RNA and subsequently to the isolation of the full-length SULT4A1 cDNA from a human brain λZap library (Falany et al., 2000). The human SULT4A1 sequence contains several conserved SULT identifier sequences and has been assigned to a separate SULT subfamily. The SULT 4A family is notable in the high degree of sequence conservation between mammalian species. Figure 5.4 shows the amino acid sequence alignment between the SULT4A1 proteins from human, rat, and rabbit brain as well as a 270 amino acid fragment of chicken brain SULT 4A1. The mouse 4A1 sequence is identical to the rat sequence. The human,

```
Rabbit SULT4A1   MAESEAETPSTPGEFESKYFEFHGVRLPPFCRGKMEEIANFPVRPSDVWIVTYPKSGTSL    60
Rat SULT4A1      MAESEAETPGTPGEFESKYFEFHGVRLPPFCRGKMEDIADFPVRPSDVWIVTYPKSGTSL    60
Human SULT4A1    MAESEAETPSTPGEFESKYFEFHGVRLPPFCRGKMEEIANFPVRPSDVWIVTYPKSGTSL    60
Chicken SULT4A1          FESKYFEYNGVRLPPFCRGKMEEIANFPVRDSDVWIVTYPKSGTGL    46
                         *******.*.*************.*********.***************
Rabbit SULT4A1   LQEVVYLVSQGADPDEIGLMNIDEQLPVLEYPQPGLDIIKELTSPRLIKSHLPYRFLPSD   120
Rat SULT4A1      LQEVVYLVSQGADPDEIGLMNIDEQLPVLEYPQPGLDIIKELTSPRLIKSHLPYRFLPSD   120
Human SULT4A1    LQEVVYLVSQGADPDEIGLMNIDEQLPVLEYPQPGLDIIKELTSPRLIKSHLPYRFLPSD   120
Chicken SULT4A1  LQEVVYLVSQGADPDEIGLMNIDEQLPVLEYPQPGLDIIKELTSPRLIKSHLPYRFLPSD   106
                 ************************************************************
Rabbit SULT4A1   LHNGDSKVIYMARNPKDLVVSYYQFHRSLRTMSYRGTFQEFCRRFMNDKLGYGSWFEHVQ   180
Rat SULT4A1      LHNGDSKVIYMARNPKDLVVSYYQFHRSLRTMSYRGTFQEFCRRFMNDKLGYGSWFEHVQ   180
Human SULT4A1    LHNGDSKVIYMARNPKDLVVSYYQFHRSLRTMSYRGTFQEFCRRFMNDKLGYGSWFEHVQ   180
Chicken SULT4A1  LHNGNSKVIYMARNPKDLVVSYYQFHRSLRTMSYRGTFQEFCRRFMNDKLGYGSWFEHVQ   166
                 ****.*******************************************************
Rabbit SULT4A1   EFWEHRMDGNVLFLKYEDMHRDLVTMVEQLARFLGVPCDKAQLESLIEHCHQLVDQCCNA   240
Rat SULT4A1      EFWEHRMDANVLFLKYEDMHRDLVTMVEQLARFLGVSCDKAQLESLIEHCHQLVDQCCNA   240
Human SULT4A1    EFWEHRMDSNVLFLKYEDMHRDLVTMVEQLARFLGVSCDKAQLEALTEHCHQLVDQCCNA   240
Chicken SULT4A1  EFWEHHMDANVLFLKYEDMHKDLATMVEQPVRFLGVSYDKAQLESMVEHCHQLIDQCCNA   226
                 *****.*.*********** .**.*****..******.******...******.******
Rabbit SULT4A1   EALPVGRGRVGLWKDIFTVSMNEKFDLVYKQKMGKCDLTFEFYL     284
Rat SULT4A1      EALPVGRGRVGLWKDIFTVSMNEKFDLVYKQKMGKCDLTFDFYL     284
Human SULT4A1    EALPVGRGRVGLWKDIFTVSMNEKFDLVYKQKMGKCDLTFDFYL     284
Chicken SULT4A1  EALPVGRGRVGLWKDIFTVSMNEKFDLVYKQKMGKCDLTFDFYL     270
                 ****************************************.***
```

FIGURE 5.4 Comparison of amino acid sequences of SULT 4A1 from human, rat, rabbit, and chicken. The translations of the SULT4A1 cDNAs were aligned using the ClustalW Alignment program in the MacVector Sequence Analysis programs. The accession numbers for the sequences are human (AF188698), rat (AF188699), and rabbit (AV196782). The mouse SULT4A1 sequence (AF059257) is identical to that of the rat enzyme and is not included in the comparison. The chicken SULT4A1 sequence represents a partial cDNA isolated from chicken brain RNA using reverse transcription-PCR and human specific SULT4A1 oligonucleotide primers.

mouse, and rabbit SULT41 sequences display 98% identity between themselves (Figure 5.4). The chicken SULT4A1 sequence is 96.7% identical to the human protein sequence, suggesting a very high degree of sequence and functional conservation even in nonmammalian species. SULT4A1 message expression in human tissues was detected only in brain and testes by Northern-blot analysis. The sequence of the testis message was identical to that of the brain. Characterization of SULT4A1 expression in different brain regions showed that message RNA levels were highest in cortical regions (Falany et al., 2000). Immunoblot analysis of rat brain cytosol showed that immunoreactive rat SULT4A1 comigrated with the expressed native protein. Falany et al. (2000) were not able to detect sulfation activity with either human or rat SULT4A1 after expression and purification from insect Sf9 cells using baculoviral vectors. Sakakibara et al. (2002) have reported trace levels of sulfation activity at millimolar concentrations of several phenolic substrates using purified mouse and human SULT4A1 expressed in *E. coli.*

The high level of sequence conservation for SULT4A1 from different mammalian species combined with selective expression in brain tissues suggests an important function for this enzyme. The marginal activity and instability of the expressed protein has limited the investigation of the physiological properties of SULT4A1.

SUMMARY

The human genome possesses the ability to express at least 11 cytosolic SULT isoforms. Whether all of these isoforms are functionally expressed in most or all humans has not been established. Characterization and cloning of many of the initial SULT cDNAs resulted from the classical approach of protein purification followed by isolation of the corresponding cDNAs using antibody screening or PCR with degenerate primers designed to peptide sequences. A number of more recently reported SULT isoforms have been identified from inspection of expressed sequence tag libraries or human genome sequences for SULT identifier sequences. The combination of these approaches has resulted in a more comprehensive understanding of the molecular characteristics of the SULTs than of their physiological properties. An excellent example of this situation occurs with SULT4A1 that is selectively localized in mammalian brain; however, the substrates and functions for this SULT remain to be elucidated. The two major current directions of human SULT research involve the pharmacogenetic/pharmacogenomic analysis of SULT variability and the characterization of the physiological functions of the individual SULT isoforms. It is interesting to reflect on the physiological function of human SULT isoforms such as SULT1A3 and SULT1E1 for which allelic variants have not been reported.

The comprehensive sequencing of the human genome has permitted the identification and characterization of the cytosolic human SULTs. It is likely that all of the human SULT isoforms have been identified. Although it is possible that other isoforms of SULT exist, their homology to the identified SULT genes is most likely very low. One major area of future SULT research will be the identification and characterization of the major allelic variants of each of the SULTs. The relative high frequency of SNPs and insertion/deletion in the genome strongly suggest that multiple allelic variants of each of the SULTs will ultimately be described. The properties and physiological consequences of the presence of these variants will be the focus of a major portion of SULT research in the next decade.

REFERENCES

Adjei, A.A., B.A. Thomae, J.L. Prondzinski, B.W. Eckloff, E.D. Wieben, and R.M. Weinshilboum, Human estrogen sulfotransferase (SULT1E1) pharmacogenomics: gene resequencing and functional genomics. *Brit. J. Pharmacol.*, 2003, 139:1373–1382.

Aksoy, I.A., D.F. Callen, S. Apostolou, C. Her, and R. Weinshilboum, Thermolabile phenol sulfotransferase gene (STM): localization to human chromosome 16p11.2. *Genomics*, 23: 275–277, 1994a.

Aksoy, I.A., T.C. Wood, and R. Weinshilboum, Human liver estrogen sulfotransferase: identification by cDNA cloning and expression. *Biochem. Biophys. Res. Comm.*, 200: 181–187, 1994b.

Arlt, W., J. Haas, F. Callies, M. Reinke, D. Hubler, M. Oettel, M. Ernst, H. Schulte, and B. Allolio, Biotransformation of oral dehydroepiandrosterone in elderly men: significant increase in circulating estrogens. *J. Clin. Endo. Metab.*, 84: 2170–2176, 1999.

Bamber, D.E., A.A. Fryer, R.C. Strange, J.B. Elder, M. Deakin, R. Rajagopal, A. Fawole, R.A. Gilissen, F.C. Campbell, and M.W. Coughtrie, Phenol sulphotransferase SULT1A1*1 genotype is associated with reduced risk of colorectal cancer. *Pharmacogenetics*, 11: 679–685, 2001.

Baulieu, E., Neurosteroids: a new function in the brain. *Biochem. J.*, 71: 3–10, 1991.

Baulieu, E., Neurosteroids: of the nervous system, by the nervous system, for the nervous system. *Recent Prog. Horm. Res.*, 52: 1–32, 1997.

Baumann, E., Ueber sulfosauren im harn. *Ber. Dtsch. Chem. Ges.*, 54., 1876.

Baxter, J.D. and J.B. Tyrrell, The adrenal cortex, in *Endocrinology and Metabolism*, P. Felig, et al., Eds., McGraw-Hill, New York, 1987, pp. 511–632.

Bernier, F., I. Lopez Solache, F. Labrie, and V. Luu-The, Cloning and expression of cDNA encoding human placental estrogen sulfotransferase. *Mol. Cell. Endocrinol.*, 99: R11–R15, 1994.

Casey, M.L., P.C. MacDonald, and E.R. Simpson, Endocrinological changes of pregnancy, in *Williams Textbook of Endocrinology*, J.D. Wilson and D.W. Foster, Eds., W.B. Saunders, Philadelphia, 1987, pp. 422–437.

Chatterjee, B., D. Majumdar, O. Ozbilen, C.V. Ramana Murty, and A.K. Roy, Molecular cloning and characterization of a cDNA for androgen-repressible rat liver protein, SMP-2. *J. Biol. Chem.*, 262: 822–825, 1987.

Comer, K.A. and C.N. Falany, Immunological characterization of dehydroepiandrosterone sulfotransferase from human liver and adrenals. *Mol. Pharm.*, 41: 645–651, 1992.

Comer, K.A., J.L. Falany, and C.N. Falany, Cloning and expression of human liver dehydroepiandrosterone sulfotransferase. *Biochem. J.*, 289: 233–240, 1993.

Dajani, R., A.M. Hood, and M.W. Coughtrie, A single amino acid, glu 146, governs the substrate specificity of a human dopamine sulfotransferase, SULT1A3. *Mol. Pharmacol.*, 54: 942–948, 1998.

Dajani, R., S. Sharp, S. Graham, S.S. Bethell, R.M. Cooke, D.J. Jamieson, and M.W. Coughtrie, Kinetic properties of human dopamine sulfotransferase (SULT1A3) expressed in prokaryotic and eukaryotic systems: comparison with the recombinant enzyme purified from *Escherichia coli. Protein Express. Purif.*, 16: 11–18, 1999.

Demyan, W.F., C.S. Song, D.S. Kim, S. Her, W. Gallwitz, T.R. Rao, M. Slomczynska, B. Chatterjee, and A.K. Roy, Estrogen sulfotransferase of the rat liver: complementary DNA cloning age- and sex-specific regulation of messenger RNA. *Mol. Endocrinol.*, 6: 589–597, 1992.

Dooley, T.P. and Z. Huang, Genomic organization and DNA sequences of two human phenol sulfotransferase genes (STP1 and STP2) on the short arm of chromosome 16. *Biochem. Biophys. Res. Comm.*, 228: 134–140, 1996.

Dooley, T.P., P. Probst, P.B. Munroe, S.E. Mole, Z. Liu, and N.A. Doggett, Genomic organization and DNA sequence of the human catecholamine-sulfating phenol sulfotransferase gene (STM). *Biochem. Biophys. Res. Comm.*, 205: 1325–1332, 1994.

Dooley, T.P., P. Probst, R.D. Obermoeller, M.J. Siciliano, N.A. Doggett, D.F. Callen, H.M. Mitchison, and S.E. Mole, Mapping of the phenol sulfotransferase gene (STP) to human chromosome 16p12.1-p11.2 and to mouse chromosome 7. *Genomics*, 18: 440–443, 1993.

Eisenhofer, G., M. Coughtrie, and D. Goldstein, Dopamine sulfate: an enigma resolved. *Clin. Exp. Pharmacol. Physiol.*, 26: S41–S53, 1999.

Engelke, C.E., W. Meinl, H.R. Boeing, and H. Glatt, Association between functional genetic polymorphisms of human sulfotransferases 1A1 and 1A2. *Pharmacogenetics*, 10: 163–169, 2000.

Falany, C.N., Enzymology of human cytosolic sulfotransferases. *FASEB J.*, 11: 206–216, 1997.

Falany, J.L. and C.N. Falany, Regulation of estrogen sulfotransferase in human endometrial adenocarcinoma cells by progesterone. *Endocrinology*, 137: 1395–1401, 1996.

Falany, J.L., R. Azziz, and C.N. Falany, Identification and characterization of the cytosolic sulfotransferases in normal human endometrium. *Chem.-Biol. Interact.*, 109: 329–339, 1998.

Falany, C.N., K.A. Comer, T.P. Dooley, and H. Glatt, Human dehydroepiandrosterone sulfotransferase: purification, molecular cloning, and characterization. *Ann. New York Acad. Sci.*, 774: 59–72, 1995.

Falany, C.N. and E.A. Kerl, Sulfation of minoxidil by human liver phenol sulfotransferase. *Biochem. Pharm.*, 40: 1027–1032, 1990.

Falany, C.N., V. Krasnykh, and J.L. Falany, Bacterial expression and characterization of a cDNA for human liver estrogen sulfotransferase. *J. Steroid Biochem. Mol. Biol.*, 52: 529–539, 1995.

Falany, C.N., M.E. Vazquez, J.A. Heroux, and J.A. Roth, Purification and characterization of human liver phenol-sulfating phenol sulfotransferase. *Arch. Biochem. Biophys.*, 278: 312–318, 1990.

Falany, C.N., M.E. Vazquez, and J.M. Kalb, Purification and characterization of human liver dehydroepiandrosterone sulfotransferase. *Arch. Biochem. Biophys.*, 260: 641–646, 1989.

Falany, C.N., J. Wheeler, T.S. Oh, and J.L. Falany, Steroid sulfation by expressed human cytosolic sulfotransferases. *J. Steroid Biochem. Mol. Biol.*, 48: 369–375, 1994.

Falany, C.N., X. Xie, J. Wang, J. Ferrer, and J.L. Falany, Molecular cloning and expression of novel sulfotransferase-like cDNAs from human and rat brain. *Biochem. J.*, 346: 857–864, 2000.

Faucher, F., L. Lacoste, and V. Luu-The, Human type 1 estrogen sulfotransferase: catecholestrogen metabolism and potential involvement in cancer promotion. *Ann. New York Acad. Sci.*, 963: 221–228, 2002.

Forbes, K.J., M. Hagen, H. Glatt, R. Hume, and M.W.H. Coughtrie, Human fetal adrenal hydroxysteroid sulfotransferase: cDNA cloning, stable expression in V79 cells and functional characterization of the expressed protein. *Mol. Cell. Endocrin.* 112: 53–60, 1995.

Freimuth, R.R., B. Eckloff, E.D. Weiben, and R.M. Weinshilboum, Human sulfotransferase SULT1C1 pharmacogenics: gene resequencing and functional genomic studies. *Pharmacogenetics*, 11: 747–756, 2001.

Freimuth, R.R., R. Raftogianis, T.C. Wood, E. Moon, U.J. Kim, J. Xu, M.J. Siciliano, and R.M. Weinshilboum, Human sulfotransferases SULT1C1 and SULT1C2: cDNA characterization, gene cloning, and chromosomal localization. *Genomics*, 65: 157–165, 2000.

Fujita, K., K. Nagata, S. Ozawa, H. Sasano, and Y. Yamazoe, Molecular cloning and characterization of rat ST1B1 and human ST1B2 cDNAs, encoding thyroid hormone sulfotransferases. *J. Biochem.*, 122: 1052–1061, 1997.

Fujita, K., K. Nagata, T. Yamazaki, E. Watanabe, M. Shimada, and Y. Yamazoe, Enzymatic characterization of human cytosolic sulfotransferases; identification of ST1B2 as a thyroid hormone sulfotransferase. *Biol. Pharm. Bull.*, 22: 446–452, 1999.

Fukuda, M., N. Hiraoka, T. Akama, and M. Fukuda, Carbohydrate-modifying sulfotransferases: structure, function, and pathophysiology. *J. Biol. Chem.*, 276: 47747–47750, 2001.

Gamage, N.U., R.G. Duggleby, A.C. Barnett, M. Tresillian, C.F. Latham, N.E. Liyou, M.E. McManus, and J.L. Martin, Structure of a human carcinogen-converting enzyme, SULT1A1: structural and kinetic implications of substrate inhibition. *J. Biol. Chem.*, 278: 7655–7662, 2003.

Ganguly, T.C., V. Krasnykh, and C.N. Falany, Bacterial expression and kinetic characterization of the human monoamine-sulfating form of phenol sulfotransferase. *Drug Metab. Disp.*, 23: 945–950, 1995.

Geese, W.J. and R. Raftogianis, Biochemical characterization and tissue distribution of human SULT2B1. *Biochem. Biophys. Res. Comm.*, 288: 280–289, 2001.

Glatt, H., Bioactivation of mutagens via sulfation. *FASEB J.*, 11: 314–321, 1997.

Glatt, H., I. Bartsch, A. Czich, A. Seidel, and C. Falany, Salmonella strains and mammalian cells genetically engineered for expression of sulfotransferases. *Tox. Lett.*, 82-83: 829–834, 1995.

Glatt, H., H. Boeing, C.E. Engelke, L. Ma, A. Kuhlow, U. Pabel, D. Pomplun, W. Teubner, and W. Meinl, Human cytosolic sulphotransferases: genetics, characteristics, toxicological aspects. *Mutation Res.*, 482: 27–40, 2001.

Glatt, H., A. Seidel, R.G. Harvey, and M.W. Coughtrie, Activation of benzylic alcohols to mutagens by human hepatic sulfotransferase. *Mutagenesis*, 9: 553–557, 1994.

Habuchi, O., Diversity and functions of glycosaminoglycan sulfotransferases. *Biochem. Biophys. Acta*, 1474: 115–127, 2000.

Hart, R.F., K.J. Renskers, E.B. Nelson, and J.A. Roth, Localization and characterization of phenol sulfotransferase in human platelets. *Life Sci.*, 24: 125–130, 1979.

Her, C., I.A. Aksoy, and R.M. Weinshilboum, Human estrogen sulfotransferase gene (STE): cloning, structure and chromosomal localization. *Genomics*, 28: 16–23, 1995.

Her, C., G.P. Kaur, R.S. Athwal, and R.M. Weinshilboum, Human sulfotransferase SULT1C1: cDNA cloning, tissue-specific expression, and chromosomal localization. *Genomics,* 41: 467–470, 1997.

Her, C., R. Raftogianis, and R.M. Weinshilboum, Human phenol sulfotransferase STP2 gene: molecular cloning, structural characterization and chromosomal localization. *Genomics*, 33: 409–420, 1996a.

Her, C., C. Szumlanski, I. Aksoy, and R. Weinshilboum, Human jejunal estrogen sulfotransferase and dehydroepiandrosterone sulfotransferase: immunological characterization of individual variation. *Drug Metab. Disp.*, 24: 1328–1335, 1996b.

Her, C., T.C. Wood, E.E. Eichler, H.W. Mohrenweiser, L.S. Ramagli, M.J. Siciliano, and R.M. Weinshilboum, Human hydroxysteroid sulfotransferase SULT2B1: two enzymes encoded by a single chromosome 19 gene. *Genomics*, 53: 284–295, 1998.

Heroux, J.A., C.N. Falany, and J.A. Roth, Immunological characterization of human phenol sulfotransferase. *Mol. Pharmacol.*, 36: 29–33, 1989.

Heroux, J.A. and J.A. Roth, Physical characterization of a monoamine-sulfating form of phenol sulfotransferase from human platelets. *Mol. Pharmacol.*, 34: 29–33, 1988.

Hornsby, P.J., Biosynthesis of DHEAS by the human adrenal cortex and its age-related decline. *Ann. New York Acad. Sci.*, 774: 29–46, 1995.

Jones, A.L., M. Hagen, M.W.H. Coughtrie, R.C. Roberts, and H. Glatt, Human platelet phenolsulfotransferases: cDNA cloning, stable expression in V79 cells and identification of a novel allelic variant of the phenol-sulfating form. *Biochem. Biophys. Res. Comm.*, 208: 855–862, 1995.

Kester, M., C. van Dijk, D. Tibboel, A. Hood, N. Rose, W. Meinl, U. Pabel, H. Glatt, C. Falany, M. Coughtrie, and T. Visser, Sulfation of thyroid hormone by estrogen sulfotransferase. *J. Clin. Metab.*, 84: 2577–2580, 1999a.

Kester, M., E. Kaptein, T. Roest, C. van Dijk, D. Tibboel, W. Meinl, H. Glatt, M. Coughtrie, and T. Visser, Characterization of human iodothyronine sulfotransferases. *J. Clin. Endocrinol. Metab.*, 84: 1357–1364, 1999b.

Kitada, H., M. Miyata, T. Nakamura, A. Tozawa, W. Honma, M. Shimada, K. Nagata, C.J. Sinal, G.L. Guo, F.J. Gonzalez, and Y. Yamazoe, Protective role of hydroxysteroid sulfotransferase in lithocholic acid-induced liver toxicity. *J. Biol. Chem.*, 278: 17838–17844, 2003.

Kong, A.-N.T., L. Yang, M. Ma, D. Tao, and T.D. Bjornsson, Molecular cloning of the alcohol/hydroxysteroid form (hSTa) of sulfotransferase from human liver. *Biochem. Biophys. Res. Comm.*, 187: 448–454, 1992.

Kotov, A., J.L. Falany, J. Wang, and C.N. Falany, Regulation of estrogen activity by sulfation in human Ishikawa endometrial adenocarcinoma cells. *J. Steroid Biochem. Mol. Biol.*, 68: 137–144, 1999.

Lewis, A., M. Kelly, K. Walle, E. Eaton, C. Falany, and T. Walle, Improved bacterial expression of the human P form phenolsulfotransferase. *Drug Metab. Disp.*, 24: 1180–1185, 1996.

Meinl, W. and H. Glatt, Structure and localization of the human SULT1B1 gene: neighborhood to SULT1E1 and a SULT1D pseudogene. *Biochem. Biophys. Res. Comm.*, 288: 855–862, 2001.

Meisheri, K.D., L.A. Cipkus, and C.J. Taylor, Mechanism of action of minoxidil sulfate-induced vasodilatation: a role for increased K^+ permeability. *J. Pharmacol. Exp. Ther.*, 245: 751–760, 1988.

Meloche, C.A. and C.N. Falany, Cloning and expression of the human 3ß-hydroxysteroid sulfotransferases (SULT 2B1a and SULT 2B1b). *J. Steroid Biochem. Mol. Biol.*, 77: 261–269, 2001.

Meloche, C., V. Sharma, S. Swedmark, P. Andersson, and C. Falany, Sulfation of budesonide by human cytosolic sulfotransferase, dehydroepiandrosterone-sulfotransferase (DHEA-ST). *Drug Metab. Disp.*, 30: 582–585, 2002.

Miller, J.A. and E.C. Miller, The metabolic activation of carcinogenic aromatic amines and amides. *Prog. Exp. Tumor Res.*, 11: 273–301, 1968.

Miller, E.C., J.A. Miller, E.W. Boberg, K.B. Delclos, C.C. Lai, T.R. Fennell, R.W. Wiseman, and A. Liem, Sulfuric acid esters as ultimate electrophilic and carcinogenic metabolites of some alkenylbenzenes and aromatic amines in mouse liver. *Carcinogenesis (London)*, 11: 93–107, 1985.

Moore, S.S., E.O.P. Thompson, and A.R. Nash, Oestrogen sulfotransferase: isolation of a high specific activity species from bovine placenta. *Austral. J. Biol. Sci.*, 41: 333–341, 1988.

Mortola, J.F. and S.S.C. Yen, The effects of oral dehydroepiandrosterone on endocrine-metabolic parameters in postmenopausal women. *J. Clin. Endo. Metab.*, 71: 696–704, 1990.

Nagata, K., S. Ozawa, M. Miyata, M. Shimada, D.-W. Gong, Y. Yamazoe, and R. Kato, Isolation and expression of a cDNA encoding a male-specific rat sulfotransferase that catalyzes activation of N-hydroxy-2-acetyl-aminofluorene. *J. Biol. Chem.*, 268: 24720–24725, 1993.

Nash, A.R., W.K. Glenn, S.S. Moore, J. Kerr, A.R. Thompson, and E.O.P. Thompson, Oestrogen sulfotransferase: molecular cloning and sequencing of cDNA for the bovine placental enzyme. *Austral. J. Biol. Sci.*, 41: 507–516, 1988.

Ogura, K., J. Kajita, H. Narihata, T. Watabe, S. Ozawa, K. Nagata, Y. Yamazoe, and R. Kato, Cloning and sequence analysis of a rat liver cDNA encoding hydroxysteroid sulfotransferase. *Biochem. Biophys. Res. Comm.*, 165: 168–174, 1989.

Ogura, K., J. Kajita, H. Narihata, T. Watabe, S. Ozawa, K. Nagata, Y. Yamazoe, and R. Kato, cDNA cloning of hydroxysteroid sulfotrasferase STa sharing a strong homology in amino acid sequence with the senescence marker protein SMP-2 in rat livers. *Biochem. Biophys. Res. Comm.*, 166: 1494–1500, 1990.

Otterness, D.M., C. Her, S. Aksoy, S. Kimura, E.D. Wieben, and R.M. Weinshilboum, Human dehydroepiandrosterone sulfotransferase gene: molecular cloning and structural characterization. *DNA Cell Biol.*, 14: 331–341, 1995b.

Otterness, D., H.W. Mohrenweiser, B.F. Brandiff, and R.M. Weinshilboum, Dehydroepiandrosterone sulfotransferase gene (STD): localization to human chromosome 19q13.3. *Cytogenet. Cell Genet.*, 70: 45–52, 1995a.

Otterness, D.M., E.D. Wieben, T.C. Wood, R.W.G. Watson, B.J. Madden, D.J. McCormick, and R.M. Weinshilboum, Human liver dehydroepiandrosterone sulfotransferase: molecular cloning and expression of the cDNA. *Mol. Pharmacol., Mol. Endocrinol.* 41: 865–872, 1992.

Ozawa, S., K. Nagata, D.-W. Gong, Y. Yamazoe, and R. Kato, Nucleotide sequence of a full-length cDNA (PST-1) for aryl sulfotransferase from rat liver. *Nucleic Acid Res.*, 18: 4001, 1990.

Ozawa, S., K. Nagata, M. Shimada, M. Ueda, T. Tsuzuki, Y. Yamazoe, and R. Kato, Primary structures and properties of two related forms of aryl sulfotransferases in human liver. *Pharmacogenetics*, 5: 135–140, 1995.

Parker, C., Jr., Dehydroepiandrosterone and dehydroepiandrosterone sulfate production in the human adrenal gland during development and aging. *Steroids*, 64: 640–647, 1999.

Parker, C.R., C.N. Falany, C.R. Stockard, A.K. Stankovic, and W.E. Grizzle, Immunohistochemical localization of dehydroepiandrosterone sulfotransferase in human fetal tissues. *J. Clin. Endo. Metab.*, 78: 234–236, 1993.

Radominska, A., K.A. Comer, P. Ziminak, J. Falany, M. Iscan, and C.N. Falany, Human liver steroid sulfotransferase sulphates bile acids. *Biochem. J.*, 273: 597–604, 1991.

Raftogianis, R., T. Wood, and R. Weinshilboum, Human phenol sulfotransferases SULT1A2 and SULT1A1: genetic polymorphisms, allozyme properties, and human genotype-phenotype correlations. *Biochem. Pharm.*, 58: 605–616, 1999.

Raftogianis, R., T. Wood, D. Otterness, J. Van Loon, and R. Weinshilboum, Phenol sulfotrans- ferase pharmacogenetics in humans: association of common SULT1A1 alleles with TS PST phenotype. *Biochem. Biophys. Res. Comm.*, 239: 298–304, 1997.

Sakakibara, N., M. Suiko, T.G. Pai, T. Nakayama, Y. Takami, J. Katafuchi, and M.C. Liu, Highly conserved mouse and human brain sulfotransferases: molecular cloning, expression, and functional characterization. *Gene*, 285: 39–47, 2002.

Sakakibara, Y., K. Yanagisawa, J. Katafuchi, D. Ringer, Y. Takami, T. Nakayama, M. Suiko, and M. Liu, Molecular cloning, expression, characterization of novel SULT1C sulfotransferases that catalyze the sulfonation of N-hydroxy-2-acetylaminofluorene. *J. Biol. Chem.*, 273: 33929–33935, 1998a.

Sakakibara, Y., Y. Takami, T. Nakayama, M. Suiko, and M.C. Liu, Localization and functional analysis of the substrate specificity/catalytic domains of human M-form and P-form phenol sulfotransferases. *J. Biol. Chem.*, 273: 6242–6247, 1998b.

Song, C.S., I. Echchgadda, B.S. Baek, S.C. Ahn, T. Oh, A.K. Roy, and B. Chatterjee, Dehydroepiandrosterone sulfotransferase gene induction by bile acid activated far- nesoid X receptor. *J. Biol. Chem.*, 276: 42549–42556, 2001.

Tashiro, A., H. Sasano, T. Nishikawa, N. Yabuki, Y. Muramatsu, M. Coughtrie, H. Nagura, and M. Hongo, Expression and activity of dehydroepiandrosterone sulfotransferase in human gastric mucosa. *J. Steroid Biochem. Mol. Biol.*, 72: 149–154, 2000.

Thomae, B., O. Rifki, M. Theobald, B. Eckloff, E. Wieben, and R. Weinshilboum, Human catecholamine sulfotransferase (SULT1A3) pharmacogenetics: common functional genetic polymorphism in African-American subjects. *Clin. Pharmacol. Ther.*, 73: P29, 2003.

Tsoi, C., C.N. Falany, R. Morgenstern, and S. Swedmark, Identification of a new subfamily of sulfotransferases: cloning and characterization of canine SULT1D1. *Biochem. J.*, 356: 891–897, 2001.

Veronese, M.E., W. Burgess, X. Zhu, and M.E. McManus, Functional characterization of two human sulphotransferase cDNAs that encode monoamine- and phenol-sulphating forms of phenol sulphotransferase: substrate kinetics, thermal-stability and inhibitor- sensitivity studies. *Biochem. J.*, 302: 497–502, 1994a.

Veronese, M.E., W. Burgess, X. Zhu, and M.E. McManus, Heterologous systems for expres- sion of mammalian sulfotransferases. *Chem.-Biol. Interact.*, 92: 77–85, 1994b.

Visser, T.J., Role of sulfation in thyroid hormone metabolism. *Chem.-Biol. Interact.*, 92: 293–303, 1994.

Visser, T.J., Role of sulfate in thyroid hormone sulfation. *Eur. J. Endocrinol.*, 134: 12–14, 1996.

Waldon, D.J., A.E. Buhl, C.A. Baker, and G.A. Johnson, Is minoxidil sulfate the active metabolite for hair growth? *J. Invest. Dermatol.*, 92: 538–545, 1989.

Wang, J., J.L. Falany, and C.N. Falany, Expression and characterization of a novel thyroid hormone-sulfating form of cytosolic sulfotransferase from human liver. *Mol. Phar- macol.*, 53: 274–282, 1998.

Wang, Y., M.R. Spitz, A.M. Tsou, K. Zhang, N. Makan, and X. Wu, Sulfotransferase (SULT) 1A1 polymorphism as a predisposition factor for lung cancer: a case control analysis. *Lung Cancer*, 35: 137–142, 2002.

Weinshilboum, R.M., Phenol sulfotransferase in humans: properties, regulation, and function. *Fed. Proc.*, 45: 2223–2228, 1986.

Weinshilboum, R., Phenol sulfotransferase inheritance. *Cell. Mol. Neurobiol.*, 8: 27–34, 1988.

Weinshilboum, R., Sulfotransferase pharmacogenetics. *Pharmac. Ther.*, 45: 93–107, 1990.

Weinshilboum, R., D. Otterness, I.A. Aksoy, T.C. Wood, C. Her, and R. Raftogianis, Sulfation and sulfotransferases 1: sulfotransferase molecular biology: cDNAs and genes. *FASEB J.*, 11: 3–14, 1997.

Wilborn, T.W., K.A. Comer, T.P. Dooley, I.M. Reardon, R.L. Heinrikson, and C.N. Falany, Sequence analysis and expression of the cDNA for the phenol-sulfating form of human liver phenol sulfotransferase. *Mol. Pharmacol.*, 43: 70–77, 1993.

Wood, T.C., I.A. Aksoy, S. Aksoy, and R.M. Weinshilboum, Human liver thermolabile phenol sulfotransferase: cDNA cloning, expression and characterization. *Biochem. Biophys. Res. Comm.*, 198: 1119–1127, 1994.

Yamazoe, Y., K. Nagata, S. Ozawa, and R. Kato, Structural similarity and diversity of sulfotransferases. *Chem.-Biol. Interact.*, 92: 107–117, 1994.

Yamazoe, Y., K. Nagata, S. Ozawa, K.-I. Fujita, and R. Kato, Primary structure and properties of ST1B forms. *ISSX Proceed.*, 8: A246, 1995.

Yoshinari, K., K. Nagata, M. Shimada, and Y. Yamazoe, Molecular characterization of ST1C1-related human sulfotransferase. *Carcinogenesis*, 19: 951–953, 1998.

Young, W.F., C.A. Gorman, and R.M. Weinshilboum, Triiodothyronine: a substrate for the thermostable and thermolabile forms of human phenol sulfotransferase. *Endocrinol.*, 122: 1816–1824, 1988.

Young, W., Jr., H. Okazaki, E. Laws Jr., and R. Weinshilboum, Human brain phenol sulfotransferase: biochemical properties and regional localization. *J. Neurochem.*, 43: 706–711, 1984.

Zhang, H., O. Varmalova, F.M. Vargas, C.N. Falany, and T.S. Leyh, Sulfuryl transfer: the catalytic mechanism of human estrogen sulfotransferase. *J. Biol Chem.*, 273: 10888–10892, 1998.

Zheng, W., D. Xie, J.R. Cerhan, T.A. Sellers, W. Wen, and A.R. Folsom, Sulfotransferase 1A1 polymorphism, endogenous estrogen exposure, well-done meat intake, and breast cancer risk. *Cancer Epidemiol. Biomarkers Prev.*, 10: 89–94, 2001.

Zhu, X., M.E. Veronese, P. Iocco, and M.E. McManus, cDNA cloning and expression of a new form of human aryl sulfotransferase. *Int. J. Biochem. Cell Biol.*, 28: 565–571, 1996.

Zhu, X., M.E. Veronese, C.A. Bernard, L.N. Sansom, and M.E. McManus, Identification of two human brain aryl sulfotransferase cDNAs. *Biochem. Biophys. Res. Comm.*, 195: 120–127, 1993.

Zou, J., R. Pentney, and J.A. Roth, Immunohistochemical detection of phenol sulfotransferase-containing neurons in human brain. *J. Neurochem.*, 55: 1154–1158, 1990.

6 Sulfotransferases in the Human Fetus and Neonate

Emma L. Stanley, Robert Hume, and Michael W.H. Coughtrie

CONTENTS

INTRODUCTION

The developing human faces numerous toxic insults from endogenous chemicals and xenobiotics *in utero* and *postpartum*. The metabolic adaptations, which occur during this critical period of development, play a key role in determining not only survival but an individual's ability to thrive. The concept of sulfation as an important metabolic reaction in the developing human fetus is one that has matured over recent years. Catalyzed by members of the sulfotransferase enzyme superfamily (SULT), sulfation is a major pathway for the metabolism and detoxification of xenobiotics and endogenous compounds including steroid hormones, iodothyronines, and catecholamines. Modulation of the biological activity of such potent molecules is of crucial importance since these compounds can dramatically influence human development. A number of SULTs are expressed at high levels in the fetus, which is in contrast to many of the enzyme systems that comprise the major adult chemical defense mechanism (i.e., Phase 2 or conjugating enzymes), which are poorly developed until after birth (e.g., UDP-glucuronosyltransferases; Coughtrie et al., 1988). The high level of expression of sulfotransferase enzymes in many human fetal tissues therefore indicates an important role during development.

SULT

A new nomenclature system for the cytosolic SULTs has recently been devised (Blanchard et al., 2004)*, and the nomenclature used here follows this new system.

Like many drug-metabolizing enzymes, cytosolic SULTs are derived from a large superfamily of genes. Full-length cDNAs encoding more than 50 mammalian and avian cytosolic SULTs have now been cloned and sequenced and many of the expressed proteins characterized. The majority of these enzymes sulfate small molecule xenobiotics or endogenous hormones and neurotransmitters and, based on amino acid sequence analysis, can be subdivided into six families (Figure 6.1). Sequence databases for other model organisms, for example the zebrafish (*Danio rerio*) and amphibian (*Xenopus laevis, Silurana tropicalis*), contain numerous cytosolic SULT sequences with orthologs in mammalian species although the *Drosophila melanogaster, Caenorhabditis elegans*, and *Saccharomyces cerevisiae* databanks contain few such sequences.

At present, the human SULT enzyme family comprises 11 isoforms, which can be subdivided based on amino acid sequence identity and enzymatic function into the SULT1 (SULTs 1A1, 1A2, 1A3, 1B1, 1C2, 1C4, and 1E1), SULT2 (SULTs 2A1, 2B1a, and 2B1b), and SULT4 (SULT4A1) families. These enzymes are produced from ten genes (SULTs 2B1a and 2B1b are generated by alternate splicing of the first exon of *SULT2B1*) that share many common structural features (reviewed in Weinshilboum et al., 1997). The SULT1 and SULT2 families are the largest and, it is assumed, the most important for xenobiotic and endobiotic metabolism. The properties of the various human SULTs are summarized in Table 6.1.

REVERSIBLE SULFATION AND HORMONAL REGULATION

Sulfation normally results in a reduction in biological activity relative to the parent molecule, and sulfate conjugates may be excreted from the body or serve as circulating or intracellular stores (or as metabolic intermediates) from which the free compound can be regenerated by the action of arylsulfatase (ARS) enzymes (Coughtrie et al., 1998). The ARS enzyme family currently comprises six members

---

* Please note, the human SULT, which is named 1C2 in this chapter, corresponds to the human isoform originally named either human SULT1C sulfotransferase 1 or human SULT1C1 (Freimuth et al., 2000; Her et al., 1997; Sakakibara et al., 1998). Human SULT1C4 referred to here was originally called SULT1C sulfotransferase 2 or SULT1C2 (Freimuth et al., 2000; Sakakibara et al., 1998).

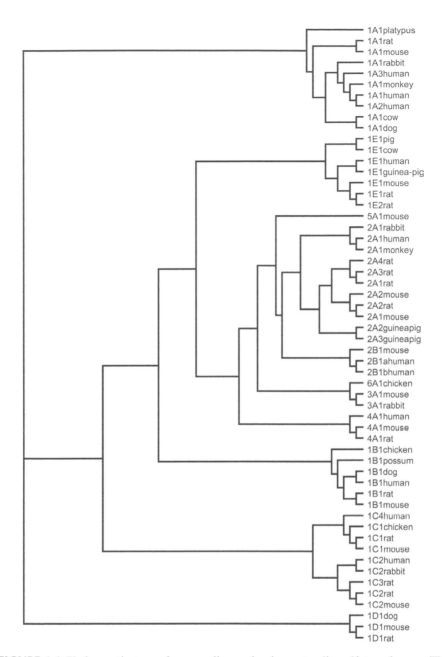

FIGURE 6.1 Phylogenetic tree of mammalian and avian cytosolic sulfotransferases. The amino acid sequence homologies between mammalian and avian cytosolic sulfotransferases. Six SULT families are represented. Clear evolutionary relationships can be seen, for example, between the 1E (estrogen) and 2 (androgen/steroid/sterol) families. Amino acid sequences were obtained from the NCBI GenBank database (http://www.ncbi.nlm.nih.gov), aligned using ClustalW from EMBL (http://www.ebi.ac.uk), and constructed using the TreeView program (http://taxonomy-zoology.gla.ac.uk/rod/treeview.html).

TABLE 6.1
Human Cytosolic Sulfotransferases

Isoform	Chromosomal Location	Substrate[a]	Principal Sites of Expression[b]
SULT1A1	16p12.1-11.2	4-Nitrophenol (diagnostic at 4μM)	Adult liver, adult GI tract, adult platelets, placenta, fetal liver
SULT1A2	16p12.1-11.2	4-Nitrophenol*	?[c]
SULT1A3	16p11.2	Dopamine	Adult GI tract, adult platelets, adult brain, placenta, fetal liver
SULT1B1	4q11-13	4-Nitrophenol*	Adult liver, adult GI tract, fetal GI tract
SULT1C2	2q11.2	4-Nitrophenol*	Fetal kidney, fetal lung, fetal GI tract
SULT1C4	2q11.2	4-Nitrophenol*	Fetal kidney, fetal lung
SULT1E1	4q13	17 β-Estradiol	Fetal liver, fetal lung, fetal kidney, adult liver, endometrium
SULT2A1	19q13.3	DHEA	Fetal adrenal, fetal liver, adult liver, adult adrenal
SULT2B1[d]	19q13.3	Cholesterol (2B1b); pregnenolone (2B1a)	Adult skin, prostate, placenta
SULT4A1	22q13.1-13.2	No substrate known	Brain

[a] Indicates substrates that can be used to measure enzyme activity of the isoform in cytosolic preparations made from human tissues. Where an asterisk (*) appears, no substrate has yet been identified that is sufficiently specific for the isoform concerned to be of value in quantifying the associated enzyme activity in tissue samples. 4-Nitrophenol may be used to follow activity of these expressed recombinant proteins *in vitro*.

[b] Expression is taken to mean protein expression. In cases where RNA expression only has been demonstrated (e.g., for SULTs 1C2 and 1C4) in certain adult tissues (Her et al., 1997; Sakakibara et al., 1998) this is not included.

[c] There is some doubt as to whether the SULT1A2 enzyme protein is expressed at significant levels in human tissues (Dooley et al., 2000).

[d] The splice variant SULT2B1b may the major form expressed at the protein level (Geese and Raftogianis, 2001; Meloche and Falany, 2001). The enzyme also sulfates DHEA although cholesterol is diagnostic (Javitt et al., 2001).

(ARSA-ARSF; Parenti et al., 1997), and a major ARS known as steroid sulfatase or ARSC is present in the endoplasmic reticulum of many tissues, including the placenta where its main function is in estrogen biosynthesis (Kung et al., 1988). ARSC has been demonstrated to hydrolyze iodothyronine sulfates, whereas ARSA and ARSB do not (Kester et al., 2002). The expression of particular SULT and ARS enzymes in individual cells and tissues can therefore provide a sensitive and responsive means of controlling the activity of potent hormones in target and nontarget cells.

In human liver, the expression of the major hydroxysteroid SULT enzyme (DHEA sulfotransferase, SULT2A1; Barker et al., 1994) increases during fetal development, whereas in the adrenal gland the expression is high (more than five-fold higher than in liver) from at least 12 weeks gestation, consistent with the critical role of the fetal adrenal in the production of DHEA sulfate to serve as substrate for placental estrogen biosynthesis. The fetus is therefore protected from the (albeit weak) androgenic effect of DHEA as the steroid is circulated as nonbiologically active DHEA sulfate.

Catecholamines are important mediators of physiological stresses such as hypoxia and hypoglycemia, and catecholamine secretion is believed to play an important role in immediate adaptation to extrauterine life, including metabolic, respiratory, and cardiovascular events (Sperling et al., 1984). Animal studies, mainly in sheep, indicate that stresses such as hypoxia and birth produce dramatic increases in fetal catecholamine production (Padbury et al., 1989). Indeed, the dramatic surge in catecholamine secretion at birth may be a key event since cate-cholamine infusion to the fetal sheep mimics the immediate postnatal profiles of glucagon and insulin and consequently gluconeogenesis. In adult humans, the vast majority of catecholamines such as dopamine, noradrenaline, and adrenaline are present in the circulation in the form of their respective sulfate conjugates (Kopin, 1985) — e.g., plasma concentrations of free dopamine (<0.1 pmol/mL) represent less than 1% of the total combined free and sulfate-conjugated dopamine. These sulfate conjugates may be thought of as biologically inactive storage and transport forms of these biogenic amines, which provide a source of the active, free compounds in target tissues through enzymatic hydrolysis. Catecholamines are important counter-regulatory hormones in the response to hypoglycemia. Recent unpublished work by us has shown that in hypoglycaemic infants, total adrenaline, free adrena-line, and total dopamine concentrations are significantly greater, and the proportion of adrenaline as sulfate conjugates is significantly lower in hypoglycemic infants. This suggests that acute control of hypoglycemia in infants is regulated in part by the balance between sulfated and nonsulfated catecholamines.

THE IMPORTANCE OF THYROID HORMONE SULFATION DURING FETAL DEVELOPMENT

Thyroid hormone is essential for normal human development, and this is most evident in the central nervous system, where thyroid hormone deficiency in the fetal and neonatal periods can result in severe and prolonged neurodevelopmental deficit. A number of factors are involved in regulating the bioavailability of receptor active thyroid hormone (T_3). Of particular importance are the deiodinase and sulfotrans-ferase enzyme systems governing the interconversion (and thus activation or deac-tivation) of the various iodothyronines.

Deiodination constitutes one of the principal pathways of thyroid hormone metabolism. Under normal conditions, the prohormone thyroxine (T_4) is the pre-dominant secretory product of the thyroid gland. T_4 is peripherally converted to receptor active T_3 (3,3′,5-triiodothyronine) by outer ring deiodination (ORD) or to receptor inactive reverse T_3 (rT_3) by inner ring deiodination (IRD). Both T_3 and rT_3 are further deiodinated, and therefore deactivated (via IRD and ORD, respectively) to T_2 (3,3′-diiodothyronine). The bioactivity of thyroid hormone is therefore

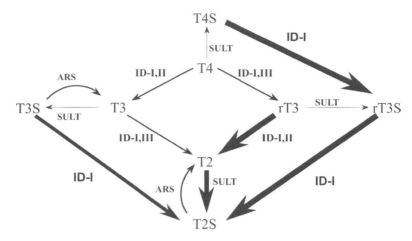

FIGURE 6.2 Model of reversible sulfation and deiodination of iodothyronines. The interaction between the deiodination and sulfation pathways. Sulfation strongly facilitates (up to 200-fold) the IRD (inactivation) of T_4 and T_3 by ID-I (indicated by thick arrows), thereby resulting in the irreversible degradation of thyroid hormone. The sulfotransferase isoenzymes involved at different stages in the pathway are currently unknown. Abbreviations: T_4 — thyroxine, T_3 — triiodiodothyronine, rT_3 — reverse triiodothyronine, T_2 — diiodothyronine, SULT — sulfotransferase, ARS — arylsulfatase, ID-I — iodothyronine deiodinase type I, ID-II — iodothyronine deiodinase type II, ID-III — iodothyronine deiodinase type III.

controlled by competing activation (ORD) and inactivation (IRD) pathways (Visser 1996; Figure 6.2). Three enzymes, known as the iodothyronine deiodinases (types I, II, and III), catalyze the reductive deiodination of T_4 and its metabolites, either by 5′-deiodination at the phenolic ring or by 5-deiodination at the tyrosyl ring (Kohrle, 1996). Factors governing the fate of the T_4 pathway therefore have a crucial role in thyroid function.

Sulfation has profound effects on iodothyronine metabolism and homeostasis. Once sulfated, T_4 is exclusively metabolized to the rT_3 and cannot be converted to receptor active T_3 (Kohrle, 1996). Sulfation also accelerates the degradation (deiodination) of T_3, and the major metabolic product of iodothyronine metabolism 3,3′-diiodothyronine (T_2) is also extensively sulfated (Kohrle, 1996; Figure 6.2). Sulfation is therefore likely to play a key role in regulating the amount of receptor active thyroid hormone in target tissues. Iodothyronine sulfates are present at very high levels in fetal (umbilical cord) blood and persist well into postnatal life, supporting the importance of sulfation as a means of protecting the fetus from excessive T_3 (Chopra et al., 1992; Hume et al, unpublished observations; Santini et al., 1999). Microsomal 3,3′-T_2S sulfatase activity is also present in developing human liver and to a lesser extent in fetal lung and brain (Richard et al., 2001) and thus has the potential to generate free iodothyronines from their respective sulfate conjugates.

TECHNIQUES USED IN SULT RESEARCH

To further our understanding of the role of sulfation during human development, researchers have studied the expression and relative activities of the major SULT isoforms in human adult and fetal tissues using a variety of biochemical and molecular techniques, which are outlined below.

A vital feature of the human SULT enzyme family is the degree of substrate specificity demonstrated by the individual isoforms, and this feature has been exploited in enzyme activity assays. SULT activity can be determined in tissue cytosol using specific substrates at concentrations selective for each isoform. SULT1A1 can be measured by virtue of its high affinity (low K_m) for 4-nitrophenol compared to other SULTS. At low micromolar concentrations, SULT1A1 selectively sulfates 4-nitrophenol, whereas at similar concentrations SULT1A3, which has a major role in the sulfation of catecholamines, preferentially sulfates dopamine. SULT1E family members preferentially sulfate endogenous (e.g., estrone and 17β-estradiol) and xenobiotic (e.g., 17α-ethinylestradiol) estrogens, generally with very high affinity (Aksoy et al., 1994; Coughtrie et al., 1994, 1998; Falany 1997; Falany et al., 1994). Although SULT1A1 and SULT2A members also sulfate estrogens, the affinity for endogenous estrogens is typically several orders of magnitude lower than the affinity of SULT1E1 for those substrates. The SULT2 family (SULTs 2A and 2B) can catalyze sulfation of steroids and sterols, such as dehydroepiandrosterone (DHEA), pregnenolone, and cholesterol. Sulfation of 4-nitrophenol and dopamine can be determined using PAPS^{35}S as originally described by Foldes and Meek (1973). Sulfation of estrone, 17β-estradiol, 17α-ethinylestradiol, and DHEA can be determined using ^{3}H-labelled substrates as previously described (Borthwick et al., 1993; Sharp et al., 1993). Sulfation of iodothyronines T_2, T_3, rT_3, and T_4 can be measured using ^{125}I-labelled substrates (as detailed in Kaptein et al., 1997). At the time of writing, substrates selective for SULTs 1B1, 1C2, and 1C4 have not yet been identified.

To determine SULT enzyme protein expression and, more importantly, to assess the expression of SULTs 1B1 and 1C2 (for which no specific substrates are available), western-immunoblot analysis can be used. Cytosolic fractions can be resolved by SDS-PAGE and transferred to nitrocellulose as first described by Laemmli (1970) and Towbin et al. (1979), respectively. Antibodies against purified, recombinant human SULT isoforms can be raised in sheep by immunization, using an adaptation of the method of Vaitukaitis (1981). Immunological detection of SULT proteins can be achieved using the enhanced chemiluminescence method, which is commercially available and has been described by several manufacturers. Our research group has successfully produced anti-SULT1A3, which has been described previously (Dajani et al., 1998; Richard et al., 2001; Stanley et al., 2001). This preparation cross-reacts with SULTs 1A1 and 1A2 on immunoblot and ELISA analysis (these isoforms share >93% amino acid sequence identity), thus it is possible to use this preparation to assess the expression of all three isoforms. We have also made anti-SULT1B1 (Kester et al., 1999a), 1C2 (Stanley, 2001), and 1E1 (Rubin et al., 1999) preparations.

To localize the sites of SULT expression, immunohistochemical analysis can be performed using the SULT antibody preparations detailed above. Tissue sections can be cut from formalin-fixed, paraffin-embedded material, and a standard perox-idase–antiperoxidase technique can be used incorporating 3,3'-diaminobenzidine as a developing agent (Sternberger et al., 1990). For more complex tissues, such as placenta, immunofluorescence can be performed using frozen sections (Stanley et al., 2001). The use of frozen sections generally does not provide as high resolution staining as formalin-fixed, paraffin-embedded material. Control sections with replacement of the primary antibody with nonimmune IgG should be performed in tandem for both immunohistochemical techniques.

To rapidly screen a large panel of human tissues for SULT expression, mRNA dot-blot analysis can be conducted. Dot-blot analysis is performed using poly(A)+RNA isolated from a number of human tissues, and expression is detected using SULT cDNA as a probe. The probe can be labeled with, for example, $[\alpha\text{-}^{32}\text{P}]$-dCTP by random priming using a commercially available kit. This technique is described in further detail in Her et al. (1997).

To detect expression of SULT genes, a PCR-based approach, also known as expression profiling, can be adopted. Total RNA can be extracted from tissue using a commercially available kit and cDNA generated by RT-PCR. Alternatively, cDNA can be purchased directly from a commercial supplier. PCR is performed on cDNA using isoform-specific primer pairs, designed specifically to avoid amplification of other SULT isoforms (as detailed in Dooley et al., 2000). Newer applications, such as real-time PCR analysis with the TaqMan® 5'-nuclease system (PE Applied Bio-systems) and with the Light Cycler™ system (Roche), can be utilized for a more quantitative estimation of SULT expression levels.

RESULTS

The results of biochemical and molecular investigations highlight substantial inter-individual variation and significant developmental regulation of SULT expression. However, the identification of the sulfotransferase enzymes primarily responsible for iodothyronine sulfation in humans *in vivo* remains to be conclusively determined.

WHICH SULT ENZYMES ARE RESPONSIBLE FOR IODOTHYRONINE SULFATION?

The sulfotransferase enzymes primarily responsible for iodothyronine sulfation in humans *in vivo* remain to be conclusively determined. All members of the SULT1A family have been shown to catalyze the sulfation of iodothyronines; for most of these enzymes, $3,3'\text{-}T_2$ is the preferred iodothyronine substrate ($3,3'\text{-}T_2 \gg T_3 \sim rT_3 > T_4$) although SULT1E1 shows an equal preference for $3,3'\text{-}T_2$ and rT_3 ($3,3'\text{-}T_2 \sim rT_3 > T_3 \sim T_4$; Kester et al., 1999a, 1999b). Our studies suggest that SULT1A1 is probably the major enzyme involved in the sulfation of $3,3'\text{-}T_2$ in the developing human liver, brain (Richard et al., 2001), and placenta (Stanley et al., 2001). Of the major SULT isoforms, SULT1A1 has the highest specificity constant toward $3,3'\text{-}T_2$

(Kester et al., 1999a), and therefore this enzyme probably provides the majority of the sulfation capacity for this iodothyronine.

The identity of the SULT enzymes involved in the sulfation of other iodothyronines is not so clear. All the major SULTs metabolize most iodothyronines *in vitro* to varying degrees (Kester et al., 1999a). We have shown that SULT1E1 appears to be the major enzyme *in vitro* for sulfation of rT_3, T_3, and to a lesser extent T_4 (Kester et al., 1999b) although others have claimed SULT1B1 (Wang et al., 1998) and SULT1C2 (Li et al., 2000) to be involved.

DIFFERENTIAL EXPRESSION OF SULT ENZYMES IN THE DEVELOPING FETUS AND NEONATE

The human fetus expresses various SULT isoforms from very early in gestation, for example, SULT1A enzymes are detectable in mesonephric kidney at 32 postovulatory days (Hume and Coughtrie, 1994) and in lung epithelium at 56 postovulatory days (Hume et al., 1996). Many studies clearly show that human SULT expression is carefully regulated with respect to tissue type, development, and hormonal influences, supporting the proposed functional roles for some of the enzymes. The function of certain SULTs in development is fairly well established; however, the role of others is not so clear. A number of SULTs are expressed at high levels in the fetus (Table 6.1), and in fact some appear to be expressed only or primarily in the prenatal period, such as the SULT1C enzymes (Her et al., 1997; Sakakibara et al., 1998; Stanley, 2001). This can be illustrated by a number of examples, which are grouped according to SULT family.

SULT1

A comparative study of SULT expression in human fetal and adult tissues (notably, liver, lung, kidney, adrenal gland, and intestine) was one of the first to demonstrate the expression of substantial SULT in the fetus (Pacifici et al., 1988).

The catecholamine-metabolizing SULT enzyme SULT1A3 is expressed at high levels in fetal liver, in contrast to the adult liver, and the gastrointestinal tract is the major site of expression (Cappiello et al., 1991; Pacifici et al., 1993; Richard et al., 2001). This observation correlates with the dopaminergic function of the gut, where the majority of dopamine sulfate is produced (Eisenhofer et al., 1999). The physiological relevance of the high level of SULT1A3 in fetal liver is unclear. There is no apparent association between gestational age and SULT1A3 activity although reduced expression (both enzyme activity and protein) is found in postnatal liver, suggesting that in the perinatal period, SULT1A3 expression is actively (and probably rapidly) switched off (Richard et al., 2001).

The SULT1A1 enzyme is not subject to the same developmental regulation in the liver as SULT1A3 (or SULT2A1). Hepatic SULT1A1 activity is present from ten-weeks gestation (the earliest gestational age studied by Richard et al., 2001) but is extremely variable over the developmental range (Cappiello et al., 1991). Expression levels of SULT1A1 do not vary between fetal and postnatal liver although adulthood levels are higher (Cappiello et al., 1991; Pacifici et al., 1988;

Richard et al., 2001). Detailed analysis of SULT1A expression using immunohistochemistry shows that a significant portion of immunoreactivity is localized in hematopoietic cells, which make up a substantial part of the fetal liver during the second trimester (Richard et al., 2001). As the antibodies used for immunohistochemistry cross-react with both SULT1A1 and SULT1A3 (also presumably SULT1A2), it is not possible to tell from these studies which isoforms contribute to the immunoreactivity.

SULT1A1 is also expressed at low levels in most regions of the developing fetal brain, with the choroid plexus of the lateral ventricle being the major site of expression, whereas SULT1A3 is not detectable at this location. SULT1A3 is low throughout the fetal brain, with highest activity levels in the germinal eminence and cerebellum (Richard et al., 2001).

To date, there are no specific substrates known for SULTs 1B1, 1C2, and 1C4. Expression of these isoforms has been studied using approaches such as RT-PCR, RNA dot-blot, and, more recently, western immunoblotting. SULT1B1 appears to be predominantly expressed in the adult although expression has been reported in fetal small bowel (Stanley, 2001). SULT1B1 mRNA has widespread expression in many adult tissues, for example stomach, duodenum, colon, colorectal, liver, ovary, and cerebellum (Dooley et al., 2000). Conversely, SULTs 1C2 and 1C4 appear to be most highly expressed in fetal tissues (Stanley, 2001) although RNA dot-blots indicate the adult stomach and kidney (SULT1C2) and ovary (SULT1C4) may also be sites of expression (Her et al., 1997; Sakakibara et al., 1998). In adult rodent liver, SULT1C enzymes are expressed in a sexually dimorphic manner, where they seem to be involved in xenobiotic metabolism — particularly the bioactivation of procarcinogens. This may not be the case in humans, where the enzymes are thought to be involved in thyroid hormone metabolism, which is a particularly important function in the fetus where appropriate thyroid hormone homeostasis is essential for normal brain development.

Another important example involves the estrogen sulfotransferase (SULT1E1), which displays a particularly high affinity (in the low nM range) for its natural substrate 17β-estradiol, suggesting an important role for this enzyme in modulating estrogen action (Song and Melner, 2000). It is also the principal human SULT involved in the sulfation of thyroxine (Kester et al, 1999b). SULT1E1 is present in the human embryo from early in development and appears to be highly expressed in a wide range of fetal tissues (Stanley, 2001). It is also expressed in hormone-dependent tissues, such as endometrium (Buirchell and Hähnel, 1975; Rubin et al., 1999) and placenta (notably in the maternal-derived decidual component; Stanley et al., 2001). The expression of SULT1E1 varies widely in the human population (Rubin et al., 1999; Song et al, 1998) although it is not known whether this is under genetic control. It is possible that the variability in SULT1E1 expression results from different chemical influences since progesterone (and potentially other compounds) are known to induce expression levels *in vitro* and *in vivo* (Falany and Falany, 1996; Song et al., 1998; Tseng and Liu, 1981).

SULT2

The major hydroxysteroid-sulfating enzyme, SULT2A1, is not expressed at any stage in the endometrium (Rubin et al., 1999) or placenta (Stanley et al., 2001). In the fetal liver and lung, SULT2A1 expression is low in early life but increases with development (Barker et al., 1994; Hume et al., 1996). In the adrenal, SULT2A1 is highly expressed (more than 5-fold higher than the liver and 40-fold higher than in the kidney) from at least 12-weeks gestation. This is consistent with the role of the fetal adrenal in the production of DHEA sulfate for placental estrogen biosynthesis (Barker et al., 1994; Parker et al., 1994).

SULT4

The SULT4A1 proteins appear to be expressed only in the adult brain (Falany et al., 2000), and despite significant effort no natural or xenobiotic substrate (ligand) has yet been identified. No data on SULT4A1 expression in fetal tissues have been reported to date.

DISCUSSION

The expression of the various SULT enzymes appears to be under specific temporal and spatial regulation during fetal development, suggesting that these tissues have different requirements during key phases of tissue differentiation and maturation for the modulation of the chemical mediators produced by the fetus.

The human fetus produces very large amounts (compared with the adult) of iodothyronine sulfates (E. Stanley, R. Hume, and M.W.H. Coughtrie, unpublished work). The high levels of SULT1A1 in various fetal tissues (Cappiello et al., 1991; Pacifici et al., 1988; Richard et al., 2001) may provide an important chemical defense function, particularly in the absence from the liver of the other major conjugating enzymes such as UDP-glucuronosyltransferases (Coughtrie et al., 1988) although the fetal kidney does express some UDP-glucuronosyltransferase (Hume et al., 1995).

An interesting observation was the expression of SULT1A1 in fetal brain, with the choroid plexus being the major site of expression. The primary function of this brain region is the production of cerebrospinal fluid. It is also the most highly vascularized tissue of the developing brain and therefore a potential portal of entry of circulating toxins that may result from maternal exposure. Thus it seems reasonable for this tissue to have a high degree of chemical defense (Richard et al., 2001). SULT1A3 is also expressed at low levels in the developing fetal brain; however, the germinal eminence appears to be the principal site of expression. This is where the majority of neuroblast cell division occurs during mammalian brain development (Richard et al., 2001).

Substantial interindividual variation was observed in SULT1A1 expression. One possible source of this variation is a common functional polymorphism in the

SULT1A1 gene (Arg^{213} to His), which causes substantially reduced SULT1A1 enzyme activity and protein stability in platelets in individuals homozygous for the variant allele (SULT1A1*2; Coughtrie et al., 1999; Raftogianis et al., 1997). This is a likely explanation for a major component of the variation observed in SULT1A1. Another possible source of variation is difference in tissue quality resulting from the range of postmortem intervals encountered. To address this phenomenon, Richard et al. (2001) sampled freshly obtained liver over a period of 12 h postmortem and determined SULT1A1 and 1A3 enzyme activities. The results clearly show that the enzyme activities are stable at least up to 12 h postmortem. Thus suggesting that instability of enzyme protein resulting from variable postmortem intervals is unlikely to be a major factor in the observed variability in enzyme activities. The variability most likely arises from a combination of genetic and environmental influences on enzyme activity and expression.

The physiological relevance of the high level of SULT1A3 in fetal liver is unclear. It is reasonable to propose that the expression of SULT1A3 in fetal liver has a protective function against the biological activity of catecholamines. The timing of the developmental switch from liver to gastrointestinal tract as the major site of SULT1A3 expression is not known, however. The major functions of gastrointestinal SULT1A3 in the adult are likely protection against the potentially toxic effects of ingested catecholamines in the diet and production of the large amounts of dopamine sulfate present in the circulation. Humans (and presumably other higher primates) have evolved a specific SULT for this purpose, with a high degree of selectivity towards catecholamines (Dajani et al., 1998). It is likely that a similar protective function would be also required by the newborn infant, perhaps not immediately after birth (assuming that breast milk is catecholamine poor), but certainly in infancy with the introduction of a weaning diet. Clearly, further studies are required to address this issue. It is also possible that SULT1A3 has a yet unknown function in the hemopoietic cells of the fetal liver and that the reduced expression with advancing gestation parallels the disappearance of these cells from the liver (Richard et al., 2001).

In conclusion, the human fetus expresses a wide array of sulfotransferase enzymes — certainly more than the adult — suggesting an important role for sulfation of endo- and xenobiotics during human development. This is in stark contrast to the situation in most experimental animal species, where sulfation is not well developed until after birth. There are clear spatial and temporal changes in SULT expression patterns during development (e.g., SULTs 1A3, 1E1, and 1C2 display a predominantly fetal expression, whereas SULT1B1 is predominantly expressed in adult tissues). Sulfation probably represents the major detoxification system in the developing human; therefore, a thorough appreciation of the distribution, regulation, and genetics of this system is essential if we are to fully understand the role of sulfation in protecting the fetus from external and internal insult. The studies discussed here have provided us with a valuable information resource, but clearly further investigations into sulfation during development are urgently required.

ACKNOWLEDGMENTS

This work was supported by grants from Tenovus Scotland and from the Commission of the European Communities (CEC QLG3-2000-00939 EUTHYROID).

REFERENCES

Aksoy, I.A., Wood, T.C., and Weinshilboum, R., 1994, Human liver estrogen sulfotransferase: identification by cDNA cloning and expression. *Biochemical and Biophysical Research Communications*, 200, 1621–1629.

Barker, E.V., Hume, R., Hallas, A., and Coughtrie, M.W.H., 1994, Dehydroepiandrosterone sulfotransferase in the developing human fetus — quantitative biochemical and immunological characterization of the hepatic, renal, and adrenal enzymes. *Endocrinology*, 134, 982–989.

Blanchard, R.L., Freimuth, R.R., Buck, J., Weinshilboum, R.M., Coughtrie, M.W.H., 2004, A proposed nomenclature system for the cytosolic sulfotransferase (SULT) superfamily. *Pharmacogenetics*, 14, 199–211.

Borthwick, E.B., Burchell, A., and Coughtrie, M.W.H., 1993, Purification and immunochemical characterization of a male-specific rat liver estrogen sulfotransferase. *Biochemical Journal*, 289, 719–725.

Buirchell, B.J. and Hähnel, R., 1975, Metabolism of estradiol-17β in human endometrium during the menstrual cycle. *Journal of Steroid Biochemistry*, 6, 1489–1494.

Cappiello, M., Giuliani, L., Rane, A., and Pacifici, G.M., 1991, Dopamine sulfotransferase is better developed than *p*-nitrophenol sulfotransferase in the human fetus. *Developmental Pharmacology and Therapeutics*, 16, 83–88.

Chopra, I.J., Wu, S.Y., Chua Teco, G.N., and Santini, F., 1992, A radioimmunoassay for measurement of 3,5,3′-triiodothyroxine sulfate: studies in thyroidal and non-thyroidal diseases, pregnancy, and neonatal life. *Journal of Clinical Endocrinology and Metabolism*, 75, 189–194.

Coughtrie, M.W.H., Bamforth, K.J., Sharp, S., Jones, A.L., Borthwick, E.B., Barker, E.V., Roberts, R.C., Hume, R., and Burchell, A., 1994, Sulfation of endogenous compounds and xenobiotics — interactions and function in health and disease. *Chemico-Biological Interactions*, 92, 247–256.

Coughtrie, M.W.H., Burchell, B., Leakey, J.E.A., and Hume, R., 1988, The inadequacy of perinatal glucuronidation — immunoblot analysis of the developmental expression of individual UDP-glucuronosyltransferase isoenzymes in rat and human-liver microsomes. *Molecular Pharmacology*, 34, 729–735.

Coughtrie, M.W.H., Gilissen, R.A.H.J., Shek, B., Strange, R.C., Fryer, A.A., Jones, P.W., and Bamber, D.E., 1999, Phenol sulphotransferase *SULT1A1* polymorphism: molecular diagnosis and allele frequencies in Caucasian and African populations. *Biochemical Journal*, 337, 45–49.

Coughtrie, M.W.H., Sharp, S., Maxwell, K., and Innes, N.P., 1998, Biology and function of the reversible sulfation pathway catalyzed by human sulfotransferases and sulfatases. *Chemico-Biological Interactions*, 109, 3–27.

Dajani, R., Hood, A.M., and Coughtrie, M.W.H., 1998, A single amino acid, Glu$_{146}$, governs the substrate specificity of a human dopamine sulfotransferase, SULT1A3. *Molecular Pharmacology*, 54, 942–948.

Dooley, T.P., Haldeman-Cahill, R., Joiner, J., and Wilborn, T.W., 2000, Expression profiling of human sulfotransferase and sulfatase gene superfamilies in epithelial tissues and cultured cells. *Biochemical and Biophysical Research Communications*, 277, 236–245.

Eisenhofer, G., Coughtrie, M.W.H., and Goldstein, D.S., 1999, Dopamine sulphate: an enigma resolved. *Clinical and Experimental Pharmacology and Physiology*, 26, S41–S53.

Falany, C.N., 1997, Enzymology of human cytosolic sulfotransferases. *FASEB Journal*, 11, 206–216.

Falany, C.N., Wheeler, J., Oh, T.S., and Falany, J.L., 1994, Steroid sulfation by expressed human cytosolic sulfotransferases. *Journal of Steroid Biochemistry and Molecular Biology*, 48, 369–375.

Falany, C.N., Xie, X.W., Wang, J., Ferrer, J., and Falany, J.L., 2000, Molecular cloning and expression of novel sulphotransferase-like cDNAs from human and rat brain. *Biochemical Journal*, 346, 857–864.

Falany, J.L. and Falany, C.N., 1996, Regulation of estrogen sulfotransferase in human endometrial adenocarcinoma cells by progesterone. *Endocrinology*, 137, 1395–1401.

Foldes, A. and Meek, J.L., 1973, Rat brain phenolsulfotransferase — partial purification and some properties. *Biochimica et Biophysica Acta*, 327, 365–374.

Freimuth, R.R., Raftogianis, R.B., Wood, T.C., Moon, E., Kim, U.J., Xu, J., Sicilaiono, M.J., and Weinshilboum, R.M., 2000, Human sulfotransferases SULT1C1 and SULT1C2: cDNA characterization, gene cloning, and chromosomal localization. *Genomics*, 65, 157–165.

Geese, W.J. and Raftogianis, R.B., 2001, Biochemical characterization and tissue distribution of human SULT2B1. *Biochemical and Biophysical Research Communications*, 288, 280–289.

Her, C., Kaur, G.P., Athwal, R.S., and Weinshilboum, R.M., 1997, Human sulfotransferase SULT1C1: cDNA cloning, tissue-specific expression and chromosomal localization. *Genomics*, 41, 467–470.

Hume, R., Barker, E.V., and Coughtrie, M.W.H., 1996, Differential expression and immunohistochemical localization of the phenol and hydroxysteroid sulfotransferase enzyme families in the developing lung. *Histochemistry and Cell Biology*, 105, 147–152.

Hume, R. and Coughtrie, M.W.H., 1994, Phenolsulfotransferase — localization in kidney during human embryonic and fetal development. *Histochemical Journal*, 26, 850–855.

Hume, R., Coughtrie, M.W.H., and Burchell, B., 1995, Differential localization of UDP-glucuronosyltransferase in kidney during human embryonic and fetal development. *Archives of Toxicology*, 69, 242–247.

Javitt, N.B., Lee, Y.C., and Strott, C.A., 2001, Sulfotransferase enzymes hHST 2A1 and 2B1a/b: unmasking their substrate specificity as DHEA and cholesterol sulfotransferases respectively. *FASEB Journal*, 15, A30.

Kaptein, E., Vanhaasteren, G.A.C., Linkels, E., Degreef, W.J., and Visser, T.J., 1997, Characterization of iodothyronine sulfotransferase activity in rat liver. *Endocrinology*, 138, 5136–5143.

Kester, M.H.A., Kaptein, E., Roest, T.J., Van Dijk, C.H., Tibboel, D., Meinl, W., Glatt, H., Coughtrie, M.W.H., and Visser, T.J., 1999a, Characterization of human iodothyronine sulfotransferases. *Journal of Clinical Endocrinology and Metabolism*, 84, 1357–1364.

Kester, M.H.A., Kaptein, E., Van Dijk, C.H., Roest, T.J., Tibboel, D., Coughtrie, M.W.H., and Visser, T.J., 2002, Characterization of iodothyronine sulfatase activities in human and rat liver and placenta. *Endocrinology*, 143, 814–819.

Kester, M.H.A., Vandijk, C.H., Tibboel, D., Hood, A.M., Rose, N.J.M., Meinl, W., Pabel, U., Glatt, H., Falany, C.N., Coughtrie, M.W.H., and Visser, T.J., 1999b, Sulfation of thyroid hormone by estrogen sulfotransferase. *Journal of Clinical Endocrinology and Metabolism*, 84, 2577–2580.

Kohrle, J., 1996, Thyroid hormone deiodinanses — a selenoenzyme family acting as gate keepers to thyroid hormone action. *Acta Medica Austriaca*, 1/2, 17–30.

Kopin, I.J., 1985, Catecholamine metabolism: basic aspects and clinical significance. *Pharmacological Reviews*, 37, 333–364.

Kung, M.P., Spaulding, S.W., and Roth, J.A., 1988, Desulfation of 3,5,3′-triiodothyronine sulfate by microsomes from human and rat tissues. *Endocrinology*, 122, 1195–1200.

Laemmli, U.K., 1970, Cleavage of structural proteins during the assembly of the head of bacteriophage T_4. *Nature*, 227, 680–685.

Li, X.Y., Clemens, D.L., and Anderson, R.J., 2000, Sulfation of iodothyronines by human sulfotransferase 1C1 (SULT1C1). *Biochemical Pharmacology*, 60, 1713–1716.

Meloche, C.A. and Falany, C.N., 2001, Expression of SULT2B1b in hormonally regulated human tissues. *Drug Metabolism Reviews*, 33 (Suppl. 1), 230.

Pacifici, G.M., Franchi, M., Colizzi, C., Giuliani, L., and Rane, A., 1988, Sulfotransferase in humans: development and tissue distribution. *Pharmacology*, 36, 411–419.

Pacifici, G.M., Kubrich, M., Giuliani, L., de Vries, M., and Rane, A., 1993, Sulfation and glucuronidation of ritodrine in human fetal and adult tissues. *European Journal of Clinical Pharmacology*, 44, 259–264.

Padbury, J.F., Martinez, A.M., Thio, S.L., Burnell, E.E., and Humme, J.A., 1989, Free and sulfoconjugated catecholamine responses to hypoxia in fetal sheep. *American Journal of Physiology*, 257, E198–E202.

Parenti, G., Meroni, G., and Ballabio, A., 1997, The sulfatase gene family. *Current Opinion in Genetics & Development*, 7, 386–391.

Parker, C.R., Falany, C.N., Stockard, C.R., Stankovic, A.K., and Grizzle, W.E., 1994, Immunohistochemical localization of dehydroepiandrosterone sulfotransferase in human fetal tissues. *Journal of Clinical Endocrinology and Metabolism*, 78, 234–236.

Raftogianis, R.B., Wood, T.C., Otterness, D.M., Van Loon, J.A., and Weinshilboum, R.M., 1997, Phenol sulfotransferase pharmacogenetics in humans: association of common *SULT1A1* alleles with TS PST phenotype. *Biochemical and Biophysical Research Communications*, 239, 298–304.

Richard, K., Hume, R., Kaptein, E., Stanley, E.L., Visser, T.J., and Coughtrie, M.W.H., 2001, Sulfation of thyroid hormone and dopamine during human development — ontogeny of phenol sulfotransferases and arylsulfatase in liver, lung and brain. *Journal of Clinical Endocrinology and Metabolism*, 86, 2734–2742.

Rubin, G.L., Harrold, A.J., Mills, J.A., Falany, C.N., and Coughtrie, M.W.H., 1999, Regulation of sulfotransferase expression in the endometrium during the menstrual cycle, by oral contraceptives and during early pregnancy. *Molecular Human Reproduction*, 5, 995–1002.

Sakakibara, Y., Yanagishita, M., Katafuchi, J., Ringer, D.P., Takami, Y., Nakayama, T., Suiko, M., and Liu, M.C., 1998, Molecular cloning, expression, and characterization of novel human SULT1C sulfotransferases that catalyze sulfonation of N-hydroxy-2-acetylaminofluorene. *Journal of Biological Chemistry*, 273, 33929–33935.

Santini, F., Chiovato, L., Ghirri, P., Lapi, P., Mammoli, C., Montanelli, L., Scartabelli, G., Ceccarini, G., Coccoli, L., Chopra, I.J., Boldrini, A., and Pinchera, A., 1999, Serum iodothyronines in the human fetus and the newborn: evidence for an important role of placenta in fetal thyroid hormone homeostasis. *Journal of Clinical Endocrinology and Metabolism*, 84, 493–498.

Sharp, S., Barker, E.V., Coughtrie, M.W.H., Lowenstein, P.R., and Hume, R., 1993, Immunochemical characterization of a dehydroepiandrosterone sulfotransferase in rats and humans. *European Journal of Biochemistry*, 211, 539–548.

Song, W.C., Qian, Y.M., and Li, A.P., 1998, Estrogen sulfotransferase expression in the human liver: marked interindividual variation and lack of gender specificity. *Journal of Pharmacology and Experimental Therapeutics*, 284, 1197–1202.

Song, W.C. and Melner, M.H., 2000, Editorial: Steroid transformation enzymes as critical regulators of steroid action *in vivo*. *Endocrinology*, 141, 1587–1589.

Sperling, M.A., Ganguli, S., Leslie, N., and Landt, K., 1984, Fetal-perinatal catecholamine secretion: role in perinatal glucose homeostasis. *American Journal of Physiology*, 247, 69–74.

Stanley, E.L., 2001, Expression and Function of Iodothyronine Metabolising Enzymes during Human Placental and Fetal Development. Ph.D. thesis, University of Dundee, Scotland.

Stanley, E.L., Hume, R., Visser, T.J., and Coughtrie, M.W.H., 2001, Differential expression of sulfotransferase enzymes involved in thyroid hormone metabolism during human placental development. *Journal of Clinical Endocrinology and Metabolism*, 86, 5944–5955.

Sternberger, L.A., Hardy, P.H., Cuculis, J., and Meyer, H.G., 1990, The unlabeled antibody method of immunohistochemistry; preparation and properties of soluble antigen–antibody complex (horseradish peroxidase–anti-horseradish peroxidase) and its use in identification of spirochaetes. *Journal of Histochemistry & Cytochemistry*, 18, 315–333.

Towbin, H., Stehilin, T., and Gordon, J., 1979, Electrophoretic transfer of proteins from polyacrylamide gels to nitrocellulose sheets: procedures and some applications. *Proceedings of the National Academy of Sciences USA*, 76, 4045–4049.

Tseng, L. and Liu, H.C., 1981, Stimulation of arylsulfotransferase activity by progestins in human endometrium *in vitro*. *Journal of Clinical Endocrinology and Metabolism*, 53, 418–421.

Vaitukaitis, J.L., 1981, Production of antisera with small doses of immunogen: multiple intradermal injections. *Methods in Enzymology*, 73, 46–52.

Visser, T.J., 1996, Pathways of thyroid hormone metabolism. *Acta Medica Austriaca*, 1/2, 10–16.

Wang, J., Falany, J.L., and Falany, C.N., 1998, Expression and characterization of a novel thyroid hormone-sulfating form of cytosolic sulfotransferase from human liver. *Molecular Pharmacology*, 53, 274–282.

Weinshilboum, R.M., Otterness, D.M., Aksoy, I.A., Wood, T.C., Her, C., and Raftogianis, R.B., 1997, Sulfotransferase molecular biology: cDNAs and genes. *FASEB Journal*, 11, 3–14.

7 Sulfation of Thyroid Hormones

Monique H.A. Kester, Michael W.H. Coughtrie, and Theo J. Visser

CONTENTS

INTRODUCTION

Deiodination is a key process in the regulation of thyroid hormone bioavailability. The prohormone thyroxine (T4) is converted by outer ring deiodination to the receptor-active hormone 3,3′,5-triiodothyronine (T3) or by inner ring deiodination (IRD) to the receptor-inactive metabolite 3,3′,5′-triiodothyronine (rT3; Kohrle, 1999). T3 is further metabolized by IRD and rT3 by outer ring deiodination to the common metabolite 3,3′-diiodothyronine (3,3′-T2). Deiodination is catalyzed by three iodothyronine deiodinases (D1-D3), which have a selenocysteine residue in their catalytic center (Bianco et al., 2002; Kohrle, 1999). D1 catalyzes both outer ring deiodination and IRD and is abundantly expressed in liver, kidney, and thyroid (Bianco et al., 2002; Kohrle, 1999). D1 is the main site for the production of plasma T3 and the clearance of plasma rT3 in adults, but its expression is lower in fetal human tissues (Polk, 1995; Richard et al., 1998). D2 has only outer ring deiodinase activity and, thus, converts T4 solely to active T3. It is mainly expressed in brain, pituitary, brown adipose tissue, and, in humans, in thyroid and skeletal muscle (Bianco et al., 2002; Croteau et al., 1996; Kohrle, 1999). The physiological role of D2 largely lies in the local production of T3 in these tissues. D3 only catalyzes IRD, with preference for T3 to 3,3′-T2 over T4 to rT3 conversion. It is predominantly expressed in brain, skin, placenta, pregnant uterus, and fetal tissues and is important

121

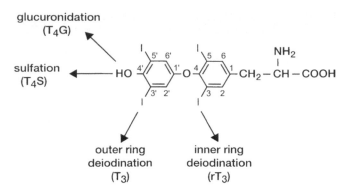

FIGURE 7.1 Pathways of thyroid hormone metabolism.

for the inactivation of tissue and plasma T3 as well as for the production of plasma rT3 (Bianco et al., 2002; Kohrle, 1999).

In addition to deiodination, thyroid hormone is importantly metabolized by the conjugation of the hydroxyl group with glucuronic acid or sulfate (Figure 7.1; Visser, 1996a). Normally, these conjugation reactions facilitate the biliary excretion of their substrates by increasing their water solubility. However, subsequent to the excretion of the glucuronides in the bile, at least part of the iodothyronines are reabsorbed after hydrolysis by β-glucuronidases present in the intestine (Visser, 1996a). Iodo-thyronine glucuronidation is carried out by multispecific UDP-glucuronyltrans-ferases (UGTs) located in the endoplasmic reticulum of tissues such as the liver, which catalyze the transfer of glucuronic acid from the cofactor UDP-glucuronic acid to the hydroxyl group of iodothyronines. Based on sequence homology, two families of UGTs have been identified (Mackenzie et al., 1997; Radominska-Pandya et al., 1999). So far, the bilirubin UGT1A1 and phenol UGT1A9 are known to be involved in the glucuronidation of T4 in humans (Findlay et al., 2000; Mackenzie et al., 1997; Radominska-Pandya et al., 1999). Whereas in the rat androsterone UGT seems largely responsible for the glucuronidation of T3, no human UGT preferring T3 over T4 has yet been identified (Beetstra et al., 1991; Findlay et al., 2000; Radominska-Pandya et al., 1999). Since expression of the UGTs catalyzing the glucuronidation of thyroid hormone starts around or after birth, they do not play an important role in thyroid hormone metabolism in the developing fetus (Coughtrie et al., 1988).

Like glucuronidation, sulfation is an important detoxification mechanism for exogenous chemicals (Coughtrie, 2002; Glatt et al., 2001). In addition, sulfation is involved in the regulation of the biological activity of endogenous compounds such as monoamines, steroids, and thyroid hormone (Coughtrie, 2002; Falany, 1997). This chapter will focus on thyroid hormone sulfation and its role in the regulation of thyroid hormone bioavailability.

THE IMPORTANCE OF SULFATION IN THYROID HORMONE METABOLISM

The role of sulfation in thyroid hormone metabolism is intriguing. Sulfated iodo-thyronines do not bind to the thyroid hormone receptors nor are they substrates for D2 and D3, but their deiodination by D1 is strongly augmented. Figure 7.2 shows the effects of sulfation on the deiodination of different iodothyronines by D1 (Visser, 1996b). IRD (inactivation) of T4 sulfate (T4S) and T3 sulfate (T3S) is markedly facilitated, whereas outer ring deiodination (activation) of T4S is blocked (Visser, 1994, 1996b; Visser et al., 1988). However, outer ring deiodination of other substrates is not inhibited by sulfation; in fact, deiodination of rT3 sulfate (rT3S) is not affected and that of 3,3'-T2 sulfate (T2S) is even greatly stimulated (Visser, 1994; Visser et al., 1988). Under normal conditions, therefore, the main function of sulfation is to induce the irreversible degradation of thyroid hormone.

Normally, as a result of the very rapid IRD of T4S and T3S and outer ring deiodination of rT3S and T2S, the plasma concentrations of these sulfated iodothy-ronines are low (Chopra et al., 1992, 1993; Eelkman Rooda et al., 1989; Wu et al., 1993). However, in conditions in which D1 activity is low, e.g., in selenium defi-ciency and after administration of iopanoic acid or propylthiouracil, iodothyronine sulfate levels are increased (Chopra et al., 1992; Rooda et al., 1989; Wu et al., 1995). Elevated serum T3S/T3 ratios have also been reported in patients with nonthyroidal illness and in hypothyroid patients, in whom D1 activity is known to be decreased (Chopra et al., 1992).

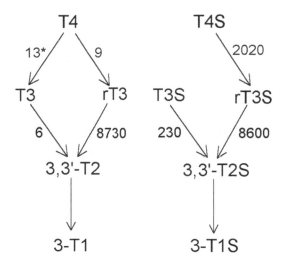

FIGURE 7.2 Efficiency of deiodination of iodothyronines and their sulfates by rat liver D1.* V_{max} (pmol/min/mg protein)/K_m (μM) ratio.

Furthermore, in human fetal serum and amniotic fluid, high iodothyronine sulfate concentrations are found (Chopra et al., 1992, 1993; Santini et al., 1993; Wu et al., 1992, 1993). Originally, these high levels were also believed to be due to low hepatic D1 expression in the human fetus. However, although D1 activity in rat fetal liver is low, increasing just before birth (Bates et al., 1999; Galton et al., 1991; Huang et al., 1988; Ruiz de Ona et al., 1991), significant D1 activity is already present in the fetal human liver in the second trimester of pregnancy (Richard et al., 1998). Little is known about the ontogeny of transporters such as Na-taurocholate cotransporting polypeptide (NTCP) and different Na-independent organic anion transporter polypeptides (OATPs), which mediate hepatic uptake of iodothyronine sulfates (Abe et al., 2002; Friesema et al., 1999) in the developing human fetus. Low expression of these transporters would be an alternative explanation for the high iodothyronine sulfate levels in the human fetus (see Discussion).

IODOTHYRONINE SULFOTRANSFERASES

Iodothyronine sulfation is catalyzed by cytosolic sulfotransferases, soluble enzymes located in different tissues such as liver, kidney, and brain, which catalyze the sulfation of the hydroxyl group of the different compounds using 3′-phosphoadenosine-5′-phosphosulfate (PAPS) as the sulfate donor (Coughtrie, 2002; Falany, 1997; Glatt et al., 2001). The sulfotransferases have a molecular weight of about 34 kDa and exist predominantly as homodimers (Matsui and Homma, 1994). The sulfotransferases are classified on the basis of amino acid sequence homology. The human SULT family consists of 11 enzymes, which can be subdivided into three families: the SULT1 family of phenol sulfotransferases (hSULT1A1, 1A2, 1A3, 1B1, 1C2, 1C4, 1E1), the SULT2 family of hydroxysteroid sulfotransferases (hSULT2A1, 2B1a, 2B1b), and the SULT4 family of sulfotransferase-like proteins (hSULT4A1), for which no enzyme activities have yet been reported (Falany et al., 2000; Glatt et al., 2001).

CHARACTERIZATION OF IODOTHYRONINE SULFOTRANSFERASES

The sulfotransferases show overlapping substrate specificity. Sulfation of the iodothyronines T4, T3, rT3, and 3,3′-T2 can be measured by incubating ^{125}I-labeled substrates with appropriate concentrations of the cofactor PAPS and enzyme. Alternatively, nonradioactive iodothyronines and [35S]PAPS are used to analyze iodothyronine sulfotransferase activity. It has thus been shown that all members of the hSULT1 family, i.e., hSULT1A1-3, 1B1, 1C2, 1C4, and 1E1, catalyze the sulfation of iodothyronines (Kester et al., 1999a, 1999b; Li et al., 2000; Wang et al., 1998). hSULT1A1-3, 1B1, and 1C4*, but also native enzymes in human kidney and liver cytosol, have a substrate preference for 3,3′-T2, which is catalyzed orders of magnitude faster than T3 and rT3, whereas sulfation of T4 is negligible (Kester et al.,

---

* M.H.A. Kester, M.W.H. Coughtrie, and T.J. Visser, unpublished observations.

1999a, 1999b). Interestingly, for crude cytosols of hSULT1C2-expressing cells, T4 is the preferred substrate, and T3, rT3, and 3,3′-T2 show successively lower sulfation rates. It should be noted, however, that this unique apparent preference of hSULT1C2 for T4 has not yet been confirmed in purified hSULT1C2.* hSULT1E1 also shows a pattern of substrate preference that differs from that of most other isoenzymes: hSULT1E1 equally prefers 3,3′-T2 and rT3 over T3 and T4 (Kester et al., 1999b).

To increase our understanding of the relative importance of the different enzymes, sulfation of iodothyronines has been characterized using purified sulfotransferases. Figure 7.3 compares sulfation of the different iodothyronines by purified hSULT1A1, 1A3, 1B1, and 1E1. 3,3′-T2 is shown to be a better substrate for hSULT1A1 than for the other enzymes, T3 is sulfated at similar rates by the different enzymes, and hSULT1E1 is the most effective in catalyzing rT3 sulfation (Kester et al., 1999b). The values for the kinetic parameters obtained by Lineweaver-Burk analysis of the sulfation of the iodothyronines by the purified sulfotransferases are presented in Table 7.1.* Although V_{max} values for 3,3′-T2 are similar for hSULT1A1 and hSULT1E1, the apparent K_m value is >10-fold lower for hSULT1A1 than for hSULT1E1, indicating a higher affinity of 3,3′-T2 for hSULT1A1. For rT3, the V_{max}/K_m ratio, which reflects the catalytic efficiency, is higher for hSULT1E1 compared to hSULT1A1. This again shows that hSULT1E1 is the most effective isoenzyme for rT3 sulfation. Furthermore, hSULT1E1 is the only enzyme catalyzing T4 sulfation. For this isoenzyme the V_{max}/K_m ratio is in the same order of magnitude for

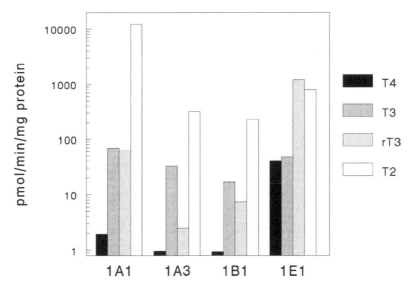

FIGURE 7.3 Sulfation of iodothyronines by purified human sulfotransferases. Reaction conditions: 0.1 μM iodothyronines, appropriate concentrations of enzymes, 50 μM PAPS, and 30 min incubation at 37°C (adapted from Kester et al., 1999b).

---

* M.H.A. Kester, M.W.H. Coughtrie, and T.J. Visser, unpublished observations.

TABLE 7.1
Kinetic Parameters of Purified Human Iodothyronine
Sulfotransferases

	K_m (µM)	V_{max} (nmol/min/mg)	V_{max}/K_m
Substrate: 3,3′-T2			
hSULT1A1	0.1–0.2	39.6–92.9	282.61–466.20
hSULT1A3	37.0–39.9	77.7–109.6	1.95–2.96
hSULT1B1	29.0–46.8	3.6–5.5	2.22–2.44
hSULT1E1	10.3–15.1	34.5–106.9	3.35–7.06
Substrate: T3			
hSULT1A1	10.2–18.0	6.9–11.5	282.61–466.20
hSULT1A3	30.9–80.2	12.2–36.5	0.39–0.46
hSULT1B1	29.0–46.8	3.6–5.5	0.12–0.12
hSULT1E1	25.0–29.5	7.2–10.9	0.29–0.37
Substrate: rT3			
hSULT1A1	4.8–5.2	2.6–3.2	0.56–0.61
hSULT1E1	1.0–3.1	14.3–30.3	9.67–14.52
Substrate: T4			
hSULT1A1	>300[a]	N.D.	
hSULT1E1	30.7[a]	7.9[a]	0.258[a]

Note: Data are presented as the range of values from two experiments.
N.D. = not determined.

[a] Value from a single experiment.

T4 as for T3. Although, compared to the other iodothyronine sulfotransferases, hSULT1E1 is clearly very efficient in catalyzing iodothyronine sulfation, it should be noted that this enzyme has much higher affinity for estrogens. Apparent K_m values of estrone (E1) and estradiol (E2) for hSULT1E1 are around 5 nM (Kester et al., 1999b), whereas apparent K_m values of 3,3′-T2 and rT3 are in the low micromolar range (Table 7.1).

To find out more about the structural requirements for iodothyronine sulfation by hSULT1E1, we also analyzed the sulfation of various unlabeled iodothyronines at a concentration of 1 µM in incubations with 0.3 µM [^{35}S] PAPS and hSULT1E1. Product formation was analyzed using the method of Foldes and Meek (1973), which is based on the precipitation of unreacted [^{35}S]PAPS with $BaSO_4$, leaving the radio-active sulfated products in solution. Besides 3,3′-T2 and rT3, 3′,5′-T2 and 3′-T1 are efficiently sulfated by hSULT1E1 (Figure 7.4). Sulfation efficiency decreases in the order 3′-T1 >> 3,3′-T2 ~ 3′,5′-T2 > rT3 >> T3 ~ T4 ~ 3-T1 > T0. This indicates a preference for substrates with 1 > 2 > 0 iodine substituents in the outer ring and with 0 > 1 > 2 iodine substituents in the inner ring.

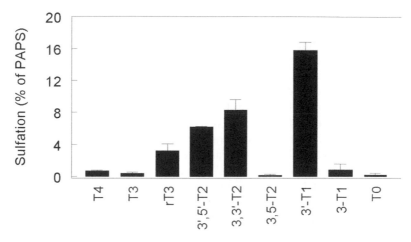

FIGURE 7.4 Sulfation of iodothyronines by hSULT1E1. Reaction conditions: 1 μM substrate, 25 μg of total cytosolic protein/ml, 0.3 μM [³⁵S]PAPS, 30 min incubation at 37°C. Results are the means and range of two experiments.

ONTOGENY AND TISSUE DISTRIBUTION OF IODOTHYRONINE SULFOTRANSFERASES

The expression of the different sulfotransferases is tissue- and developmental stage-dependent. Richard et al. (2001) studied the ontogeny of SULT1A1 and 1A3 in human tissues. They found high but variable hSULT1A1 expression in the fetal and postnatal human liver, which reached half the expression level of adult liver. hSULT1A3 is differently regulated: highest hSULT1A3 expression in the liver was found early in development, decreasing in the late fetal and early neonatal period, being absent in the adult liver (Richard et al., 2001). hSULT1A1 expression was also found in the fetal human brain, especially in the choroid plexus. In contrast, hSULT1A3 expression was low in the developing brain, with relatively the highest expression in cerebellum and germinal eminence (Richard et al., 2001).

RT-PCR and immunoblotting data have shown that hSULT1B1 is predominantly expressed in a wide range of adult tissues such as liver, stomach, duodenum, colon, colorectal, ovary, and cerebellum (Dooley et al., 2000); in the fetus, hSULT1B1 is only observed in the small bowel (Stanley, 2001). In contrast, although hSULT1C2 and hSULT1C4 mRNA have been found in adult kidney and ovary, respectively (Her et al., 1997; Sakakibara et al., 1998), hSULT1C2 and hSULT1C4 are mainly expressed in fetal tissues, such as small bowel, kidney, and liver (Stanley, 2001).

hSULT1E1 is expressed in the liver, mammary gland, uterus, placenta, and in a wide range of fetal tissues, such as lung, liver, adrenal, kidney, small bowel, thyroid, heart, and brain (Rubin et al., 1999; Stanley, 2001; Stanley et al., 2001; Strott, 1996). It is possible that the hSULT1E1 in the endometrium contributes to the high iodothyronine sulfate levels in the fetal serum and amniotic fluid (see Discussion).

DISCUSSION

Thyroid hormone metabolism is an important process in the regulation of thyroid hormone homeostasis. The prohormone T4 can be activated by outer ring deiodination to T3 or inactivated by IRD to rT3. T3 and rT3 are further metabolized to 3,3'-T2. In addition, iodothyronines are also metabolized by sulfotransferase-catalyzed conjugation of the phenolic hydroxyl group (Visser, 1996b).

The human fetus and adult express a wide range of sulfotransferases. In this chapter we assessed the role of sulfotransferases in iodothyronine metabolism. All members of the hSULT1 family catalyze iodothyronine sulfation. Whereas hSULT1A1 is expressed in fetal as well as in adult tissues such as liver, kidney, and brain, hSULT1A3, 1E1, and 1C2 are predominantly expressed in fetal tissues, and hSULT1B1 is mainly present in adult tissues (Dooley et al., 2000; Her et al., 1997; Sakakibara et al., 1998; Stanley, 2001; Stanley et al., 2001). The kinetic characterization of purified sulfotransferases revealed that hSULT1A1 and hSULT1E1 are the most potent iodothyronine sulfotransferases (Kester et al., 1999b).

Still, the tissue-specific expression of the sulfotransferases makes the understanding of the physiological importance of the different sulfotransferases rather complex. We showed a strong correlation between substrate specificities of hSULT1A1 and those of native iodothyronine sulfotransferase activities in human liver and kidney, which suggests an important role for hSULT1A1 in thyroid hormone sulfation in these tissues (Figure 7.5; Kester et al., 1999a). Richard et al. (2001) studied hSULT1A1 and hSULT1A3 expression and 3,3'-T2 sulfation in the

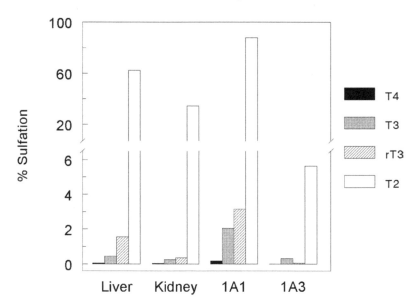

FIGURE 7.5 Sulfation of iodothyronines by human liver and kidney cytosol, SULT1A1 and SULT1A3. Reaction conditions: 0.1 μM ^{125}I-labeled iodothyronines, 0.1 mg protein/ml, 50 μM PAPS, and 30 min incubation at 37°C (adapted from Kester et al., 1999a).

developing liver, lung, and brain. They found a strong correlation between 3,3'-T2 sulfation and hSULT1A1 expression in the different tissues.

hSULT1E1 may be especially important for iodothyronine sulfation in the uterus. As was mentioned in the introduction, the iodothyronine sulfate levels in the human fetal serum and amniotic fluid are very high (Chopra et al., 1992, 1993; Santini et al., 1993; Wu et al., 1992, 1993). Also, T3 is low and rT3 is high in fetal serum compared to the adult (Burrow et al., 1994; Polk, 1995). These low T3 and high rT3 levels have been demonstrated to be due at least in part to high placental D3 activities, which in this way limit transplacental passage of maternal T4 and T3 to the fetus (Mortimer et al., 1996; Roti et al., 1981; Santini et al., 1999). However, at least in the rat, the pregnant uterus also expresses high D3 activity, and hSULT1E1 is expressed in the human endometrium and in the pregnant mouse uterus. Therefore, the uterus may play an additional role in protecting the fetus from excessive thyroid hormone by catalyzing the reversible (hSULT1E1) and irreversible (D3) inactivation of the hormone. Furthermore, in addition to the placenta, a role for the uterus in the supply of thyroid hormone (sulfates) from mother to fetus via the fetal membranes and the amniotic fluid is not excluded. It is remarkable that the products of thyroid hormone deiodination by D3, i.e., rT3 and 3,3'-T2, are also the preferred substrates for hSULT1E1, suggesting that T4 and T3 are metabolized in the uterus by successive deiodination and sulfation. Since rT3, but not T3, has profound and acute effects on the cytoskeleton in brain cells (Leonard and Farwell, 1997), it is possible that rT3 has an important function in fetal brain development. The production of rT3S by the uterus may thus be an important means to provide the developing brain with (reversibly inactivated) rT3. In addition, 3,3'-T2 has acute effects on mitochondrial respiration independent of the nuclear T3 receptor (Moreno et al., 1997). Sulfation may thus also be a route for the reversible inactivation of this metabolite.

Clearly, understanding the physiological role of the different iodothyronine sulfotransferases remains complex. Although hSULT1A1 and hSULT1E1 seem the most potent iodothyronine sulfotransferases, the other sulfotransferases from the hSULT1 family contribute to iodothyronine sulfation in a tissue-specific and developmental stage-dependent manner. The polymorphic variation of the different iodothyronine sulfotransferases (Glatt et al., 2001; Weinshilboum et al., 1997) and the finding that, at least in the rat, the different isoforms can also form heterodimers (Kiehlbauch et al., 1995) add to the complexity.

Sulfation is a reversible inactivation pathway. The iodothyronine sulfates (such as the above-mentioned rT3S) in the human fetal circulation may represent a reservoir of inactive thyroid hormone from which active thyroid hormone is liberated by action of arylsulfatases expressed in a tissue-specific and developmental stage-dependent manner (Darras et al., 1999; Santini et al., 1992; Visser, 1994). Multiple arylsulfatases have been identified now (Parenti et al., 1997). So far, arylsulfatase C (ARSC, also called steroid sulfatase) is the only member of the arylsulfatase family known to catalyze iodothyronine sulfate hydrolysis (Kester et al., 2002). Human ARSC is predominantly expressed in tissues such as placenta, liver, and brain (Coughtrie et al., 1998). In the placenta, the enzyme is principally involved in estrogen biosynthesis (Kuss, 1994). We characterized iodothyronine sulfatase activities of human ARSC and of human placenta and liver and demonstrated that ARSC is the main sulfatase for iodothyronine sulfate hydrolysis in the placenta, whereas in the liver, at least in the adult, other, still unidentified sulfatases are also involved (Kester et al., 2002).

The reasons for the high iodothyronine sulfate levels in the fetus are not completely clear. Certain iodothyronine sulfotransferases are expressed at a higher level in the fetal stage. Indeed, it has been demonstrated in sheep that the production of iodothyronine sulfates is increased in the fetus compared to the newborn (Polk, 1995). In addition, the clearance of iodothyronine sulfates may be decreased in the fetus. Clearance of iodothyronine sulfates involves three important processes: tissue uptake by transporters, deiodination by D1, and hydrolysis by sulfatases.

In the fetal rat, hepatic D1 activity is low until around after birth (Bates et al., 1999; Galton et al., 1991; Huang et al., 1988; Ruiz de Ona et al., 1991). Since D1 very efficiently degrades iodothyronine sulfates, the high iodothyronine sulfate levels in the human fetus were believed to be due to low hepatic fetal D1 activity. However, our studies have shown that D1 is already present in the human fetal liver at the end of the first trimester (Richard et al., 1998). Alternatively, the iodothyronine sulfates may accumulate because of the absence of hepatic transporters, which mediate the removal of iodothyronine sulfates from the plasma. In the rat, it has been demonstrated that NTCP and different members of the Na-independent OATP family mediate iodothyronine sulfate transfer over the plasma membrane (Abe et al., 1998; Friesema et al., 1999). Studies in rats have shown that the expression of NTCP starts on the last gestational day (Boyer et al., 1993) and OATP1 is not significantly expressed until after weaning (Dubuisson et al., 1996). Little is known about the ontogeny of the expression of these transporters in human tissues. The identification of the iodothyronine sulfatases and in particular their temporal and spatial profiles remain to be studied in detail.

In conclusion, under normal conditions sulfation is a primary step in the irreversible degradation of T4 and T3 by D1 (Visser, 1994, 1996b; Visser et al., 1988). In conditions of high iodothyronine sulfate levels, such as during fetal development, the inactivation of thyroid hormone may be reversible due to the expression of arylsulfatases in different tissues (Kester et al., 2002; Santini et al., 1992). Possibly, in the fetus the iodothyronine sulfates form a reservoir from which active thyroid hormone is released when needed. However, it should be realized that the hydrolysis of iodothyronine sulfates does not only depend on the presence of arylsulfatases but also on the presence of iodothyronine sulfate transporters. Therefore, further research on the ontogeny and tissue distribution of not only the iodothyronine sulfotransferases but also the iodothyronine sulfatases and iodothyronine sulfate transporters is needed to further elucidate the role of sulfation/desulfation in iodothyronine metabolism.

ACKNOWLEDGMENTS

This work was supported by NWO grant 903-40-204, EC grant QLG-2000-00930, and by Tenovus Scotland/Leng Trust.

REFERENCES

Abe, T., Kakyo, M., Sakagami, H., Tokui, T., Nishio, T., Tanemoto, M., Nomura, H., Hebert, S.C., Matsuno, S., Kondo, H., and Yawo, H., 1998, Molecular characterization and tissue distribution of a new organic anion transporter subtype (oatp3) that transports thyroid hormones and taurocholate and comparison with oatp2, *J Biol Chem* 273:22395–22401.

Abe, T., Suzuki, T., Unno, M., Tokui, T., and Ito, S., 2002, Thyroid hormone transporters: recent advances, *Trends Endocrinol Metab* 13:215–220.

Bates, J.M., St. Germain, D.L., and Galton, V.A., 1999, Expression profiles of the three iodothyronine deiodinases, D1, D2, and D3, in the developing rat, *Endocrinology* 140:844–851.

Beetstra, J.B., van Engelen, J.G., Karels, P., van der Hoek, H.J., de Jong, M., Docter, R., Krenning, E.P., Hennemann, G., Brouwer, A., and Visser, T.J., 1991, Thyroxine and 3,3′,5-triiodothyronine are glucuronidated in rat liver by different uridine diphosphate-glucuronyltransferases, *Endocrinology* 128:741–746.

Bianco, A.C., Salvatore, D., Gereben, B., Berry, M.J., and Larsen, P.R., 2002, Biochemistry, cellular and molecular biology, and physiological roles of the iodothyronine selenodeiodinases, *Endocr Rev* 23:38–89.

Boyer, J.L., Hagenbuch, B., Ananthanarayanan, M., Suchy, F., Stieger, B., and Meier, P.J., 1993, Phylogenic and ontogenic expression of hepatocellular bile acid transport, *Proc Natl Acad Sci USA* 90:435–438.

Burrow, G.N., Fisher, D.A., and Larsen, P.R., 1994, Maternal and fetal thyroid function, *N Engl J Med* 331:1072–1078.

Chopra, I.J., Santini, F., Hurd, R.E., and Chua Teco, G.N., 1993, A radioimmunoassay for measurement of thyroxine sulfate, *J Clin Endocrinol Metab* 76:145–150.

Chopra, I.J., Wu, S.Y., Teco, G.N., and Santini, F., 1992, A radioimmunoassay for measurement of 3,5,3′-triiodothyronine sulfate: studies in thyroidal and nonthyroidal diseases, pregnancy, and neonatal life, *J Clin Endocrinol Metab* 75:189–194.

Coughtrie, M.W., 2002, Sulfation through the looking glass — recent advances in sulfotransferase research for the curious, *Pharmacogenomics J* 2:297–308.

Coughtrie, M.W., Burchell, B., Leakey, J.E., and Hume, R., 1988, The inadequacy of perinatal glucuronidation: immunoblot analysis of the developmental expression of individual UDP-glucuronosyltransferase isoenzymes in rat and human liver microsomes, *Mol Pharmacol* 34:729–735.

Coughtrie, M.W., Sharp, S., Maxwell, K., and Innes, N.P., 1998, Biology and function of the reversible sulfation pathway catalysed by human sulfotransferases and sulfatases, *Chem Biol Interact* 109:3–27.

Croteau, W., Davey, J.C., Galton, V.A., and St Germain, D.L., 1996, Cloning of the mammalian type II iodothyronine deiodinase: a selenoprotein differentially expressed and regulated in human and rat brain and other tissues, *J Clin Invest* 98:405–417.

Darras, V.M., Hume, R., and Visser, T.J., 1999, Regulation of thyroid hormone metabolism during fetal development, *Mol Cell Endocrinol* 151:37–47.

Dooley, T.P., Haldeman-Cahill, R., Joiner, J., and Wilborn, T.W., 2000, Expression profiling of human sulfotransferase and sulfatase gene superfamilies in epithelial tissues and cultured cells, *Biochem Biophys Res Commun* 277:236–245.

Dubuisson, C., Cresteil, D., Desrochers, M., Decimo, D., Hadchouel, M., and Jacquemin, E., 1996, Ontogenic expression of the Na(+)-independent organic anion transporting polypeptide (oatp) in rat liver and kidney, *J Hepatol* 25:932–940.

Eelkman Rooda, S.J., Kaptein, E., and Visser, T.J., 1989, Serum triiodothyronine sulfate in man measured by radioimmunoassay, *J Clin Endocrinol Metab* 69:552–556.

Falany, C.N., 1997, Enzymology of human cytosolic sulfotransferases, *Faseb J* 11:206–216.

Falany, C.N., Xie, X., Wang, J., Ferrer, J., and Falany, J.L., 2000, Molecular cloning and expression of novel sulphotransferase-like cDNAs from human and rat brain, *Biochem J* 346:857–864.

Findlay, K.A., Kaptein, E., Visser, T.J., and Burchell, B., 2000, Characterization of the uridine diphosphate-glucuronosyltransferase-catalyzing thyroid hormone glucuronidation in man, *J Clin Endocrinol Metab* 85:2879–2883.

Foldes, A., and Meek, J.L., 1973, Rat brain phenolsulfotransferase: partial purification and some properties, *Biochim Biophys Acta* 327:365–374.

Friesema, E.C., Docter, R., Moerings, E.P., Stieger, B., Hagenbuch, B., Meier, P.J., Krenning, E.P., Hennemann, G., and Visser, T.J., 1999, Identification of thyroid hormone trans-porters, *Biochem Biophys Res Commun* 254:497–501.

Galton, V.A., McCarthy, P.T., and St. Germain, D.L., 1991, The ontogeny of iodothyronine deiodinase systems in liver and intestine of the rat, *Endocrinology* 128:1717–1722.

Glatt, H., Boeing, H., Engelke, C.E., Ma, L., Kuhlow, A., Pabel, U., Pomplun, D., Teubner, W., and Meinl, W., 2001, Human cytosolic sulphotransferases: genetics, characteris-tics, toxicological aspects, *Mutat Res* 482:27–40.

Her, C., Kaur, G.P., Athwal, R.S., and Weinshilboum, R.M., 1997, Human sulfotransferase SULT1C1: cDNA cloning, tissue-specific expression, and chromosomal localization, *Genomics* 41:467–470.

Huang, T.S., Chopra, I.J., Boado, R., Soloman, D.H., and Chua Teco, G.N., 1988, Thyroxine inner ring monodeiodinating activity in fetal tissues of the rat, *Pediatr Res* 23:196–199.

Kester, M.H., Kaptein, E., Roest, T.J., van Dijk, C.H., Tibboel, D., Meinl, W., Glatt, H., Coughtrie, M.W., and Visser, T.J., 1999a, Characterization of human iodothyronine sulfotransferases, *J Clin Endocrinol Metab* 84:1357–1364.

Kester, M.H., Kaptein, E., Van Dijk, C.H., Roest, T.J., Tibboel, D., Coughtrie, M.W., and Visser, T.J., 2002, Characterization of iodothyronine sulfatase activities in human and rat liver and placenta, *Endocrinology* 143:814–819.

Kester, M.H., van Dijk, C.H., Tibboel, D., Hood, A.M., Rose, N.J., Meinl, W., Pabel, U., Glatt, H., Falany, C.N., Coughtrie, M.W., and Visser, T.J., 1999b, Sulfation of thyroid hormone by estrogen sulfotransferase, *J Clin Endocrinol Metab* 84:2577–2580.

Kiehlbauch, C.C., Lam, Y.F., and Ringer, D.P., 1995, Homodimeric and heterodimeric aryl sulfotransferases catalyze the sulfuric acid esterification of N-hydroxy-2-acetylami-nofluorene, *J Biol Chem* 270:18941–18947.

Kohrle, J., 1999, Local activation and inactivation of thyroid hormones: the deiodinase family, *Mol Cell Endocrinol* 151:103–119.

Kuss, E., 1994, The fetoplacental unit of primates, *Exp Clin Endocrinol* 102:135–165.

Leonard, J.L. and Farwell, A.P., 1997, Thyroid hormone-regulated actin polymerization in brain, *Thyroid* 7:147–151.

Li, X., Clemens, D.L., and Anderson, R.J., 2000, Sulfation of iodothyronines by human sulfotransferase 1C1 (SULT1C1)*, *Biochem Pharmacol* 60:1713–1716.

Mackenzie, P.I., Owens, I.S., Burchell, B., Bock, K.W., Bairoch, A., Belanger, A., Fournel-Gigleux, S., Green, M., Hum, D.W., Iyanagi, T., Lancet, D., Louisot, P., Magdalou, J., Chowdhury, J.R., Ritter, J.K., Schachter, H., Tephly, T.R., Tipton, K.F., and Nebert, D.W., 1997, The UDP glycosyltransferase gene superfamily: recommended nomen-clature update based on evolutionary divergence, *Pharmacogenetics* 7:255–269.

Matsui, M. and Homma, H., 1994, Biochemistry and molecular biology of drug-metabolizing sulfotransferase, *Int J Biochem* 26:1237–1247.

Moreno, M., Lanni, A., Lombardi, A., and Goglia, F., 1997, How the thyroid controls metabolism in the rat: different roles for triiodothyronine and diiodothyronines, *J Physiol* 505:529–538.

Mortimer, R.H., Galligan, J.P., Cannell, G.R., Addison, R.S., and Roberts, M.S., 1996, Maternal to fetal thyroxine transmission in the human term placenta is limited by inner ring deiodination, *J Clin Endocrinol Metab* 81:2247–2249.

Parenti, G., Meroni, G., and Ballabio, A., 1997, The sulfatase gene family, *Curr Opin Genet Dev* 7:386–391.

Polk, D.H., 1995, Thyroid hormone metabolism during development, *Reprod Fertil Dev* 7:469–477.

Radominska-Pandya, A., Czernik, P.J., Little, J.M., Battaglia, E., and Mackenzie, P.I., 1999, Structural and functional studies of UDP-glucuronosyltransferases, *Drug Metab Rev* 31:817–899.

Richard, K., Hume, R., Kaptein, E., Sanders, J.P., van Toor, H., de Herder, W.W., den Hollander, J.C., Krenning, E.P., and Visser, T.J., 1998, Ontogeny of iodothyronine deiodinases in human liver, *J Clin Endocrinol Metab* 83:2868–2874.

Richard, K., Hume, R., Kaptein, E., Stanley, E.L., Visser, T.J., and Coughtrie, M.W., 2001, Sulfation of thyroid hormone and dopamine during human development: ontogeny of phenol sulfotransferases and arylsulfatase in liver, lung, and brain, *J Clin Endocrinol Metab* 86:2734–2742.

Rooda, S.J., Kaptein, E., Rutgers, M., and Visser, T.J., 1989, Increased plasma 3,5,3'-triiodothyronine sulfate in rats with inhibited type I iodothyronine deiodinase activity, as measured by radioimmunoassay, *Endocrinology* 124:740–745.

Roti, E., Fang, S.L., Green, K., Emerson, C.H., and Braverman, L.E., 1981, Human placenta is an active site of thyroxine and 3,3'5-triiodothyronine tyrosyl ring deiodination, *J Clin Endocrinol Metab* 53:498–501.

Rubin, G.L., Harrold, A.J., Mills, J.A., Falany, C.N., and Coughtrie, M.W., 1999, Regulation of sulphotransferase expression in the endometrium during the menstrual cycle, by oral contraceptives and during early pregnancy, *Mol Hum Reprod* 5:995–1002.

Ruiz de Ona, C., Morreale de Escobar, G., Calvo, R., Escobar del Rey, F., and Obregon, M.J., 1991, Thyroid hormones and 5'-deiodinase in the rat fetus late in gestation: effects of maternal hypothyroidism, *Endocrinology* 128:422–432.

Sakakibara, Y., Takami, Y., Nakayama, T., Suiko, M., and Liu, M.C., 1998, Localization and functional analysis of the substrate specificity/catalytic domains of human M-form and P-form phenol sulfotransferases, *J Biol Chem* 273:6242–6247.

Santini, F., Chiovato, L., Ghirri, P., Lapi, P., Mammoli, C., Montanelli, L., Scartabelli, G., Ceccarini, G., Coccoli, L., Chopra, I.J., Boldrini, A., and Pinchera, A., 1999, Serum iodothyronines in the human fetus and the newborn: evidence for an important role of placenta in fetal thyroid hormone homeostasis, *J Clin Endocrinol Metab* 84:493–498.

Santini, F., Chopra, I.J., Wu, S.Y., Solomon, D.H., and Chua Teco, G.N., 1992, Metabolism of 3,5,3'-triiodothyronine sulfate by tissues of the fetal rat: a consideration of the role of desulfation of 3,5,3'-triiodothyronine sulfate as a source of T3, *Pediatr Res* 31:541–544.

Santini, F., Cortelazzi, D., Baggiani, A.M., Marconi, A.M., Beck-Peccoz, P., and Chopra, I.J., 1993, A study of the serum 3,5,3'-triiodothyronine sulfate concentration in normal and hypothyroid fetuses at various gestational stages, *J Clin Endocrinol Metab* 76:1583–1587.

Stanley, E.L., 2001, Expression and Function of Iodothyronine Metabolising Enzymes during Human Placental Development, Ph.D. thesis, University of Dundee, Scotland.

Stanley, E.L., Hume, R., Visser, T.J., and Coughtrie, M.W., 2001, Differential expression of sulfotransferase enzymes involved in thyroid hormone metabolism during human placental development, *J Clin Endocrinol Metab* 86:5944–5955.

Strott, C.A., 1996, Steroid sulfotransferases, *Endocr Rev* 17:670–697.

Visser, T.J., 1994, Role of sulfation in thyroid hormone metabolism, *Chem Biol Interact* 92:293–303.

Visser, T.J., 1996a, Role of sulfate in thyroid hormone sulfation, *Eur J Endocrinol* 134:12–14.

Visser, T.J., 1996b, Pathways of thyroid hormone metabolism, *Acta Med Austriaca* 23:10–16.

Visser, T.J., Kaptein, E., Terpstra, O.T., and Krenning, E.P., 1988, Deiodination of thyroid hormone by human liver, *J Clin Endocrinol Metab* 67:17–24.

Wang, J., Falany, J.L., and Falany, C.N., 1998, Expression and characterization of a novel thyroid hormone-sulfating form of cytosolic sulfotransferase from human liver, *Mol Pharmacol* 53:274–282.

Weinshilboum, R.M., Otterness, D.M., Aksoy, I.A., Wood, T.C., Her, C., and Raftogianis, R.B., 1997, Sulfation and sulfotransferases 1: sulfotransferase molecular biology: cDNAs and genes, *Faseb J* 11:3–14.

Wu, S.Y., Huang, W.S., Chopra, I.J., Jordan, M., Alvarez, D., and Santini, F., 1995, Sulfation pathway of thyroid hormone metabolism in selenium-deficient male rats, *Am J Physiol* 268:E572–E579.

Wu, S.Y., Huang, W.S., Polk, D., Chen, W.L., Reviczky, A., Williams, J., III, Chopra, I.J., and Fisher, D.A., 1993, The development of a radioimmunoassay for reverse triiodo-thyronine sulfate in human serum and amniotic fluid, *J Clin Endocrinol Metab* 76:1625–1630.

Wu, S.Y., Huang, W.S., Polk, D., Florsheim, W.H., Green, W.L., and Fisher, D.A., 1992, Identification of thyroxine-sulfate (T4S) in human serum and amniotic fluid by a novel T4S radioimmunoassay, *Thyroid* 2:101–105.

8 Estrogen Sulfotransferase in Breast Cancer

Jorge R. Pasqualini and Gerard S. Chetrite

CONTENTS

INTRODUCTION

Most breast cancers (95-97%) are initially hormone-dependent, where the hormone estradiol plays an important role in their development and progression. Recent information suggests that this pathogenic action can be indirect. The estradiol-receptor complex can mediate the activation of proto-oncogenes, oncogenes (e.g., c-myc, c-fos), and nuclear proteins, as well as other genes. Consequently, factors that modulate the tissular concentration of active estrogens are important to the etiology of this disease.

The "intracrinology concept," where a hormone can have its biological response in the same organ as it is produced, is applied to breast carcinoma tissue because the enzyme systems for the bioformation and metabolic transformation of estrogens

135

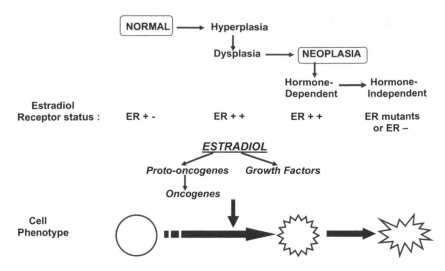

FIGURE 8.1 Evolutive transformation of the breast cell from normal to carcinogen. ER+: Estrogen receptor positive (detectable and functional); ER mutants: Estrogen receptor detectable but nonfunctional; ER−: Estrogen receptor negative (not detectable).

are present in this tissue. In addition, breast cancer tissue accumulates huge quantities of estrogens (unconjugated and in the form of sulfoconjugates), particularly in postmenopausal patients.

The transformation from normal to carcinogenic breast cell is a long process, probably lasting decades, which involves a complex mechanism. A hypothetical scheme of the various potential steps in this evolutive transformation of the breast cell is indicated in Figure 8.1.

Two main enzymatic pathways are implicated in the intratumoral biosynthesis of estrogens: the "aromatase pathway," which converts androgens into estrogens, and the "sulfatase pathway," which transforms estrone sulfate (E_1S) to estrone. Estrone is then reduced to estradiol by the 17β-hydroxysteroid dehydrogenase type 1. Data indicate that the activity of estrone sulfatase in breast tumors is 10- to 500-fold higher than aromatase activity (Pasqualini et al., 1996)

Conversion to the sulfates through the action of sulfotransferases (SULTs), which are present in huge quantities in both the normal and cancerous breast tissues, can inactivate estrogens. Figure 8.2 gives a schematic representation of the enzymatic process involved in the formation and transformation of estrogens in breast cancer tissue.

This chapter concerns the role of the estrogen sulfotransferase (EC 2.8.2.4) activity in breast cancer, its control by different substances, and its possible clinical applications.

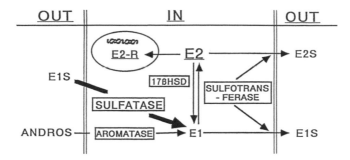

FIGURE 8.2 Origin and intracellular regulation of estradiol in breast cancer. out = extracellular compartment; in = intracellular compartment. ANDROS = androstenedione; E_1 = estrone; E_2 = estradiol; E_1S = estrone sulfate; E_2S = estradiol sulfate; E_2-R = estrogen receptor–estradiol complex; 17β-HSD = 17β-hydroxysteroid dehydrogenase.

CONCENTRATION OF ESTROGENS AND THEIR SULFATES IN NORMAL AND PATHOLOGICAL BREAST

Estrogen sulfotransferase activity is very high in both physiological and pathophysiological conditions. It is well established that E_1S is quantitatively the most important form of circulating estrogen during the cycle (Honjo et al., 1987; Nunez et al., 1977) as well as in postmenopausal women (Noel et al., 1981; Roberts et al., 1980).

Different studies agree that the plasma levels of unconjugated estrone and estradiol are similar in normal women and in breast cancer patients (for a review, see Pasqualini and Chetrite, 1996), although Thomas et al. (1997) reported a strong positive relationship between breast cancer susceptibility and increased serum estradiol concentration. However, the concentration of E1S is significantly higher in the follicular phase of premenopausal breast cancer patients than in normal women (Pasqualini et al., 1996).

In breast cancer tissues, most authors agree that the concentrations of unconjugated estrogens are found at high levels (Bonney et al., 1983; Pasqualini et al., 1996; van Landeghem et al., 1985). Data on the values of E_1S in this tissue are limited, but information from two laboratories indicated high concentrations, particularly in postmenopausal patients (Pasqualini et al., 1996; Vermeulen et al., 1986). Important information is obtained by comparing estrogen levels in the tumor and plasma, where it is observed that the tumor/plasma ratio for E_1S and for estradiol increases very significantly in postmenopausal breast cancer patients (Pasqualini et al., 1996; see Table 8.1). Comparative studies on the evaluation of estrone and estradiol and their sulfates show that the levels of these estrogens are significantly higher in the tumoral tissue than in the area of the same breast considered normal (Chetrite et al., 2000).

In a series of studies in patients with breast fibroadenoma, similar information was obtained: the tissular concentrations of estrone, estradiol, and E_1S were significantly higher than in the breast considered normal (Pasqualini et al, 1997).

TABLE 8.1
Ratio Concentration in the Tumor Tissue and Plasma of Estrone (E_1), Estradiol (E_2), and Their Sulfates (E_1s, E_2s) in Human Breast Cancer

Patients	E_1	E_2	E_1S	E_2S
Premenopausal	7	5	0.3	2
Postmenopausal	6	22	9.0	4

Note: The ratio corresponds to values obtained with the tissue concentration of each estrogen (pmol/g) divided by the plasma concentration of the respective estrogen (pmol/ml). The data represent the average values obtained with 10 to 15 patients.

Quoted from Pasqualini JR, Chetrite G, Blacker M-C, Feinstein C, Delalonde L, Talbi M, Maloche C, *J Clin Endocrinol Metab* 81: 1460–1464, 1996.

ESTROGENS AND SULT ACTIVITY

The family of human cytosolic SULTs includes two major subfamilies: I) the Phenol SULT family, which includes the SULTs 1A1, 1A2, 1A3, 1B1, 1C2, 1C4 and the estrogen sulfotransferase, SULT1E1 and II) the hydroxysteroid SULTs, which include dehydroepiandrosterone sulfotransferase (SULT2A1) and the two SULT2B1 forms that sulfate cholesterol, pregnenolone and other steroids (Blanchard et al., 2004).

SULT1E1 conjugates 3-hydroxy-estrogens by transfer of the sulfuryl group from the co-substrate 3'-phosphoadenosine 5'-phosphosulfate (PAPS). The SULT1E1 enzyme is important in the inactivation of estradiol during the luteal phase of the menstrual cycle. SULT1E1 has a significantly higher activity/affinity for the sulfation of E_2 (K_m value of 4 nM) and 17α-ethinylestradiol (EE$_2$) than for other potent estrogens, such as diethylstilbestrol and equine estrogens. Consequently, the ability of SULT1E1 to sulfate estrogens at physiological concentrations is important in regulating their activation of estrogen receptor (ER) in target tissues of estrogens.

The human SULT1E1 enzyme contains 294 amino acids and was first cloned by Weinshilboum's group (Aksoy et al., 1994). The native enzyme is a dimer of 35 kDa subunits, and kinetic studies indicate that two E_2 are bound per subunit (Zhang et al., 1998). Human endometrial Ishikawa adenocarcinoma cells demonstrate high levels of estrogen sulfotransferase activity (Chetrite and Pasqualini, 1997; Hata et al., 1987).

SULT1E1 is not the only enzyme to sulfate estrogens, but it is distinguishable from the other enzymes involved by the very high affinity for E_2 (e.g. Zhang et al., 1998). Members of the SULT1A family, particularly SULT1A1, are able to sulfate estrogens at μM concentrations (e.g. Falany et al., 1994). SULT1E1 is present mainly in the normal breast cell. However, in the breast carcinoma cell lines, estrone and estradiol could be sulfoconjugated by the action of SULT1A1 and possibly SULT1A3, and SULT1E appears to be expressed at low levels in breast cancer cells (Falany and Falany, 1996a).

Pedersen et al. (2002) obtained the crystal structure of the human EST–PAPS complex. The authors suggested that in the PAPS-bound structure, the side chain nitrogen of the catalytic Lys (47) interacts with the side chain hydroxyl of Ser (137) and not with the bridging oxygen between the 5′-phosphate and sulfate groups of the PAPS molecule as is seen in the PAP-bound structures.

Another important aspect of the metabolic transformation of estrogens is its conversion to catecholestrogens by hydroxylations in C-2 and C-4 to form, respectively, 2-hydroxyestrone (2-OHE$_1$), 2-hydroxyestradiol (2-OHE$_2$), 4-hydroxyestrone (4-OHE$_1$), and 4-hydroxyestradiol (4-OHE$_2$). These catecholestrogens can undergo further metabolism to form quinones that interact with DNA and could be involved in carcinogenesis, and the sulfonation of catecholestrogens could impede this process. Adjei and Weinshilboum (2002) show that of all SULTs, EST has the lowest K_m values, with 0.31, 0.18, 0.27, and 0.22 μM for 4-OHE$_1$, 4-OHE$_2$, 2-OHE$_1$, and 2-OHE$_2$, respectively.

SULTS IN THE BREAST CELLS

NORMAL BREAST

High levels of estrone sulfotransferase were detected in a normal breast cell line, the Huma 7 obtained from reduction mammoplasty (Wild et al., 1991). These authors observed that EST activity in this cell line far exceeded that in either MCF-7 or ZR-75-1 breast cancer cells. In a study, after 24-hours incubation, the normal cell sulfated 50% of estrogens compared with less than 10% in the malignant cells. The data were confirmed by Anderson and Howell (1995) using two normal breast epithelial cells: the MTSV 1-7 and the MRSV 4-4 produced by simian virus 40 immortalization cells obtained from human milk.

In these normal breast cells, SULT1E1 has the affinity for estradiol sulfation in the nanomolar concentration range. Consequently, SULT1E1 may be active in altering the levels of unconjugated estrogens in the cell and thus cellular responsiveness to estrogens, as estrogens in the nanomolar concentration range interact with the ER. Estrogen-dependent breast cells with high SULT1E1 levels grow more slowly than cells with low or no detectable estrogen sulfotransferase. Metabolic evidence

TABLE 8.2
Sulfotransferase[a] Activity in Human Breast

	Experimental Condition	References
A) Normal Breast		
Epithelial cell (Huma 7)	Cell culture	Wild et al., 1991
Human mammary epithelium (HME)	Cell culture	Malet et al., 1991
Breast tissue	Immunochemistry	Sharp et al., 1994
MRSV 4-4 and MTSV 1-7	Cell culture	Anderson and Howell, 1995
HME	Cell culture	
HME	Cell culture	Falany and Falany, 1996a
HME	Cell culture	Otake et al., 2000
B) Cancerous Breast		
Primary and metastatic		
Tumor	Homogenate	Dao et al., 1974
Tumor	Cytosol	Godefroi et al., 1975
Primary tumor	Cytosol	Adams et al., 1979
Tumor	700 g, supernatant	Raju et al., 1980
Primary tumor	Cytosol	Pewnim et al., 1982
Primary tumor	Cytosol	Tseng et al., 1983
Tumor	Cytosol	Adams and Phillips, 1990
MCF-7	Cell culture	Godefroi et al., 1975
MCF-7, BT-20	Cell culture	Raju et al., 1980
MCF-7	Cell culture	Rozhin et al., 1986
MCF-7, T-47D, MDA-MB-361, ZR-75-1	Cell culture	Adams et al., 1989
MDA-MB-468	Cell culture	Pasqualini, 1992
MCF-7	Cell culture	Falany et al., 1993

[a]Including all isoforms of SULTs.

indicates that this is due to the ability of SULT1E1 to render estrogens physiologically inactive via sulfoconjugation (Falany and Falany, 1996a, 1996b; Qian et al., 1998, 2001). Table 8.2 summarizes the presence of SULTs in the normal and carcinoma breast demonstrated by different authors.

BREAST CANCER

There are some discrepancies in the various reports of SULT activities in breast cancer: some authors found significant amounts of phenol sulfotransferase, but only trace levels of hydroxysteroid and estrogen sulfotransferase activities in several hormone-dependent breast cancer cells (Falany et al., 1993; Falany and Falany, 1996a). However, others report EST and HST activity in MCF-7 and ZR-75 cells and in mammary tumors. An interesting observation was made by Falany and Falany (1996a) who felt that human SULT1E1 is not detectable in most breast cancer cell lines and suggested that the sulfoconjugated activity in these cells is mainly due to the human phenol sulfotransferase SULT1A1, an enzyme that is more efficient with estrogens at micromolar than at nanomolar concentrations. SULT1A1 has an affinity for estrogen sulfation about 300-fold lower than that of human SULT1E1 (Falany et al., 1993, 1994).

Comparative studies using normal human mammary epithelial (HME) and MCF-7 breast cancer cells showed that after incubation with 20 nM E_2, the level of sulfated E_2 detected in the medium of HME was 10 times that found in the medium of MCF-7 cells (Falany and Falany, 1996a). The data indicate that HME cells secreted E_2-sulfate into the medium at a significantly higher rate than did MCF-7 breast cancer cells. As estrogen sulfates do not bind to the ER, factors that modify estrogen sulfotransferase levels and consequently affect estrogen metabolism may be important in controlling hormone-dependent cellular growth. The data suggested that in normal breast tissue, estrogen stimulation of growth and differentiation is carefully controlled, contrasting markedly with the abnormal proliferation of breast cancer cells (Zajchowski et al., 1993).

Normal HME cells possess endogenous SULT1E1 at physiological levels and are not present in MCF-7 and some other breast cancer cells (e.g., T47-D, BT-20, ZR75-1, and MDA-MB-231; Falany et al., 1993; Falany and Falany, 1996a). These authors suggested that in the breast cancer cells, estradiol or estrone is sulfated by the SULT1A1, an enzyme that only acts preferentially at micromolar concentrations of estrogens. The loss of SULT1E1 expression during the process of breast cancer oncogenesis may be critical because this enzyme inactivates estradiol, suggesting that the inability of the breast cell to block E_2 could be an important mechanism in contributing to abnormal growth of these cells through the presence of this hormone.

To explore the possibility that SULT1E1 disappears during the process of tumorigenesis, Falany and Falany (1996a, 1997) transfected MCF-7 cells with a SULT1E1-expression vector and observed that after incubation of 20 nM of estradiol, sulfation occurs significantly more rapidly with the transfected MCF-7 cells than in control cells, thereby rendering E_2 physiologically inactive. In addition, SULT1E1/MCF-7 cells require a higher concentration of estradiol to stimulate growth than do the MCF-7 control cells.

This observation was confirmed by Qian et al. (1998), who evaluated the physiological significance of SULT1E1 expression by cDNA transfection using MCF-7 cells and observed that in these transformed cells, the response to physiological concentrations of E2 (10 nM) is reduced by up to 70%, as determined in an estrogen-responsive reporter gene assay.

The physiological importance of SULT1E1 was also largely demonstrated in other estrogen-responsive tissue such as the endometrium. Falany et al. (1994) have reported that SULT1E1 is present at significant levels in human endometrial tissues during the secretory phase of the cycle, but this enzyme was not detectable during the proliferative phase.

Using another estrogen-dependent cell line, the human Ishikawa endometrial carcinoma cells (ISH), Kotov et al. (1999) found that these ISH cells, transformed with a SULT1E1 expression vector, were 200-fold less sensitive to E_2 and EE_2 than were control cells in their ability to activate the estrogen receptor. Consequently, in the SULT1E1/ISH cells, E_2 and EE_2 are rapidly sulfated at concentrations where these estrogens bind the ER, suggesting that this mechanism plays an important physiological role in modulating the response of ER-positive tissue to estrogenic stimulation. However, the same authors demonstrated that potent estrogens, such as diethylstilbestrol or equine estrogens, are not sulfated by the SULT1E1/ISH cells and the biological responses are similar to the control cells. These results indicate that SULT1E1 has a role in the selective inactivation of estradiol but is significantly less effective in inhibiting the ER binding of other potent estrogens. It is well known that estrogens play an important role in regulating the proliferation of breast tumors via the induction or suppression of growth regulatory factors (Molis et al., 1995). As an interesting effect, estrogens inhibit expression of the potent growth factor repressor, transforming growth factor (TGF)-β1 (Cho et al., 1994; Eckert and Katzenellenbogen, 1982; Knabbe et al., 1987). Also, it was observed that MCF-7 cells expressing SULT1E1 activity did not show a decrease in ER-α levels, an increase in progesterone receptor, or a decrease in transforming growth factor-β expression upon exposure to 100 pM or 1 nM of estradiol, which is suggested due to the rapid sulfation and inactivation of the unconjugated estradiol by SULT1E1 (Falany et al., 2002).

In conclusion, knowledge of the expression and regulation of the different SULTs is of extreme importance in understanding the changes in the normal breast cell during the process of carcinogenesis as well as the hormonal implication in this mechanism.

CONTROL OF SULT ACTIVITIES IN NORMAL AND CANCEROUS BREAST

Comparative studies of the quantitative evaluation of SULT activity in various breast cancer cells show significantly higher levels in the hormone-dependent (e.g., MCF-7, T47-D) than in the hormone-independent (e.g., MDA-MB-231) cells (Chetrite et al., 1999a; see Table 8.3).

TABLE 8.3
Transformation of Estrone to Estrogen Sulfates in the Cell Compartment and Culture Medium after Incubation with the Hormone-Dependent (MCF-7, T47-D) and Hormone-Independent (MDA-MB-231) Human Mammary Cancer Cells

Cell Lines	Estrogen Sulfates (in pmol/mg DNA)	
	In the Cells	In the Culture Medium
MCF-7	N.D.	14.90 ± 3.15
T-47D	N.D.	17.30 ± 2.80
MDA-MB-231	N.D.	2.01 ± 0.45

Quoted from Chetrite GS, Kloosterboer HJ, Philippe J-C, Pasqualini JR, *Anticancer Research* 19: 269–276, 1999b. With permission.

The control of the formation of estrogen sulfoconjugates represents an important mechanism to modulate the biological effect of the hormone in breast tissue as it is well established that estrogen sulfates are biologically inactive. Here we summarize the action of various substances that may inhibit or stimulate SULTs in breast tissue.

EFFECT OF PROGESTINS

Different progestins have been tested on the effect of the SULT activities in breast cancer cells. Medrogestone, a synthetic pregnane derivative that is used in the treatment of pathological deficiency of the natural progesterone, has a biphasic effect on SULT activity in MCF-7 and T47-D breast cancer cells: at a low concentration $(5 \times 10^{-8} \text{ mol/L})$, it stimulates the formation of estrogen sulfates, whereas at a high concentration $(5 \times 10^{-5} \text{ mol/L})$, the SULT activity is not modified in the MCF-7 cells or inhibited in T47-D cells. Other progestins, promegestone (R-5020), nomegestrol acetate at low concentration, can also increase SULT activity in breast cancer cells (Chetrite et al., 1999a; Figure 8.3). In relation to these findings, it is interesting to mention that the natural progesterone can induce SULT1E1 activity in the Ishikawa human endometrial adenocarcinoma cells, as well as in the excretory endometrial tissue (Clarke et al., 1982; Falany and Falany, 1996b; Tseng and Liu, 1981).

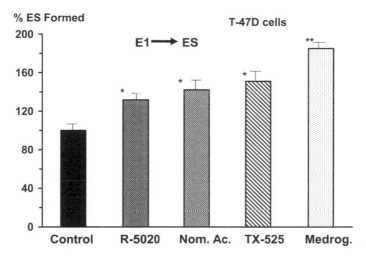

FIGURE 8.3 Comparative effects of various progestins on the conversion of estrone (E_1) to estrogen sulfates (ES) in the hormone-dependent T-47D human breast cancer cell line. Results (pmol of ES formed in culture medium per mg DNA from E_1) are expressed in percent (%) of control value considered as 100%. The data represent the mean ± S.E.M. of duplicate determinations of three to six independent experiments. R-5020: promegestone; Nom.Ac.: nomegestrol acetate; TX-525: a 19-nor progestin of Theramex Laboratories; Medrog.: medrogestone. * $P \leq 0.05$ vs control value; ** $P \leq 0.01$ vs control value.

EFFECT OF TIBOLONE AND ITS METABOLITES

Tibolone (the active substance in Livial®) is a 19-nortestosterone derivative with estrogenic, androgenic, and progestagenic properties used to prevent climacteric symptoms and postmenopausal bone loss (Bjarnason et al., 1996; Kicovic et al., 1982).

Tibolone is largely metabolized in three main derivatives: the 3α-and-3β hydroxy, which are estrogenic, and the 4-en isomer, which is progestagenic. These compounds also provoke a dual effect on SULT activity in breast cancer cells: stimulatory at low doses (5×10^{-8} mol/L), inhibitory at high doses (5×10^{-5} mol/L). The 3β-hydroxy derivative is the most potent compound in the stimulatory effect of the SULT activity in both the MCF-7 and T-47D breast cancer cells (Chetrite et al., 1999b; Figure 8.4).

As the apparent affinities of SULT1E1 for estrogens are in the same order as those of Kd for the ER (nanomolar concentrations), it was postulated that SULT1E1 can compete with ER for estradiol binding and abolish the steroid action after processing of ligand-charged ER (Anderson and Howell, 1995; Hobkirk, 1993; Hobkirk et al., 1985; Roy, 1992; Saunders et al., 1989). In support of this hypothesis, it is interesting to remark that a significant sequence homology was observed between the ligand domain of the ER and putative estrogen-binding domain deduced from bovine placental SULT1E1 cDNA (Nash et al., 1988).

It was demonstrated in previous studies in this laboratory that medrogestone, as well as tibolone and its metabolites, at low doses can block the sulfatase activity in the conversion of E1S to estradiol. As these compounds also stimulate SULT1E1 activity in the same concentration range, this dual effect can contribute to decreasing

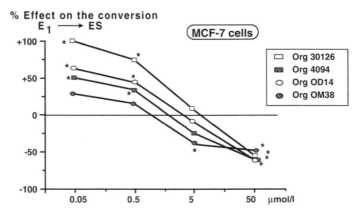

FIGURE 8.4 Comparative effects of tibolone (Org OD14; active substance of Livial®) and of its main metabolites on the conversion of estrone (E_1) to estrogen sulfates (ES) in the hormone-dependent MCF-7 human breast cancer cell line. Results (pmol of ES formed in culture medium per mg DNA from E_1) are expressed in percent (%) of control value considered as 100%. The data represent the mean ± S.E.M. of duplicate determinations of three to four experiments. Org OM38: 4-en isomer of tibolone; Org 4094: 3α-hydroxy derivative of tibolone; Org 30126: 3β-hydroxy derivative of tibolone. $* P \leq 0.05$ vs control value [3H]-E_1 alone. Quoted from Chetrite, GS, Kloosterboer, HJ, Philippe, J-C, Pasqualini, JR, *Anticancer Research*, 19: 269–276, 1999b.

the estrogenic stimulation by encouraging the excretion of estrogens to the sulfate form. If a similar action can operate *in vivo*, we have a new possibility to block estradiol with interesting clinical applications.

The mechanism implicated for the different dose-response effects observed with medrogestone or tibolone and its metabolites remains to be elucidated. However, there are a substantial number of examples where a hormone or antihormone produced an opposite effect according to its concentration.

EFFECT OF QUERCETIN AND RESVERATROL

Epidemiological studies have suggested that dietary phytoestrogens (e.g., soya products, tea, fruit, etc.) rich in flavonoids, isoflavonoids, and other phenolic compounds can protect against hormone-dependent breast cancer. One of the mechanisms implicated for this chemoprotective effect is the ability of phytoestrogens to inhibit human cytosolic SULTs as the sulfation process is a key step in the metabolic activation of some dietary or environmental procarcinogens and promutagens in mammary tissues (Banoglu, 2000; Kirk et al., 2001; Pai et al., 2001).

Quercetin and resveratrol are dietary flavonoids that can inhibit estrogen sulfatase activity (Huang et al., 1997). It was demonstrated that these flavonoids are also potent inhibitors of the human SULT1A1 (Eaton et al., 1996; Walle et al., 1995)

Otake et al. (2000) observed that quercetin and resveratrol are substrates for EST in the normal HME cells, with K_m values similar to their K_i values for inhibition of estradiol sulfation. Quercetin is 25 times more potent in inhibiting SULT1E1 in the HME cells than in inhibiting SULT1A1 activity in the intact human hepatoma cell line Hep G2, which has SULT1A1 expression levels similar to the human liver (Shwed et al., 1992). The mechanism for this potent inhibition is unclear. Otake et al. (2000) proposed that it could involve 1) a mechanism concentrating quercetin inside the breast cell; 2) bioactivation to a more potent form, e.g., by *O*-methylation; 3) inhibition of synthesis of the cosubstrate PAPS; or 4) inhibition of some factor involved in the regulation of EST expression.

The IC_{50} of 0.1 μM corresponds to a quercetin concentration of about 30 ng/mL, which is 5 to 10 times lower than concentrations in plasma reported in humans after consuming common foodstuffs rich in quercetin, such as onions and apples (Hollman et al., 1997). Inhibition of SULT1E1 by quercetin resulted in elevated estradiol levels in the normal breast cell, which can be a potentially harmful effect. However, it is interesting to note that in the HME cells, SULT1E1 could catalyze the bioactivation of the cooked food mutagen and procarcinogen *N*-hydroxy-2-amino-1-methyl-6-phenylimidazol (4,5-b)pyridine (*N*-OH-PhIP) and its subsequent binding to genomic DNA (Lewis et al., 1998). It has been reported that resveratrol (50 μM) leads to a decrease in PhIP–DNA adducts from 31 to 69% in primary cultures of HME cells. In breast cancer cell lines (MCF-7, ZR-75-1), resveratrol suppresses *O*-acetyltransferase and SULT activities (Dubuisson et al., 2002).

SULFOTRANSFERASE EXPRESSION AND ITS
CONTROL IN BREAST CANCER

SULT1E1 is the only sulfotransferase that displays affinity for 17β-estradiol in a physiological (nanomolar) concentration range (Zhang et al., 1998). Other SULTs, including SULT1A1 and SULT1A3, are able to sulfate estrogens *in vitro* however. For example, cells transfected with the cDNA coding for the enzyme originally called placental hEST 1 (Bernier et al., 1994b; Luu-The et al., 1996), now identified as SULT1A3, was able to transform estrone to E_1S at nanomolar concentrations (Bernier et al., 1994b).

Using reverse transcriptase–polymerase chain reaction amplification, the expression of SULT1A3 mRNA was detected in the hormone-dependent MCF-7 and T47-D, as well as in hormone-independent MDA-MB-231 and MDA-468, human breast cancer cells. An interesting correlation of the relative SULT activity and the SULT1A3 mRNA expression was found in various breast cancer cells studied (Chetrite et al., 1998; Figure 8.5).

Qian et al. (1998) demonstrated that the restoration of SULT1E1 expression in MCF-7 cells by cDNA transfection could significantly attenuate the response on both gene activity and DNA synthesis, and cell numbers were used as markers of estrogen-stimulated cell growth and proliferation. These authors suggest that loss or down regulation of SULT1E1 expression may enhance the growth-stimulating effect of estrogens and can contribute to the process of tumor initiation.

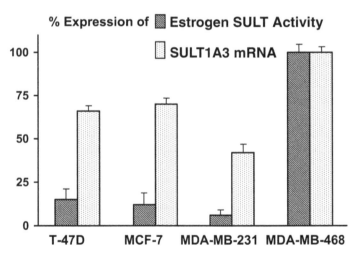

FIGURE 8.5 Relative SULT activity and SULT1A3 mRNA expression in the hormone-dependent (MCF-7, T-47D) and hormone-independent (MDA-MB-231, MDA-MB-468) human breast cancer cell lines. Results were expressed in percent (%), and the value of 100% was assigned to the activity and the mRNA expression of SULT1A3 in the MDA-MB-468 cells. The data represent the mean ± S.E.M. of three to five experiments. Quoted from Chetrite G, Le Nestour E, and Pasqualini JR, *J Steroid Biochem Molec Biol* 66: 295–302, 1998. With permission.

A study on the effects of the progestin promegestone (R-5020) on SULT1A3 mRNA and the formation of estrogen sulfates in the T-47D and MCF-7 cells showed that at low doses of R-5020, there was a significant increase in the levels of the mRNA in these breast cancer cell lines. This was accompanied by an increased formation of estrogen sulfates in these cell lines following the treatment. However, at high doses of this progestin, an inhibitory effect was observed on SULT1A3 mRNA and the formation of estrogen sulfates (Chetrite et al., 1998; Figure 8.6).

HYPOTHETICAL CORRELATION OF PROLIFERATION OF THE BREAST CANCER CELL AND SULT ACTIVITY

Maximal epithelial mitosis of the normal breast cell is found between 22 and 26 days of the cycle, which corresponds to the high levels of estradiol and progesterone (Longacre and Bartow, 1986). During pregnancy, it is suggested that the elevated values of circulating progesterone are responsible for the induction of lobular–alveolar development, to prepare the breast for lactation (Russo and Russo, 2002; Topper and Freeman, 1980). The data on the effect of progesterone on breast epithelial proliferation are contradictory. It has been found that progesterone can increase DNA synthesis in normal breast epithelium in organ culture (Laidlaw et al., 1995).

Using normal human breast epithelial cells, it was demonstrated that the progestin promegestone can decrease cell proliferation (Gompel et al., 1986; Malet et al., 2000). These authors also found that progestins can inhibit the proliferative effect

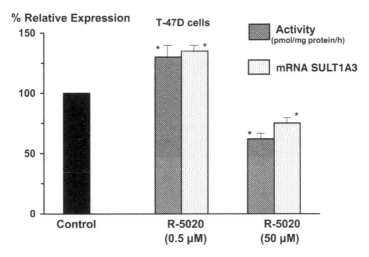

FIGURE 8.6 Effects of the progestin R-5020 (promegestone) on the SULT activity and the mRNA expression of SULT1A3 in the hormone-dependent T-47D human breast cancer cell line. Relative expression of the mRNA (using RT-PCR amplification) and the estrogen sulfation activity (in pmol/mg protein/h) in T-47D cells nontreated (control) and treated with R-5020 at the concentration of 5×10^{-5} or $5 \times 1 0^{-7}$ mol/L. The control value is assigned 100%. The data represent the mean ± S.E.M. of two to three experiments. Quoted from Chetrite G, Le Nestour E, Pasqualini JR, *J Steroid Biochem Molec Biol* 66: 295–302, 1998. With permission.

provoked by estradiol; however, McManus and Welsch (1984) and Longman and Buehring (1987) demonstrate no effect.

The proliferative effect of progestins using various isolated breast cancer models — cell lines, organ culture, or transplantation of breast cancer cells — in nude mice is contradictory as it was reported that these compounds can either inhibit (Botella et al., 1994; Horwitz and Freidenberg, 1985; Musgrove et al., 1991; Vignon et al., 1983), stimulate (Catherino and Jordan, 1995; Jeng et al., 1992; Kalkhoven et al., 1994), or have no effect (Schatz et al., 1985).

The biological action of progestins is derived from many factors: structure, affinity for the progesterone receptor or for other steroid receptors, the experimental conditions (source of the cell lines, media, sera, presence of phenol red, insulin, duration of treatment), the dose concentrations, and their metabolic transformation (Braunsberg et al., 1987; Schoonen et al., 1995a, 1995b; van der Burg et al., 1992).

Interesting information was obtained with nomegestrol acetate (Lutenyl®), a 19-nor-progestin derivative. This compound does not possess estrogenic activity and is exclusively antiproliferative in MCF-7 and T-47D breast cancer cells. Recently, it was demonstrated also that another progestin, medrogestone, can inhibit proliferation of the T-47D cells as well as block the proliferative effect provoked by estradiol (Pasqualini and Chetrite, 2002). Since estradiol plays an important role in regulating the proliferation of breast cancer cells via the induction or suppression of growth factors, the control of the hormone by SULT1E1 activity could be an important aspect in the mechanism of cell growth and proliferation.

As was indicated in preceding sections, SULT1E1 is present mainly in the normal breast cell and is active at nanomolar concentrations of E_2; consequently, it can block cell proliferation by formation of the inactive E_2S. However, in the breast cancer cells, SULT1A1 is present and is active only at micromolar concentrations; most of the E_2 remains in an unconjugated form and can be involved in cell proliferation. As the progestin medrogestone can stimulate SULT1E1 in breast cancer cells and as this compound can block the proliferation of T-47D cells (Figure 8.7), it is suggested that the antiproliferative action of medrogestone is correlated with the stimulatory effect of SULT1E1 in this breast cancer cell. This hypothetical mechanism of the correlation of SULT1E1 and proliferation in breast cancer is schematically represented in Figure 8.8 (A, B, and C).

DISCUSSION AND CONCLUSIONS

The intracellular metabolism of estrogens within breast cancer cells is important for the bioformation and activity of estradiol as well as to stimulate breast cancer cell growth. Cellular levels of E_2 are the result of a balance between A) its synthesis via the sulfatase pathway, and the aromatase pathway and B) its transformation via SULT1E1, which converts E_2 to the biologically inactive estradiol sulfate.

An intriguing question is the site of SULT action in breast cancer cells. Most studies indicate that SULT activity is highest in cytosol (Adams and Phillips, 1990; Evans et al., 1993; Rozhin et al., 1986). However, in all breast cancers studied, very little or no estrogen sulfates could be localized inside the cells after incubation with

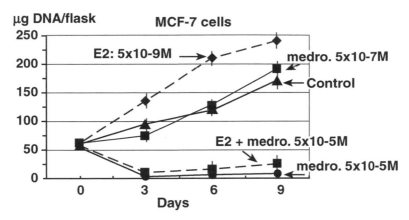

FIGURE 8.7 Effect of medrogestone (Medro) alone or in the presence of estradiol (E_2) on proliferation of MCF-7 breast cancer cells. MCF-7 cells were grown in 75 cm^2 flasks in MEM culture medium containing 5% FCS treated with dextran-coated charcoal. The cells were treated with medrogestone alone (5×10^{-5} M, 5×10^{-7} M), E_2 alone (5 nM), or Medro (5×10^{-5} M) + E_2 (5 nM; Day 0). DNA content was evaluated after 3, 6, and 9 days of culture. * $P \le 0.05$ vs. control value; ** $P \le 0.01$ vs. control value.

E_1 or E_2. In addition, when these cells were incubated with estrogen sulfates, only unconjugated estrogens were detected inside the cell. Similar findings were observed in the Ishikawa human endometrial adenocarcinoma cells (Chetrite and Pasqualini, 1997). The data suggest that both SULT and sulfatase are present inside the cell but that sulfated estrogens are rapidly exported from the cells. These compounds are substrates for members of the MRP family of efflux pumps, which may well explain these findings (Keppler et al., 2000; Leslie et al., 2001).

The finding that some progestins (e.g., promegestone, nomegestrol acetate, medrogestone) as well as tibolone and its metabolites at physiological concentrations can stimulate sulfotranferase activity in hormone-dependent breast cancer cells is an important point in the physiopathology of this disease as it is well known that estrogen sulfates do not bind to the ER.

As different substances, including antiestrogens, various progestins, as well as tibolone and its metabolites, can block very significantly sulfatase and 17β-hydroxysteroid dehydrogenase, the exploration of various progestins or other molecules in trials with breast cancer patients, showing an inhibitory effect on sulfatase and 17β-HSD and a stimulatory effect on SULTs will, in combination with antiaromatase agents, provide interesting and attractive possibilities in hormone replacement therapy, as well as in the treatment of breast cancer.

For the inhibitory or stimulatory effects on the control of the enzymes involved in the formation and transformation of estrogens in breast cancer, we propose the concept of "selective estrogen enzyme modulators" (SEEM), which is schematically represented in Figure 8.9.

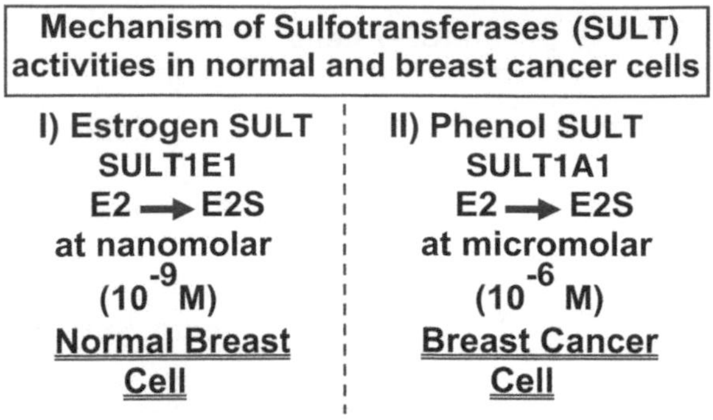

**Mechanism of Sulfotransferases (SULT)
activities in normal and breast cancer cells**

I) Estrogen SULT	II) Phenol SULT
SULT1E1	SULT1A1
E2 → E2S	E2 → E2S
at nanomolar	at micromolar
$(10^{-9}$ M)	$(10^{-6}$ M)
Normal Breast Cell	**Breast Cancer Cell**

FIGURE 8.8A Mechanism of SULT activity in normal and breast cancer cells. In normal breast cells, it is suggested that the action of SULT1E1 works at physiological (nanomolar) concentrations of estradiol to form estradiol sulfate, which is biologically inactive. This enzyme is absent from breast cancer cells where the PST activity acts only at micromolar (nonphysiological) concentrations.

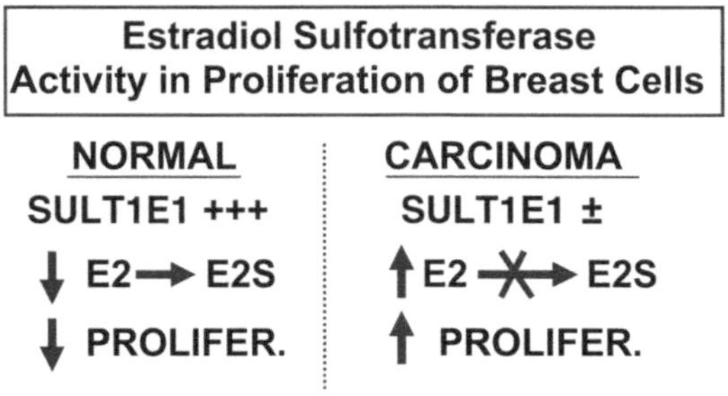

**Estradiol Sulfotransferase
Activity in Proliferation of Breast Cells**

NORMAL	CARCINOMA
SULT1E1 +++	SULT1E1 ±
↓ E2 → E2S	↑ E2 ⤢ E2S
↓ PROLIFER.	↑ PROLIFER.

FIGURE 8.8B Effects of SULT1E1 activity on the proliferation of breast cancer cells. In normal breast cells, as a consequence of the SULT1E1 activity, proliferation is inhibited as estradiol sulfate (E₂S) is biologically inactive. In opposition to breast cancer cells, SULT1E1 activity is very low or nonexistent as E₂S is not formed and E₂ can stimulate proliferation.

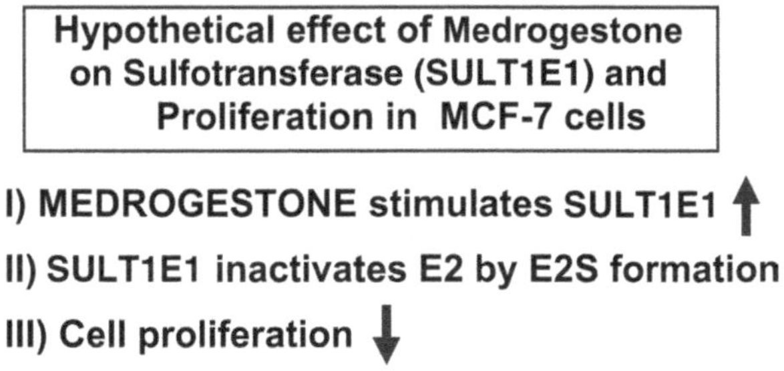

**Hypothetical effect of Medrogestone
on Sulfotransferase (SULT1E1) and
Proliferation in MCF-7 cells**

I) MEDROGESTONE stimulates SULT1E1 ↑

II) SULT1E1 inactivates E2 by E2S formation

III) Cell proliferation ↓

FIGURE 8.8C Hypothetical effects of medrogestone on human sulfotransferase (SULT1E1) and proliferation in T-47D and MCF-7 breast cancer cells. As medrogestone can stimulate SULT1E1 in the cancer cell, the effect of estradiol becomes inactive by the formation of estradiol sulfate and consequently cell proliferation is inhibited.

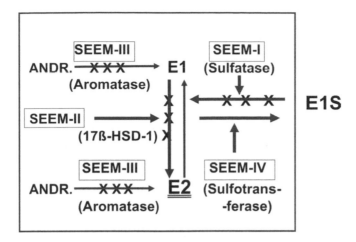

FIGURE 8.9 The selective estrogen enzyme modulator (SEEM) concept in human hormone-dependent breast cancer cells. The SEEM can control the enzymatic mechanisms involved in the formation and transformation of estrogens in breast cancer cells, where the sulfatase pathway is quantitatively higher than the aromatase. SEEM-I inhibits the estrone sulfatase; SEEM-II the 17β-hydroxysteroid dehydrogenase type 1; SEEM-III the aromatase activities; and SEEM-IV stimulates the estrone SULT activity. It is suggested that E^1S is present in the tumor outside the cell and reaches the cell membrane where it is in contact with the intracellular estrone sulfatase. ANDR: androgens; E_1: estrone; E_2: estradiol; E_1S: estrone sulfate. Quoted from Chetrite G, Pasqualini JR, *J Steroid Biochem Molec Biol* 76: 95–104, 2001. With permission.

REFERENCES

Adams JB, Phillips NS, Properties of estrogen and hydroxysteroid sulphotransferases in human mammary cancer. *J Steroid Biochem* 36: 695–701, 1990.

Adams JB, Pewnim T, Chandra DP, Archibald L, Foo MS, A correlation between estrogen sulfotransferase levels and estrogen receptor status in human primary breast carcinoma. *Cancer Res* 39: 5124–5126, 1979.

Adams JB, Phillips NS, Pewnim T, Expression of hydroxysteroid sulphotransferase is related to estrogen receptor status in human mammary cancer. *J Steroid Biochem* 33: 637–642, 1989.

Adjei AA, Weinshilboum RM, Catecholestogen sulfation: possible role in carcinogenesis. *Biochem Biophys Res Comm* 292: 402–408, 2002.

Aksoy IA, Wood R, Weinshilboum R, Human liver estrogen sulfotransferase: identification by cDNA cloning and expression. *Biochem Biophys Rec Commun* 200: 1621–1629, 1994.

Anderson E, Howell A, Oestrogen sulphotransferases in malignant and normal human breast tissue. *Endocr-Relat Cancer* 2: 227–233, 1995.

Banoglu E, Current status of the cytosolic sulfotransferases in the metabolic activation of promutagens and procarcinogens. *Current Drug Metab* 1: 1–30, 2000.

Bernier F, Leblanc G, Labrie F, Luu-The V, Structure of human estrogen and aryl sulfotransferase gene. *J Biol Chem* 269: 28200–28205, 1994a.

Bernier F, Lopez-Solache I, Labrie F, Luu-The V, Cloning and expression of cDNA encoding human placental estrogen sulfotransferase. *Molec Cell Endocr* 99: R11–R15, 1994b.

Bjarnason NH, Bjarnason K, Haarbo J, Christiansen C, Tibolone: prevention of bone loss in late postmenopausal women. *J Clin Endocrinol Metab* 81: 2419–2422, 1996.

Blanchard RB, Freimuth RR, Buck J, Weinshilboum RM, Coughtrie MWH, A proposed nomenclature system for the cytosolic sulfotransferase (SULT) superfamily. *Pharmacogenetics* 14: 199–211, 2004.

Bonney RC, Reed MJ, Davidson K, Beranek PA, James VHT, The relationship between 17ß-hydroxysteroid dehydrogenase activity and oestrogen concentrations in human breast tumours and in normal breast tissue. *Clin Endocrinol (Oxf)* 19: 727–739, 1983.

Botella J, Duranti E, Duc I, Cognet AM, Delansorne R, Paris J, Inhibition by nomegestrol acetate and other synthetic progestins on proliferation and progesterone receptor content of T47-D human breast cancer cells. *J Steroid Biochem Molec Biol* 50: 41–47, 1994.

Braunsberg H, Coldham NG, Leake RE, Cowan SK, Wong W, Actions of the progestagen on human breast cancer cells: mechanisms of growth stimulation and inhibition. *Europ J Clin Oncol* 23: 563–571, 1987.

Catherino WH, Jordan VC, Nomegestrol acetate, a clinically useful 19-norprogesterone derivative which lacks estrogenic activity. *J Steroid Biochem Molec Biol* 55: 239–246, 1995.

Chetrite G, Pasqualini JR, Steroid sulphotransferase and 17β-hydroxysteroid dehydrogenase activities in Ishikawa human endometrial adenocarcinoma cells. *J Steroid Biochem Molec Biol* 61: 27–34, 1997.

Chetrite G, Pasqualini JR, The selective estrogen enzyme modulator (SEEM) in breast cancer. *J. Steroid Biochem Molec Biol* 76: 95–104, 2001.

Chetrite G, Le Nestour E, Pasqualini JR, Human estrogen sulfotransferase (hEST1) activities and its mRNA in various breast cancer cell lines. Effect of the progestin, promegestone (R-5020). *J Steroid Biochem Biol* 66: 295–302, 1998.

Chetrite G, Cortes-Prieto J, Philippe JC, Wright F, Pasqualini JR, Comparison of estrogen concentrations, estrone sulfatase and aromatase activities in normal, and in cancerous, human breast tissues. *J Steroid Biochem Molec Biol* 72: 23–27, 2000.

Chetrite GS, Ebert C, Wright F, Philippe J-C, Pasqualini JR, Control of sulfatase and sulfotransferase activities by medrogestone in the hormone-dependent MCF-7 and T-47D human breast cancer cells lines. *J Steroid Biochem Molec Biol* 70: 39–45, 1999a.

Chetrite GS, Kloosterboer HJ, Philippe J-C, Pasqualini JR, Effect of Org OD14 (Livial®) and its metabolites on human estrogen sulphotransferase activity in the hormone-dependent MCF-7 and T-47D, and the hormone-independent MDA-MB-231, breast cancer cells lines. *Anticancer Research* 19: 269–276, 1999b.

Cho H, Aronica SM, Katzenellenbogen BS, Regulation of progesterone receptor gene expression in MCF-7 breast cancer cells: a comparison of the effects of cyclic adenosine 3′,5′-monophosphate, estradiol, insulin-like growth factor-I, and serum factors. *Endocrinology* 134: 658–664, 1994.

Clarke CL, Adams JB, Wren BG, Induction of oestrogen sulfotransferase activity in the human endometrium by progesterone in organ culture. *J Clin Endocrinol Metab* 55: 70–75, 1982.

Dao TL, Hayes C, Libby PR, Steroid sulfatase in human breast tumors. *Proc Soc Exp Biol Med* 146: 381–384, 1974.

Dubuisson JG, Dyess DL, Gaubatz JW, Resveratrol modulates human mammary epithelial cell O-acetyltransferase, sulfotransferase, and kinase activation of the heterocyclic amine carcinogen N-hydroxy-PhIP. *Cancer Lett* 182: 27–32, 2002.

Eaton EA, Walle UK, Lewis AJ, Hudson T, Wilson AA, Walle T, Flavonoids, potent inhibitors of the human P-form phenolsulfotransferase: potential role in drug metabolism and chemoprevention. *Drug Metab Dispos* 24: 232–237, 1996.

Eckert RL, Katzenellenbogen BS, Effects of estrogens and antiestrogens on estrogen receptor dynamics and the induction of progesterone receptor in MCF-7 human breast cancer cells. *Cancer Res* 42: 139–144, 1982.

Evans T, Rowlands M, Silva M, Law M, Coombes R, Prognostic significance of aromatase and estrone sulfatase enzymes in human breast cancer. *J Steroid Biochem Mol Biol* 44: 583–587, 1993.

Falany JL, Falany CN, Expression of cytosolic sulfotransferases in normal mammary epithelial cells and breast cancer cell lines. *Cancer Res* 56: 1551–1555, 1996a.

Falany JL, Falany CN, Regulation of oestrogen sulfotransferase in human endometrial adenocarcinoma cells by progesterone. *Endocrinology* 137: 1395–1401, 1996b.

Falany JL, Falany CN, Regulation of estrogen activity by sulfation in human MCF-7 human breast cancer cells. *Oncology Research* 9: 589–596, 1997.

Falany JL, Lawing L, Falany CN, Identification and characterization of cytosolic sulfotransferase activities in MCF-7 human breast carcinoma cells. *J Steroid Biochem Molec Biol* 46: 481–487, 1993.

Falany JL, Macrina N, Falany CN, Regulation of MCF-7 breast cancer cell growth by ß-estradiol sulfation. *Breast Cancer Res Treat* 74: 167–176, 2002.

Falany CN, Wheeler J, Oh TS, Falany JL, Steroid sulfation by expressed human cytosolic sulfotransferases. *J Steroid Biochem Molec Biol* 48: 369–375, 1994.

Godefroi VC, Locke ER, Singh DV, Brooks SC, Boch H, The steroid alcohol and estrogen sulfotransferase in rodent and human mammary tumors. *Cancer Res* 35: 1791–1798, 1975.

Gompel A, Malet C, Spritzer P, Lalardrie J-P, Kuttenn F, Mauvais-Jarvis P, Progestin effect on cell proliferation and 17ß-hydroxysteroid dehydrogenase activity in normal human breast cells in culture. *J Clin Endocrinol Metab* 63: 1174–1180, 1986.

Hata H, Holinka CF, Pahuja SL, Hochberg RB, Kuramoto H, Gurpide E, Estradiol metabolism in Ishikawa endometrial cancer cells. *J Steroid Biochem Molec Biol* 26: 699–704, 1987.

Hobkirk R, Steroid sulfation — current contents. *Trends Endocr Metab* 4: 69–74, 1993.

Hobkirk R, Girard LR, Durham NJ, Khalil MW, Behavior of mouse placental and uterine estrogen sulfotransferase during chromatography and other procedures. *Biochem Biophys Acta* 828: 123–129, 1985.

Hollman PCH, van Trijp JMP, Buysman MNCP, Gaag MS, Mengelers MJB, De Vries JHM, Katan MB, Relative bioavailability of the antioxidant flavonoid quercetin from various foods in man. *FEBS Letters* 418: 152–156, 1997.

Honjo H, Kitawaki J, Itoh M, Yasuda J, Iwasaku K, Urabe M, Naitoh K, Yamamoto T, Okada H, Ohkubo T, Nambara T, Serum and urinary estrone sulfate during the menstrual cycle, measured by a direct radio-immunoassay, and fate of exogenously injected estrone sulfate. *Horm Res* 27: 61–68, 1987.

Horwitz KB, Freidenberg GR, Growth inhibition and increase of insulin receptors in antiestrogen-resistant T-47Dco human breast cancer cells by progestins: implication for endocrine therapies. *Cancer Res* 45: 167–173, 1985.

Huang Z, Fasco MJ, Kaminsky LS, Inhibition of estrone sulfatase in human liver microsomes by quercetin and other flavonoids. *J Steroid Biochem Mol Biol* 63: 9–15, 1997.

Jeng M-H, Parker CJ, Jordan VC, Estrogenic potential in oral contraceptives to stimulate human breast cancer cell proliferation. *Cancer Res* 52: 6539–6546, 1992.

Kalkhoven E, Kwakkenbos-Isbrücker L, De Latt SW, van der Saag PT, van den Burg B, Synthetic progestins induce proliferation of breast tumor cell lines via the progesterone or estrogen receptor. *Molec Cell Endocr* 102: 45–52, 1994.

Keppler D, Kamisako T, Leier I, Cui YH, Nies AT, Tsujii H, Konig J, Localization, substrate specificity, and drug resistance conferred by conjugate export pumps of the MRP family. *Adv Enzyme Reg* 40: 339–349, 2000.

Kicovic PM, Cortes-Prieto J, Luisi M, Franchi F, Placebo-controlled cross-over study of the effects of Org OD 14 in menopausal women. *Reproduccion* 6: 81–91, 1982.

Kirk CJ, Harris RM, Wood DM, Waring RH, Hughes PJ, Do dietary phytoestrogens influence susceptibility to hormone-dependent cancer by disrupting the metabolism of endogenous oestrogens? *Biochem Soc Trans* 29: 209–216, 2001.

Knabbe C, Lippman ME, Wakefield LM, Flanders KC, Kasid A, Derynck R, Dickson RB, Evidence that transformation growth factor-beta is a hormonally regulated negative growth factor in human breast cancer cells. *Cell* 48: 417–428, 1987.

Kotov A, Falany JL, Wang J, Falany CN, Regulation of estrogen activity by sulfation in human Ishikawa endometrial adenocarcinoma cells. *J Steroid Biochem Mol Biol* 52: 137–144, 1999.

Laidlaw IJ, Clarke RB, Howell A, Owen AWMC, Potten CS, Anderson E, The proliferation of normal human breast tissue implanted into athymic nude mice is stimulated by estrogen but not progesterone. *Endocrinology* 136: 164–171, 1995.

Leslie EM, Deeley RG, Cole SPC, Toxicological relevance of the multidrug resistance protein 1, MRP1 (ABCC1) and related transporters. *Toxicology* 161: 3–23, 2001.

Lewis AJ, Walle UK, King RS, Kadlubar FF, Fanaly CN, Walle T, Bioactivation of the cooked food mutagen N-hydroxy-2-amino-1-methyl-6-phenylimidazol [4,5-b]pyridine by estrogen sulfotransferase in cultured human mammary epithelial cells. *Carcinogenesis* 19: 2049–2053, 1998.

Longacre TA, Bartow SA, A correlative morphologic study of human breast and endometrium in the menstrual cycle. *Am J Surgical Pathol* 10: 382–393, 1986.

Longman SM, Buehring GC, Oral contraceptives and breast cancer: *in vitro* effect of contraceptive steroids on human mammary cell growth. *Cancer* 59: 281–287, 1987.

Luu-The V, Bernier F, Dufort I, Steroid sulfotransferases. *J Endocrinol* 150: S87–S97, 1996.

Malet C, Gompel A, Yaneva H, Cren H, Fidji N, Mowszowicz I, Kuttenn F, Mauvais-Jarvis P, Estradiol and progesterone receptors in culture normal breast epithelial cells and fibroblasts: immunocytochemical studies. *J Clin Endocrinol Metab* 73: 8–17, 1991.

Malet C, Spritzer P, Guillaumin D, Kuttenn F, Progesterone effect on cell growth, ultrastructural aspect and estradiol receptors of normal human breast epithelial (HBE) cells in culture. *J Steroid Biochem Molec Biol* 73: 171–181, 2000.

McManus MJ, Welsch CW, The effect of estrogen, progesterone, thyroxine, and human placental lactogen on DNA synthesis of human breast ductal epithelium maintained in athymic nude mice. *Cancer* 54: 1920–1927, 1984.

Molis TG, Spriggs LL, Jupiter Y, Hill SM, Melatonin modulation of estrogen-regulated proteins, growth factors, and proto-oncogenes in human breast cancer. *J Pineal Res* 18: 93–103, 1995.

Musgrove EA, Lee CS, Sutherland RL, Progestins both stimulate and inhibit breast cancer cell cycle progression while increasing expression of transforming growth factor alpha, epidermal growth factor receptor, c-fos and c-myc genes. *Molec Cell Biol* 11: 5032–5043, 1991.

Nash AR, Glenn WK, Moore SS, Kerr J, Thompson AR, Thompson EOP, Oestrogen sufotransferase: molecular cloning and sequencing of cDNA for the bovine placenta enzyme. *Aust J Biol Sci* 41: 507–516, 1988.

Noel CT, Reed MJ, Jacobs HS, James VHT, The plasma concentration of oestrone sulphate in postmenopausal women: lack of diurnal variation, effect of ovariectomy, age and weight. *J Steroid Biochem Molec Biol* 14: 1101–1105, 1981.

Nunez M, Aedo A-R, Landgren B-M, Cekan SZ, Dicszfalusy E, Studies of the pattern of circulating steroids in the normal menstrual cycle. *Acta Endocrinol (Kbh)* 86: 621–633, 1977.

Otake Y, Nolan AL, Walle UK, Walle T, Quercetin and resveratrol potently reduce estrogen sulfotransferase activity in normal human mammary epithelial cells. *J Steroid Biochem Molec Biol* 73: 265–270, 2000.

Pai TG, Suiko M, Sakakibara Y, Liu M, Sulfation of flavonoids and other phenolic dietary compounds by the human cytosolic sulfotransferases. *Biochem Biophys Res Comm* 285: 1175–1179, 2001.

Pasqualini JR. Steroid sulphotransferase activity in the human hormone-independent MDA-MB-468 mammary cancer cells. *Eur J Cancer* 28A: 758–762, 1992.

Pasqualini JR, Chetrite G, Activity, regulation and expression of sulfatase, sulfotransferase, and 17ß-hydroxysteroid-dehydrogenase in breast cancer, in *Hormone-Dependent Cancer*, Pasqualini JR, Katzenellenbogen BS, Eds., New York, 1996, pp. 25–80.

Pasqualini JR, Chetrite GS, The selective estrogen enzyme modulators (SEEM) in breast cancer, in *Breast Cancer: Prognosis, Treatment and Prevention*, Pasqualini JR, Ed., Marcel Dekker, New York, 2002, pp. 187–249.

Pasqualini JR, Chetrite G, Blacker M-C, Feinstein C, Delalonde L, Talbi M, Maloche C, Concentrations of estrone, estradiol, estrone sulfate and evaluation of sulfatase and aromatase activities in pre- and post-menopausal breast cancer patients. *J Clin Endocrinol Metab* 81: 1460–1464, 1996.

Pasqualini JR, Cortes-Prieto J, Chetrite G, Talbi M, Ruiz A, Concentrations of estrone, estradiol and their sulfates, and evaluation of sulfatase and aromatase activities in patients with breast fibroadenoma. *Int J Cancer* 70: 639–643, 1997.

Pedersen LC, Petrotchenko E, Shevtsov S, Negishi M, Crystal structure of the human estrogen sulfotransferase-PAPS complex: evidence for catalytic role of Ser137 in the sulfuryl transfer reaction. *J Biol Chem* 277: 17928–17932, 2002.

Pewnim T, Adams JB, Ho KP, Estrogen sulfurylation as an alternative indicator of hormone dependence in human breast cancer. *Steroids* 39: 47–52, 1982.

Qian Y, Deng C, Song W-C. Expression of estrogen sulfotransferase in MCF-7 cells by cDNA transfection suppresses the estrogen response: potential role of the enzyme in regulating estrogen-dependent growth of breast epithelial cells. *J Pharmacol Exp Ther* 286: 555–560, 1998.

Qian Y-M, Sun XJ, Tong MH, Li XP, Richa J, Song W-C, Targeted disruption of the mouse estrogen sulfotransferase gene reveals a role of estrogen metabolism in intracrine and paracrine estrogen regulation. *Endocrinology* 142: 5342–5350, 2001.

Raju U, Blaustein A, Levitz M, Conjugation of androgens and estrogens by human breast tumors in vitro. *Steroids* 35: 685–695, 1980.

Roberts KD, Rochefort JG, Bleau G, Chapdelaine A, Plasma estrone sulfate levels in post-menopausal women. *Steroids* 35: 179–187, 1980.

Roy AK, Regulation of steroid hormone action in target cells by specific hormone-inactivating enzymes. *Proc Soc Experim Biol Med* 199: 265–272, 1992.

Rozhin J, Corombos JD, Horwitz JP, Brooks SC, Endocrine steroid sulfotransferases: steroid alcohol sulfotransferase from human breast carcinoma cell line MCF-7. *J Steroid Biochem* 25: 973–979, 1986.

Russo J, Russo IH, Mechanisms involved in carcinogenesis of the breast, in *Breast Cancer: Prognosis, Treatment and Prevention*, Pasqualini JR, Ed., Marcel Dekker, New York, 2002, pp. 1–18.

Saunders DE, Lozon MM, Corombos JD, Brooks SC, Role of porcine endometrial estrogen sulfotransferase in progesterone mediated downregulation of estrogen receptor. *J Steroid Biochem Molec Biol* 32: 749–757, 1989.

Schatz RW, Soto AM, Sonnenschein C, Effects of interaction between estradiol-17ß and progesterone on the proliferation of cloned breast tumor cells (MCF-7 and T-47D). *J Cell Physiol* 124: 386–390, 1985.

Schoonen WGEJ, Joosten JWH, Kloosterboer HJ, Effects of two classes of progestagens, pregnane and 19-nortestosterone derivatives, on cell growth of human breast tumor cells. I. MCF-7 cell lines. *J Steroid Biochem Molec Biol* 55: 423–437, 1995a.

Schoonen WGEJ, Joosten JWH, Kloosterboer HJ, Effects of two classes of progestagens, pregnane and 19-nortestosterone derivatives, on cell growth of human breast tumor cells. II. T-47D cell lines. *J Steroid Biochem Molec Biol* 55: 439–444, 1995b.

Sharp S, Anderson JM, Coughtrie MWH, Immunohistochemical localisation of hydroxysteroid sulphotransferase in human breast carcinoma tissue: a preliminary study. *Eur J Cancer* 30A: 1654–1658, 1994.

Shwed JA, Walle UK, Walle T, Hep G2 cell line as a human model for sulphate conjugation of drugs. *Xenobiotica* 22: 973–982, 1992.

Thomas HV, Key TJ, Allen DS, Moore JW, Dowsett M, Fentiman IS, Wang DY, A prospective study of endogenous serum hormone concentrations and breast cancer risk in post-menopausal women on the island of Guernsey. *Br J Cancer* 76: 401–405, 1997.

Topper YJ, Freeman CS, Multiple hormone interactions in the developmental biology of the mammary gland. *Physiol Rev* 60: 1049–1106, 1980.

Tseng L, Liu HC, Stimulation arylsulfotransferase activity by progestins in human endo-metrium in vitro. *J Clin Endocrinol Metab* 53: 418–421, 1981.

Tseng L, Mazella J, Lee LY, Stone ML, Estrogen sulfatase and estrogen sulfotransferase in human primary mammary carcinoma. *J Steroid Biochem* 19: 1413–1417, 1983.

van der Burg B, Kalkhoven E, Isbrücker L, de Laat SW, Effects of progestins on the prolifer-ation of œstrogen-dependent human breast cancer cells under growth factor-defined conditions. *J Steroid Biochem Molec Biol* 42: 457–469, 1992.

van Landeghem AAJ, Poortman J, Nabuurs M, Thijssen JHH, Endogenous concentration and subcellular distribution of estrogens in normal and malignant human breast tissue. *Cancer Res* 45: 2900–2906, 1985.

Vermeulen A, Deslypere JP, Paridaens R, Leclercq G, Roy F, Heuson JC, Aromatase, 17ß-hydroxysteroid dehydrogenase and intratissular sex hormone concentrations in can-cerous and normal glandular breast tissue in postmenopausal women. *Eur J Cancer Clin Oncol* 22: 515–525, 1986.

Vignon F, Bardon S, Chalbos D, Rochefort H. Antiestrogenic effect of R5020, a synthetic progestin in human breast cancer cells in culture. *J Clin Endocrinol Metab* 56: 1124–1130, 1983.

Walle T, Eaton EA, Walle UK, Quercetin, a potent and specific inhibitor of the human P-form phenolsulfotransferase. *Biochem Pharmacol* 50: 731–734, 1995.

Wild MJ, Rudland PS, Back DJ, Metabolism of the oral contraceptive steroids ethynylestradiol and norgestimate by normal (Huma 7) and malignant (MCF-7 and ZR 75-1) human breast cells in culture. *J Steroid Biochem Molec Biol* 39: 535–543, 1991.

Zajchowski DA, Sager R, Webster L. Estrogen inhibits the growth of estrogen receptor-negative, but not estrogen receptor-positive, human mammary epithelial cells express-ing a recombinant estrogen receptor. *Cancer Res* 53: 5004–5011, 1993.

Zhang HP, Varlamova O, Vargas FM, Falany CN, Leyh TS, Varmalova O, Sulfuryl transfer: the catalytic mechanism of human estrogen sulfotransferase. *J Biol Chem* 273: 10888–10892, 1998.

9 Sulfation of Drugs

Gian Maria Pacifici

CONTENTS

INTRODUCTION

Sulfotransferase (EC 2.8.2) is an important enzyme that catalyzes the sulfation of neurotransmitters, hormones, drugs, and chemicals (Falany, 1991; Mulder and Jakoby, 1990; Pacifici and De Santi, 1995; Pacifici and Rossi, 2001). Drugs that possess a hydroxy group or an amine group are often sulfated (Pacifici and De Santi, 1995). Sulfation usually determines inactivation of the drug effects, with the exception of minoxidil, which requires conjugation with sulfate to relax smooth muscles (Meisheri et al., 1988).

Sulfotransferase is a family of different forms with different substrate specificities and regulation (Coughtrie and Johnston, 2001; Falany, 1991; Weinshilboum, 1986). SULT1A1 and SULT1A3 develop before the 14th week of gestation in the human fetal liver (Cappiello et al., 1991; Pacifici et al., 1993a). Adenosine-3'-phosphate-5'-phosphosulfate (PAPS), the endogenous substrate of sulfotransferase, is present in the mid-gestation human fetal liver (Cappiello et al., 1990), and sulfation should take place in the human fetus.

The rate of a drug sulfation varies from three- to eight-fold in human adult liver. This information derives from studies with ethinyloestradiol (Temellini et al., 1991), ritodrine (Pacifici et al., 1998), minoxidil (Pacifici et al., 1993b), (−)-salbutamol (Pacifici et al., 1997a), testosterone (Pacifici et al., 1997b), the eterocyclic amine 1,2,3,4-tetrahydroisoquinoline (TIQ; Pacifici et al., 1997c), and budesonide (Pacifici et al., 1994a).

In the presence of a chiral atom, drug molecules are constituted by two enantiomers that have different biological properties. As an example, terbutaline (Walle and Walle, 1990), 4-hydroxypropranolol (4-OHP; Walle and Walle, 1991), and salbutamol (Walle et al., 1993a, 1993b) consist of two enantiomers that are sulfated at different rates or have different affinities for sulfotransferase.

Finally, the sulfation rate of a drug may be inhibited by drugs and dietary chemicals. Mefenamic acid and salicylic acid, nonsteroidal antiinflammatory drugs (NSAIDs), are potent inhibitors of human liver SULT1A1 and SULT1A3 activities in human liver (Vietri et al., 2000a, 2000b). Quercetin, a natural flavonoid, is a potent inhibitor of SULT1A1 activity and the sulfation rate of minoxidil, (−)-salbutamol, and paracetamol in human liver and duodenum (Marchetti et al., 2001). This article reviews the interindividual variability, the stereoselectivity, and the inhibition of sulfotransferases in human tissues.

SULFATION OF DRUGS *IN VIVO*

Although the entire book is devoted to the *in vitro* studies of cytosolic sulfotransferases, a review on the sulfation of drugs requires a brief consideration of the *in vivo* sulfation of the therapeutic agents. When salbutamol is administered orally, the plasma levels of salbutamol sulfate are higher than the plasma levels of salbutamol itself (Morgan, 1986). An early study performed with radiolabeled salbutamol revealed that, after inhalation, the radioactivity concentration of salbutamol metabolites was higher than those of the authentic salbutamol (Evans et al., 1973). Salbutamol is mainly eliminated by sulfation.

Paracetamol was sulfated *in vivo* in man, and paracetamol sulfate excreted in the urine was 35% of the paracetamol dose (Reither and Weinshilboum, 1982). Parkinson's disease and Huntington's disease do not impair the sulfation rate of paracetamol (Roos et al., 1993).

The sulfation of the NSAID diflunisal was studied in young healthy volunteers (Verbeeck et al., 1990). The recovery of diflunisal sulfate in the urine was 6.1% and 9.1% after a single oral dose of 250 and 500 mg, respectively. After two daily doses of diflunisal for several days, the urinary recovery of diflunisal sulfate was 10.9% (250 mg) and 15.9% (500 mg). The sulfation of diflunisal is dose dependent.

Fenoldopam is a selective dopamine (DA-1) agonist with pronounced cardiovascular and renal effects in humans. This drug undergoes sulfation *in vivo* in man (Ziemniak et al., 1987). After a single dose of fenoldopam (100 mg), the area under the curve (AUC) values (ng × h/ml) were 37.8 (fenoldopam) and 2177 (fenoldopam sulfate at position 8).

Brashear et al. (1988) administered ritodrine intravenously to seven pregnant women before delivery, and the urine concentration of ritodrine and ritodrine sulfate was measured in the mother and newborn pairs at several times after delivery. In the mother, the concentration of ritodrine sulfate was two- to three-fold higher than that of ritodrine itself, and in the newborn, the concentration of ritodrine sulfate was three- to sevenfold higher than authentic ritodrine. Pacifici et al. (1993a) have observed that the rate of ritodrine sulfation was higher in the mid-gestation human fetal liver than in adult liver.

SULFATION OF DRUGS IN HUMAN FETAL TISSUES

The metabolism of mid-gestation human fetus is a useful model to predict the metabolism of the premature infant. The expression of the various drug-metabolizing enzymes in human fetal liver varies as compared with those of the adult liver. Ritodrine sulfation was measured with 20 mM ritodrine, and the ritodrine sulfation rate (median) was 204 pmol/min/mg (human fetal liver) and 133 pmol/min/mg (human adult liver; Pacifici et al., 1993a). To sulfate a drug, the presence of the enzyme and the endogenous substrate, PAPS are necessary. The endogenous concentration of PAPS was measured in the human fetal liver, in human adult liver, and in placenta, and it was (nmol/g wet tissue) 10.1, 23.4, and 3.6, respectively (Cappiello et al., 1990). PAPS is present in human fetal liver at half the concentration of adult liver, and sulfation should occur in human fetus.

SULT1A1 and SULT1A3 activities and the sulfation rate of ritodrine were measured with 4 μM 4-nitrophenol, 60 μM dopamine, and 20 mM ritodrine, respectively. Ritodrine sulfation rate and the activities of SULT1A1 and SULT1A3 were measured in the livers obtained from 48 human fetuses aged between 14 and 27 weeks with a mean ±SD of 20±3 weeks (Pacifici et al., 1993a). The rate of ritodrine sulfation varied from 7.8 to 1050 pmol/min/mg between the 5th and 95th percentiles with a variation of 134-fold. The rate of ritodrine sulfation did not correlate with the fetal age ($r = -0.024$; $p = 0.872$) and had a median value of 204 pmol/min/mg. Figure 9.1 shows the probit plot, and Table 9.1 summarizes the descriptive statistical analysis of ritodrine sulfation rate. In 9 fetal livers, aged between 17 and 27 weeks, the rate of ritodrine sulfation was lower than 40 pmol/min/mg, one fifth of the median value.

SULT1A3 activity ranged between 1.4 and 166 pmol/min/mg between the 5th and 95th percentiles with a median of 41 pmol/min/mg (Pacifici et al., 1993a). There was an intense interindividual variability in the activity of SULT1A3 with a 118-fold variation between the 5th and 95th percentiles (Figure 9.1 and Table 9.1). SULT1A3 activity did not correlate with the fetal age ($r=-0.057$; $p=0.699$), and in 14 livers, aged from 15 to 27 weeks, the activity of SULT1A3 was lower than 8 pmol/min/mg, one fifth of the median value.

SULT1A1 activity ranged between 6.0 and 244 pmol/min/mg between the 5th and 95th percentiles, with a median of 60 pmol/min/mg (Pacifici et al., 1993a). There was intense interindividual variability in the activity of SULT1A1 (Figure 9.1 and Table 9.1), and SULT1A1 activity did not correlate with the fetal age ($r=-0.181$; $p=0.219$). There was 41-fold variation of SULT1A1 activity between the 5th and 95th percentiles. In 3 livers, aged between 17 and 23 weeks, SULT1A3 activity was lower than 8 pmol/min/mg, one fifth of the median value. These results indicate that SULT1A3 and SULT1A1 activities develop in the fetal liver before the 14th week of gestation and are subject to considerable interindividual variability.

The activity of SULT1A3 is better developed than SULT1A1 in mid-gestation human fetal liver but decreases after birth (Richard et al., 2001). The activities of these two enzymes were measured in 8 human fetal livers, aged from 18 to 25 weeks, and 6 human adult livers, aged from 22 to 76 years (Cappiello et al., 1991). The activity of SULT1A3 (mean ± SD) was 97 ± 53 (fetal liver) and 29 ± 17 (adult liver;

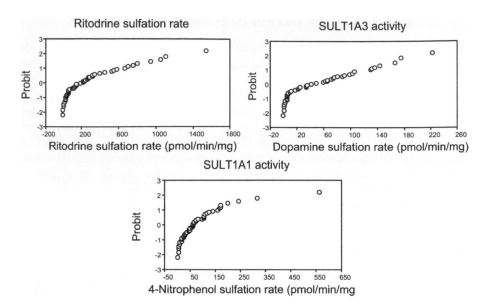

FIGURE 9.1 Probit plots of ritodrine sulfation rate and SULT1A3 and SULT1A1 activities in 48 samples of mid-gestation human fetal liver. The fetal age ranged between 14 and 27 weeks and did not correlate with the rate of ritodrine sulfation ($r = -0.024$; $p = 0.872$), with the activity of SULT1A3 ($r = -0.057$; $p = 0.699$), or with the activity of SULT1A1 ($r = -0.181$; $p = 0.219$). Descriptive statistical analysis is summarized in Table 9.1. The data were generated by Pacifici et al. [Pacifici, G.M., Kubrich, M., Giuliani, L., de Vries, M., and Rane, A. (1993a) *Eur. J. Clin. Pharmacol.* 44: 259–264].

$p < 0.01$) pmol/min/mg, and the activity of SULT1A1 was 98 ± 26 (fetal liver) and $1,077 \pm 293$ pmol/min/mg (adult liver; $p < 0.001$).

SULT1A3 and SULT1A1 are also present in blood platelets; the interindividual variability in the activities of these two enzymes was investigated in 100 samples of platelets isolated from term placental blood (Pacifici and Marchi, 1993). Table 9.2 summarizes the descriptive statistical analysis for SULT1A3 and SULT1A1 activities, and Figure 9.2 shows the probit plots. The frequency distribution of SULT1A3 activity did not deviate from normality, whereas the distribution of SULT1A1 was positively skewed. The fold of variation between the 5th and 95th percentiles was 7.1 for SULT1A3 and 16.7 for SULT1A1.

INTERINDIVIDUAL VARIABILITY IN THE SULFATION RATE OF DRUGS IN HUMAN ADULT LIVER AND DUODENUM

Ten years of activity of this laboratory has shown that the rate of drug sulfation varies in human liver, and variation between the 5th and the 95th percentiles ranged from three to eightfold with different drugs. Several factors, such as environmental factors and genetic factors, contribute to such variation.

TABLE 9.1
Descriptive Statistical Analysis of Ritodrine Sulfation Rate and SULT1A3 and SULT1A1 Activities in Mid-Gestation Human Fetal Liver

Enzyme	Ritodrine Sulfotransferase	SULT1A3	SULT1A1
Substrate	20 mM Ritodrine	60 μM Dopamine	4 μM 4-Nitrophenol
Number of Cases	48	48	48
Mean ±SD[d]	307	59	88
±SD[d]	341	57	96
Median[d]	204	41	60
5th Percentile[d]	7.8	1.4	6.0
95th Percentile[d]	1,050	166	244
Variation[a]	134	118	41
Skewness	1.678	0.952	2.988
W-Statistic[b]	0.81016	0.8761	0.72563
P[c]	<0.001	$P < 0.001$	$P < 0.001$

[a] The variation was calculated by dividing the 95th percentile by the 5th percentile.

[b] Critical values of the Shapiro and Wilk test.

[c] The distribution deviates from normality when $p<0.05$.

[d] pmol/min/mg. The data were generated by Pacifici et al. [Pacifici, G.M., Kubrich, M., Giuliani, L., de Vries, M., and Rane, A. (1993a) *Eur. J. Clin. Pharmacol.* 44: 259–264].

The interindividual variability of SULT1A1 and SULT1A3 activities was studied by Pacifici et al. (1994b) in human liver and by Pacifici et al. (1997a) in human duodenum. Figure 9.3 shows the probit plots of SULT1A1 and SULT1A3 activities, and Table 9.3 summarizes the descriptive statistical analysis. The fold of variation of SULT1A1 and SULT1A3 activities was smaller in human duodenum than human liver. These enzyme activities were not gender regulated, and their frequency distribution was positively skewed. It must be observed that the fold of variation of SULT1A1 and SULT1A3 activities is considerably greater in the fetal (Table 9.1) than adult liver (Table 9.3).

The interindividual variability of (−)-salbutamol (Pacifici et al., 1997a), ritodrine (Pacifici et al., 1998), and minoxidil (Pacifici et al., 1993b) was studied in human liver. The substrate final concentration in the incubation assay was 1 mM (−)-salbutamol, 20 mM ritodrine, and 14.7 mM minoxidil. Figure 9.4 shows the probit plots of (−)-salbutamol, ritodrine, and minoxidil sulfation rates, and Table 9.4 summarizes their descriptive statistical analysis. The rates of (−)-salbutamol, ritodrine, and minoxidil sulfation were not gender regulated, and their distributions were positively skewed. Budesonide (Pacifici et al., 1994a), testosterone (Pacifici et al., 1997b), and TIQ (Pacifici et al., 1997c) sulfation rates were gender regulated and were higher in males than females. Budesonide sulfation rate (median) was 37.7 (males) and 26.0 pmol/min/mg (females; $p < 0.001$). Testosterone sulfation rate (median) was 20.8 (males) and 16.8 pmol/min/mg (females; $p = 0.002$). TIQ sulfation

TABLE 9.2

Descriptive Statistical Analysis of SULT1A3 and SULT1A1 Activities in the Platelets Isolated from Term Placental Blood

Enzyme	SULT1A3 Activity	SULT1A1 Activity
Substrate	60 μM Dopamine	4 μM 4-Nitrophenol
Number of Cases	100	100
Mean[d]	13.3	4.8
±SD[d]	6.4	4.3
Median[d]	13.2	3.4
5th Percentile[d]	3.4	0.80
95th Percentile[d]	24.1	13.4
Variation[a]	7.1	16.7
Skewness	0.532	2.241
W-Statistic[b]	0.91630	0.78333
P[c]	0.177	<0.001

[a] The variation was calculated by dividing the 95th percentile by the 5th percentile.
[b] Critical values of the Shapiro and Wilk test.
[c] The distribution deviates from normality when $p < 0.05$.
[d] pmol/min/mg. The data were generated by Pacifici and Marchi [Pacifici, G.M. and Marchi, G. (1993) *Br. J. Clin. Pharmacol.* 36: 593–597].

rate (median) was 549 (males) and 456 pmol/min/mg (females; $p = 0.009$). The frequency distribution of budesonide, testosterone, and TIQ sulfation rates was positively skewed in males and females (results not shown).

STEREOSELECTIVE SULFATION OF DRUGS

It has been demonstrated that the sulfation of drugs that have a chiral atom may be stereoselective. The stereoselectivity may be due to different sulfation rates or different affinity of sulfotransferase toward the two enantiomers. Studies of pharmacokinetics have shown that the disposition of terbutaline is stereoselective: (–)-terbutaline has a higher oral bioavailability than (+)-terbutaline (Borgstrom et al., 1989a, 1989b). Walle and Walle (1990) observed that the sulfation of terbutaline, catalyzed by human liver cytosol, was stereoselective. Average sulfation rates of (+)- and (–)-terbutaline were 0.721 and 0.378 pmol/min/mg, respectively ($p < 0.05$), whereas the K_m values for the two enantiomers were virtually identical. Sulfation may take place in the intestinal mucosa and liver, and these two tissues are involved in determining the bioavailability of terbutaline. Using 50 mM terbutaline, a concentration higher than that used by Walle and Walle (1990), the sulfation rates of (+)- and (–)-terbutaline were measured in human duodenal mucosa and liver (Pacifici et al., 1993c). In duodenal mucosa, the average sulfation rate of (+)- and (–)-terbutaline was 1195 and 948 pmol/min/mg,

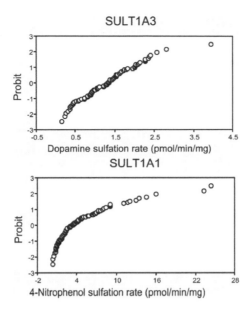

FIGURE 9.2 Probit plots of SULT1A3 and SULT1A1 activities measured in the platelets isolated from 100 samples of term placental blood. Statistical analysis is summarized in Table 9.2. The data were generated by Pacifici and Marchi [Pacifici, G.M. and Marchi, G. (1993) *Br. J. Clin. Pharmacol.* 36: 593–597].

respectively ($p < 0.001$), and in the liver, it was 45 and 33 pmol/min/mg, respectively ($p < 0.05$). Terbutaline is a β_2-adrenoceptor agonist and is used in the treatment of lung obstruction disease. The sulfation rates of enantiomers of terbutaline were measured in human lung cytosol (Pacifici et al., 1993c). The average sulfation rate of (+)- and (−)-terbutaline was 118 and 82 pmol/min/mg, respectively ($p < 0.05$). The pattern of stereoselective sulfation of terbutaline is consistent in human liver, duodenal mucosa, and lung. The sulfation rates of both enantiomers of terbutaline were sensitive to the inhibition by 2,6-dichloro-4-nitrophenol (DCNP) and to elevated temperature (Walle and Walle, 1990). These authors conclude that the sulfation of terbutaline should be catalyzed by SULT1A1 and SULT1A3.

Walle et al. (1993a) studied the sulfation of (+)- and (−)-salbutamol with partially purified human SULT1A3 and SULT1A1. SULT1A3 catalyzed salbutamol sulfation, whereas SULT1A1 was unable to catalyze the sulfation of this drug. The average K_m values obtained with (+)- and (−)-salbutamol were 1394 and 103 μM ($p < 0.001$), respectively. The V_{max} values for both enantiomers were virtually identical. Work performed with human jejunal mucosa cytosol and platelet homogenate was consistent, with the results obtained with partially purified SULT1A3 (Walle et al., 1993a). Salbutamol is a β_2-adrenoceptor agonist and the lung is the target organ. The sulfation of (+)- and (−)-salbutamol was studied in human lung cytosol, and it was observed that the average K_m values for the two enantiomers were 1198 and 190 μM ($p < 0.001$), respectively, and the V_{max} values were similar (Pacifici et al., 1996).

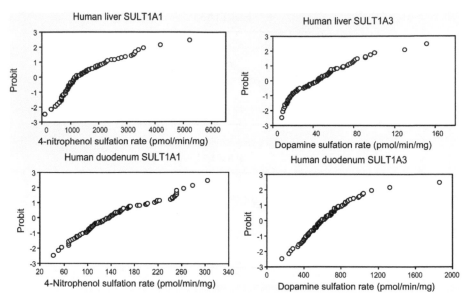

FIGURE 9.3 Probit pots of SULT1A1 and SULT1A3 measured in 100 samples of human liver and duodenum. Descriptive statistical analysis is summarized in Table 9.3. The data in liver were generated by Pacifici et al. [Pacifici, G.M., Ferroni, M.A., Temellini, A., Gucci, A., Morelli, M.C., and Giuliani, L. (1994a) *Eur. J. Clin. Pharmacol.* 46: 49–54], and those in duodenum were generated by Pacifici et al. [Pacifici, G.M., Giulianetti, B., Quilici, M.C., Spisni, R., Nervi, M., Giuliani, L., and Gomeni, R. (1997a) *Xenobiotica* 27: 279–286].

Isoproterenol, a β-adrenoceptor agonist, possesses a chiral atom that generates the (+)- and the (−)-enantiomers. The sulfation of isoproterenol is stereoselective, and the human liver and jejunum cytosols and the blood platelet homogenate catalyzed the sulfation of (+)-isoproterenol at a higher rate than (−)-isoproterenol (Pesola and Walle, 1993). The sulfation of isoproterenol was complex, and the curve of isoproterenol sulfation rate vs. isoproterenol concentration consisted of two components, one at lower concentration of isoproterenol (up to 50 to 100 μM) and another at higher concentrations (up to 1 mM). Using partially purified SULT1A3 and SULT1A1, Pesola and Walle (1993) observed that SULT1A3 has higher affinity for (+)- and (−)-isoproterenol than SULT1A1. The K_m values for SULT1A3 were 5.1 μM [(+)-isoproterenol] and 16 μM [(−)-isoproterenol], and the K_m values for SULT1A1 were 970 μM [(+)-isoproterenol] and 2900 μM [(−)-isoproterenol].

4-OHP, a metabolite of propranolol, a blocker of the $β_1$-adrenoceptor, was sulfated *in vitro* using human liver cytosol (Walle and Walle, 1991). The sulfation of 4-OHP was complex; the curve of the enzyme activity vs. 4-OHP concentration showed two peaks, one at 3 μM and another at 500 μM 4-OHP. With 3 μM 4-OHP, the sulfation of this drug was stereoselective as the sulfation rate of (+)-4-OHP is four to five times higher than that of (−)-4-OHP. In contrast, with 500 μM 4-OHP, the sulfation rate of this drug was not stereoselective. The authors conclude that the sulfation of 3 μM and 500 μM 4-OHP should be catalyzed by different enzymes.

TABLE 9.3

Descriptive Statistical Analysis of SULT1A1 and SULT1A3 Activities in the Human Liver and Duodenum

Tissue	Liver		Duodenum	
Enzyme	SULT1A1	SULT1A3	SULT1A1	SULT1A3
Substrate	4 μM 4-NP	60 μM Dopamine	4 μM 4-NP	60 μM Dopamine
Number of Cases	100	100	100	100
Mean[d]	1492	47	144	631
±SD[d]	876	29	56	234
Median[d]	1200	44.1	131	593
5th Percentile[d]	620	12	68	334
95th Percentile[d]	3310	92	252	1014
Variation[a]	5.3	7.7	3.7	3.0
Skewness	1.570	1.420	0.758	1.753
W-Statistic[b]	0.862	0.886	0.936	0.910
p[c]	<0.001	<0.001	<0.001	<0.001
References	A	A	B	B

[a] The variation was calculated by dividing the 95th percentile by the 5th percentile.

[b] Critical values of Shapiro and Wilk test.

[c] The distribution deviates from normality when $p < 0.05$.

[d] pmol/min/mg. 4-NP = 4-Nitrophenol. A: Pacifici, G.M., Temellini, A., Castiglioni, M., D'Alessandro, C., Ducci, A., and Giuliani, L. (1994b) *Chem. Biol. Interact.* 92: 219–231; B: Pacifici, G.M., Giulianetti, B., Quilici, M.C., Spisni, R., Nervi, M., Giuliani, L., and Gomeni, R. (1997a) *Xenobiotica* 27: 279–286.

Preincubation at 43°C generated a marked decrease of 3 μM 4-OHP sulfation rate, whereas it had little effect on the sulfation rate of 500 μM 4-OHP.

Antiarrhytmic propafenone is metabolized into 5-hydroxypropafenone (5-OHPPF), which is sulfated *in vivo* in man (Chen et al., 2000). Sulfation is a minor pathway of 5-OHPPF metabolism; however, it is stereoselective and the plasma levels of (−)-5-OHPPF sulfate were higher than those of (+)-5-OHPPF sulfate.

Fenoldopam has three hydroxy groups, and those in position 7 and 8 form a catechol group and are sulfated in human liver cytosol and liver slices (Klecker and Collins, 1997). This drug has a chiral atom that generates two enantiomers. (R)-Fenoldopam was preferentially sulfated at position 8, whereas (S)-fenoldopam was preferentially sulfated at position 7 both in the human liver cytosol and in liver slices.

INHIBITION OF SULFOTRANSFERASE

DCNP was introduced by Mulder and Scholtens (1977) as a selective inhibitor of rat liver SULT1A1 activity. This compound was found to be a better inhibitor of SULT1A1 activity than SULT1A3 activity. Differential inhibition of DCNP was confirmed in human frontal cortex, jejunum, and platelets (Rein et al., 1982). Consistent results have been obtained in human frontal cortex (Rein et al., 1984; Young et al., 1984),

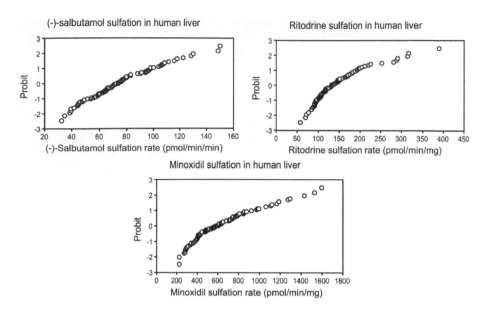

FIGURE 9.4 Probit plots of (–)-salbutamol, ritodrine, and minoxidil sulfation rates in the human liver. Sulfation rates of (–)-salbutamol and ritodrine were measured in 100 samples of human liver, and the sulfation rate of minoxidil was measured in 118 samples of human liver. Descriptive statistical analysis is summarized in Table 9.4. The data relative to (–)-salbutamol were generated by Pacifici et al. [Pacifici, G.M., Giulianetti, B., Quilici, M.C., Spisni, R., Nervi, M., Giuliani, L., and Gomeni, R. (1997a) *Xenobiotica* 27: 279–286], and those relative to ritodrine and minoxidil were generated by Pacifici et al. (1998) [and Pacifici, G.M., Bigotti, R., Marchi, G., and Giuliani, L. (1993b) *Eur. J. Clin. Pharmacol.* 45: 337–341, respectively].

human pituitary (Anderson et al., 1986), human liver (Campbell et al., 1987), and human jejunum (Sundaram et al., 1989). Campbell et al. (1987) measured the IC_{50} value of DCNP for partially purified SULT1A1 and SULT1A3, and it was 2.3×10^{-6} M and 6.6×10^{-5} M, respectively. Seah and Wong (1994) studied the inhibition type of DCNP for rat liver SULT1A1 and observed that it was the mixed noncompetitive inhibition type. DCNP was sulfated and was defined as an "alternate substrate-inhibitor"; in other words, DCNP is a dead-end inhibitor (Seah and Wong, 1994).

Hydroxylated polychlorinated biphenyls (HPCBs) are environmental pollutants that exert a variety of toxic effects in animals. Among these effects, is the disturbance of sexual development and reproductive function. Kester et al. (2000) tested the effect of 32 HPCBs on recombinant human estrogen sulfotransferase (EST) activity expressed in *Salmonella typhimurium*. Ortho-HPCBs were relatively weak inhibitors; increasing potency was observed with metahydroxylated compounds, and even higher inhibitory activity was found with parahydroxylated compounds. It was concluded that a hydroxy group in para position with two adjacent Cl substituents is required for maximal EST inhibition. The five most potent inhibitors of EST activity were 4-OH-3,3′,4′-tri-chlorinated biphenyl, 4-OH-2,3,5,2′,4′,5′-hexa-chlorinated biphenyl, 3-OH-2,4,5,2′,3′,4′,5′-hepta-chlorinated biphenyl, 4-OH 3,5,2′,3′,4′,5′-hexa-chlorinated

TABLE 9.4

Descriptive Statistical Analysis of the Sulfation Rates of (–)-Salbutamol, Ritodrine, and Minoxidil in Human Liver

Substrate	1 mM (–)-Salbutamol	20 mM Ritodrine	14.7 mM Minoxidil
Number of Cases	100	100	118
Mean[d]	76.2	146	631
± SD[d]	23.8	58.2	330
Median[d]	73.7	133	595
5th Percentile[d]	40.6	84.8	284
95th Percentile[d]	120	288	1270
Variation[a]	3.0	3.4	4.5
Skewness	0.734	1.667	0.734
W-Statistic[b]	0.955	0.869	0.911
P[c]	0.08	<0.001	<0.001
References	A	B	C

[a] The variation was calculated by dividing the 95th percentile by the 5th percentile.

[b] Critical values of Shapiro and Wilk test.

[c] The distribution deviates from normality when $p < 0.05$.

[d] pmol/min/mg. A: Pacifici, G.M., Giulianetti, B., Quilici, M.C., Spisni, R., Nervi, M., Giuliani, L., and Gomeni, R. (1997a) *Xenobiotica* 27: 279–286; B: Pacifici, G.M., Quilici, M.C., Giulianetti, B., Spisni, R., Nervi, M., Giuliani, L., and Gomeni, R. (1998) *Eur. J. Drug Metab. Pharmacokin.* 23: 67–74; C: Pacifici, G.M., Bigotti, R., Marchi, G., and Giuliani, L. (1993b) *Eur. J. Clin. Pharmacol.* 45: 337–341.

biphenyl, and 4-OH-2,3,5,6,2',3',4',5'-hepta-chlorinated biphenyl. Their IC_{50} ranged from 4.3 to 30 nM.

HPCBs were potent inhibitors of the 3,3',5-triiodothyronine sulfation rate, and the IC_{50} values were in the low μM range (Schuur et al., 1998a). HPCB inhibited the sulfation rate of 3,3'-diiodothyronine (T2) catalyzed by SULT1A1 but did not inhibit the sulfation rate of T2 catalyzed by SULT1A3 (Schuur et al., 1998b). 3-hydroxy-2,3',4,4',5-penta-chlorinated biphenyl did not inhibit the sulfation of T2 in rat brain cytosol, while it was a potent inhibitor of T2 sulfation rate in rat liver cytosol and this inhibition was competitive. The inhibitory effect of 3-hydroxy-2,3',4,4',5-penta-chlorinated biphenyl is tissue dependent.

The sulfation rate of dehydroepiandrosterone and estrone was inhibited by several drugs of common clinical use. The five most potent inhibitors of dehydroepiandrosterone were clomifene, testosterone, danazol, spironolactone, and cyproterone; and the five most potent inhibitors of estrone were cyclizine, ibuprofen, chlorpheniramine, dimenhydrinate, and tamoxifen (Bamforth et al., 1992).

Curcumin has anticarcinogen effects and is under clinical evaluation as a potential colon cancer chemopreventive agent (Cheng et al., 2001). In a study of 50 samples of human liver (Vietri et al., 2003), curcumin was found to be a potent inhibitor of human liver SULT1A1 activity, with mean $\pm SD$ and median of IC_{50} of 14.1 ± 7.3 and 12.8 nM, respectively. The IC_{50} ranged from 6.2 to 30.6 nM between the 5th and 95th percentiles, and the fold of variation was 4.9. The distribution of IC_{50} was positively skewed (skewness 1.521) and deviated from normality ($p=0.0004$).

Most of the NSAIDs have a carboxylic group, and Rao and Duffel (1991) observed that carboxylic acids inhibit rat liver AST IV sulfotransferase activity. Among the carboxylic acids used by Rao and Duffel (1991), there were salicylic acid and naproxen, and their K_i values were 67 μM and 130 μM, respectively.

Harris et al. (1998) were the first to show the inhibition of SULT1A1 activity by salicylic acid in human blood platelet and colon mucosa. These authors suggested that the prophylactic activity of salicylates against bowel cancer and other cancers (Funkhouser and Sharp, 1995; Greenberg et al., 1993; Heath et al., 1994; Rosenberg et al., 1991; Thun et al., 1991) may be due to the inhibition of sulfotransferase. Boberg et al. (1983) and Tsutsumi et al. (1995) have shown that sulfotransferase inhibitors dramatically reduce the sulfation of activated carcinogens in mice and hamsters.

The work by Harris et al. (1998) stimulated the search for potent inhibitors of human liver SULT1A1 activity among NSAIDs. Ten NSAIDs were tested as inhibitors of SULT1A1, and the IC_{50} value ranged from 0.02 μM (mefenamic acid) to 437 μM (naproxen; Vietri et al., 2000a). It was hypothesized that the intracellular drug concentration equilibrates with the unbound plasma drug concentration. When the IC_{50} values of NSAIDs were compared with their unbound plasma concentrations, it turned out that mefenamic acid and salicylic acid only had IC_{50} value greater than their unbound plasma concentrations (Vietri et al., 2000a). The IC_{50} values for mefenamic acid and salicylic acid were 0.02 μM and 30 μM, respectively (Vietri et al., 2000a). The inhibition type of salicylic acid for human liver SULT1A1 was mixed noncompetitive, with a K_i of 18 μM (Vietri et al., 2000a).

Five fenamates (mefenamic acid, tolfenamic acid, niflumic acid, meclofenamic acid, and flufenamic acid) were tested as inhibitors of human liver SULT1A1 and SULT1A3 activities, and mefenamic acid was the most potent inhibitor of SULT1A1 activity (Vietri et al., 2000b; Table 9.5). SULT1A3 was less susceptible than SULT1A1 to the inhibition by fenamates. The SULT1A3 IC_{50} to SULT1A1 IC_{50} ratio was drug dependent and it was 3800 for mefenamic acid (Table 9.5). This drug is a potent and selective inhibitor of SULT1A1 activity.

The intestinal mucosa is an important site of sulfation. The IC_{50} value of mefenamic acid and salicylic acid for SULT1A1 and SULT1A3 activities and for the sulfation rate of (−)-salbutamol was measured in human liver and duodenum (Vietri et al., 2000c). The IC_{50} values of mefenamic acid and salicylic acid for SULT1A1 and SULT1A3 activities were in the same order of magnitude in liver and duodenum, while the IC_{50} value for (−)-salbutamol was 23,000-fold higher in duodenum than liver (Table 9.6). This difference may be explained by the different forms of sulfotransferase present in these two tissues; in other words, forms of sulfotransferase

TABLE 9.5
Inhibition of SULT1A1 and SULT1A3 Activities by Drugs

Inhibitor	IC$_{50}$ SULT1A1	IC$_{50}$ SULT1A3	IC$_{50}$ SULT1A3 / IC$_{50}$ SULT1A1	Reference
Indomethacin	74	103	1.4	1
DCNP[a]	2.3[b]	44[b]	19	2
Salicylic acid	30	609	20	1
Diflunisal	3.7	79	21	1
Flufenamic acid	1.50	70	47	3
Meclofenamic acid	0.87	185	213	3
Tolfenamic acid	0.12	36	300	3
Curcumin	0.014	4.23	302	4
Niflumic acid	0.28	180	643	3
Mefenamic acid	0.02	76	3800	3

IC$_{50}$ values were measured in human liver and are expressed in µM. SULT1A1 and SULT1A3 were measured with 4 µM 4-nitrophenol and 60 µM dopamine, respectively. Modified from Vietri, M., De Santi, C., Pietrabissa, A., Mosca, F., and Pacifici, G.M. (2000b) *Xenobiotica* 30: 111–116.

[a] DCNP.
[b] IC$_{50}$ values were measured with partially purified enzyme. 1. Vietri, M., De Santi, C., Pietrabissa, A., Mosca, F., and Pacifici, G.M. (2000a) *Eur. J. Clin. Pharmacol.* 56: 81–87; 2. Campbell, N.R.C., Van Loon, J.A., and Weinsilboum R.M. (1987) *Biochem. Pharmacol.* 36: 1435–1446; 3. Vietri, M., De Santi, C., Pietrabissa, A., Mosca, F., and Pacifici, G.M. (2000b) *Xenobiotica* 30: 111–116; 4. Vietri, M., Pietrabissa, A., Mosca, F., Spisni, R., and Pacifici, G.M. (2003) *Xenobiotica* 33: 357–363.

that catalyze the sulfation of (−)-salbutamol and are resistant to the inhibition by mefenamic acid may be present in the duodenum. Mefenamic acid and salicylic acid inhibited resveratrol sulfation rate in human liver and duodenum, and mefenamic acid was a better inhibitor than salicylic acid in both tissues (De Santi et al., 2000a).

Drugs cross the placenta and act on the fetus (Pacifici and Nottoli, 1995). The possibility was explored that SULT1A1 and SULT1A3 activities and the sulfation rate of (−)-salbutamol and minoxidil are inhibited by mefenamic acid and salicylic acid in mid-gestation human fetal liver. The IC$_{50}$ value of mefenamic acid and salicylic acid for SULT1A1 and SULT1A3 were similar in human adult and fetal livers. Salicylic acid did not inhibit the sulfation rate of (−)-salbutamol and minoxidil in fetal liver, and the average IC$_{50}$ value of mefenamic acid for minoxidil sulfation rate was 1.6 µM (adult liver) and 0.15 µM (fetal liver; $p = 0.007$; Vietri et al., 2001). The lack of inhibition of (−)-salbutamol sulfation rate by salicylic acid in fetal liver indicates that the forms of sulfotransferase that catalyze the sulfation of (−)-salbutamol in the fetal liver were not susceptible to inhibition by salicylic acid.

The first observation that red wine inhibits sulfotransferase was provided by Littlewood et al. (1985). At a dilution of 1/75 from the original beverage, extract of red wine inhibited human platelet SULT1A1 activity by 99% and SULT1A3 activity

TABLE 9.6
Inhibition of Sulfotransferase Activities in Human Liver and Duodenum

Inhibitor	Substrate	Liver	Duodenum	Duodenum Liver	Reference
Salicylic acid	10 mM (−)-Salbutamol	93	705	8	1
3-OH-Flavone	2 µM 7-OH-Flavone	18	213	12	2
5-OH-Flavone	2 µM 7-OH-Flavone	3.5	69	20	2
Quercetin	60 µM Dopamine	5.7	170	30	3
Quercetin	5 mM Minoxidil	0.1	3	30	3
Quercetin	10 mM (−)-Salbutamol	0.4	16	40	3
Quercetin	50 µM (−)-Apomorphine	0.02	4.3	215	4
Quercetin	20 mM Paracetamol	0.03	35	1167	3
Mefenamic acid	50 µM (−)-Apomorphine	0.02	221	11,050	4
Mefenamic acid	10 mM (−)-Salbutamol	0.007	161	23,000	1
Salicylic acid	50 µM (−)-Apomorphine	54	n.m.[a]	—	4

The cases with duodenum/liver ratio greater than 2 are reported. IC_{50} values are expressed in µM.

[a] Not measurable. 1. Vietri, M., Pietrabissa, A., Spisni, R., Mosca, F., and Pacifici, G.M. (2000c), *Eur. J. Clin. Pharmacol.* 56: 477–479; 2. Vietri, M., Pietrabissa, A., Spisni, R., Mosca, F., Pacifici, G.M. (2002c) *Xenobiotica* 32: 563–571; 3. Marchetti, F., De Santi, C., Vietri, M., Pietrabissa, A., Spisni, F., Mosca, F., and Pacifici, G.M. (2001) *Xenobiotica* 31: 841–847; 4. Vietri, M., Vaglini, F., Pietrabissa, A., Spisni, R., Pacifici, G.M. (2002a) *Xenobiotica* 32: 587–594.

by 12%. Red wine, added at a final percentage of 5% (v/v) to the incubation mixture, decreased the rate of resveratrol sulfation by 91% (De Santi et al., 2000b).

Gibb et al. (1987) tested the effect of food and drink constituents and observed that many inhibited human blood platelet SULT1A1 activity. Among natural compounds, the most potent inhibitors were cyanidin 3-rutinoside, a simple anthocyanin, and (+)-catechin and (−)-epicatechin, two catechins. Among synthetic food colorants, the most potent inhibitors of human platelet SULT1A1 activity were carmosine (E122), erythrosine (E127), amaranth (E123), pounceau 4R (E124), and sunset yellow (E110).

Harris and Waring (1996) tested the effect of 35 substances obtained from vegetables on human platelet SULT1A1 and SULT1A3 activities and observed that some of them were potent inhibitors of both enzymes. The five most potent inhibitors of SULT1A1 and SULT1A3 activities were radish, spinach, banana, broccoli, and leek.

Walle et al. (1995) observed that quercetin, a natural flavonoid, was a potent inhibitor of human liver SULT1A1 activity ($IC_{50} = 0.1$ µM), whereas it was inactive on human liver SULT1A3 and dehydroepiandrosterone sulfotransferase activities. Kaempferol, naringenin, and naringin, natural flavonoids, inhibited human liver SULT1A1 (Walle et al., 1995). Eaton et al. (1996) observed that quercetin inhibited the sulfation rates of paracetamol and minoxidil using partial purified or recombinant SULT1A1. Quercetin also inhibited the sulfation rate of 50 µM (−)-apomorphine. The IC_{50} value was 0.02 µM (liver) and 4.3 µM (duodenum; Vietri et al., 2002a), and the duodenum IC_{50} value to the liver IC_{50} value ratio was 215. (−)-Apomorphine was sulfated in the human brain, and the sulfation rate of (−)-apomorphine was inhibited by quercetin with an IC_{50} (mean ±SD) of 16 ±2.3 nM (Vietri et al., 2002b). Quercetin was also found to inhibit 2 μM resveratrol sulfation, and the IC_{50} value was 12 pM (human liver) and 15 pM (human duodenum; De Santi et al., 2000a). Fisetin, myricetin, kaempferol, and apigenin inhibited the resveratrol sulfation rate in human liver and duodenum (De Santi et al., 2000b); the IC_{50} values were in the order of μM and were similar in liver and duodenum.

EST is the sole sulfotransferase expressed in normal human breast epithelial cells and plays an important role in determining the levels of free estrogen hormone in these cells (Otake et al., 2000). These authors observed that quercetin and resveratrol inhibited EST activity in competitive fashion with a K_i of about 1 μM both in human breast epithelial cells and in recombinant human EST.

Ghazali and Waring (1999) studied the inhibition of human platelet SULT1A1 and SULT1A3 activities by 6.7 μM quercetin, genistein, daidzein, equol, (+)-catechin, and flavone. SULT1A1 was more sensitive than SULT1A3 to the inhibition by these compounds. Quercetin and flavone were the most potent and weakest inhibitors, respectively, with an inhibition of 67% (quercetin) and 23% (flavone). The inhibition of SULT1A1 activity by these compounds was mixed noncompetitive with K_i value ranging from 0.1 μM (quercetin) to 0.94 μM (flavone). Bamforth et al. (1993) observed that (+)-catechin inhibited the sulfation rate of 17α-ethinyloestradiol, dopamine, 1-naphthol, and estrone, and the highest inhibition was found with dopamine and estrone.

Quercetin inhibited human adult and fetal liver SULT1A1. The average IC_{50} value of quercetin for SULT1A1 was 13 nM (adult liver) and 12 nM (fetal liver), and the average K_i of quercetin was 4.7 nM (adult liver) and 4.8 nM (fetal liver; De Santi et al., 2002). Human adult and fetal liver SULT1A1 were similarly susceptible to inhibition by quercetin.

The hypothesis that human duodenum sulfotransferase was less susceptible than human liver sulfotransferase to the inhibition by quercetin was tested. The following substrates of sulfotransferase were used: 4-nitrophenol, dopamine, (−)-salbutamol, minoxidil, and paracetamol. With the exception of 4-nitrophenol, whose IC_{50} value was similar in liver and duodenum, with the other substrates, quercetin was a better inhibitor of liver than duodenum sulfotransferase (Marchetti et al., 2001; Table 9.6).

Epicatechin gallate and epigallocatechin gallate are potent inhibitors of human liver SULT1A1, SULT1A2, and SULT1A3 activities (Coughtrie and Johnston, 2001). These enzymes were expressed in *Escherichia coli* or were purified recombinant preparations. There were significant differences by which these compounds exerted their effects. With SULT1A1, epicatechin gallate and epigallocatechin gallate reduced K_m and V_{max}, whereas they reduced V_{max} of SULT1A2 and SULT1A3 but did not have an effect on K_m. The inhibitory potency of the two compounds were enzyme dependent. The K_i of epicatechin gallate and epigallocatechin gallate were 60 nM and 40 nM, respectively (SULT1A1), 490 nM and 230 nM, respectively (SULT1A2), and 2.8 μM and 1.8 μM, respectively (SULT1A3). Tamura and Matsui (2000) observed that (−)-epigallocatechin was the most potent inhibitor of SULT1A1 present in green tea, with an IC_{50} value of 0.93 μM.

5-OH-flavone and 3-OH-flavone were potent inhibitors of SULT1A1 activity, whereas 7-OH-flavone was a substrate of sulfotransferase (Vietri et al., 2002c). 5-OH-flavone inhibited SULT1A1 activity, and the IC_{50} value was 0.3 μM in the human liver and duodenum. The IC_{50} values of 3-OH-flavone for SULT1A1 activity was 1 μM (liver) and 1.6 μM (duodenum). 5-OH-flavone and 3-OH-flavone also inhibited the sulfation rate of 2 μM 7-OH-flavone, and the IC_{50} values were one order of magnitude lower in liver than duodenum (Table 9.6).

CONCLUSION

Sulfotransferase is a major drug-metabolizing enzyme that catalyzes the sulfation of hormones, neurotransmitters, and drugs. Several forms of sulfotransferase exist that perform a multitude of important functions — most of which are concerned with the detoxification of endogenous compounds and drugs. SULT1A1 and SULT1A3 are the most studied forms of sulfotransferase. They develop before the 14th week of gestation in human fetal liver. In mid-gestation human fetal liver, the variation of the activities of SULT1A1 and SULT1A3 and the sulfation rate of ritodrine ranged 41-, 118-, and 134-fold, respectively. This makes it difficult to predict the sulfation capacity in the mid-gestation human fetus. In human adult liver, the activities of several sulfotransferases varied from three- to eight-fold. Substrates of sulfotransferase may have a chiral atom and consist of two enantiomers that have different biological properties. (+)-Terbutaline and (+)-4-OHP are sulfated at higher rates than the corresponding (−)-enantiomers. (−)-Salbutamol has greater affinity to sulfotransferase than (+)-salbutamol. The activities of SULT1A1 and SULT1A3 are inhibited by NSAIDs and by several natural flavonoids. SULT1A1 is more sensitive than SULT1A3 to such inhibitors. Among NSAIDs, the most potent inhibitor of SULT1A1 is mefenamic acid; the IC50 value is 0.02 μM, and the IC50 value of SULT1A3 is 3800-fold greater than that of SULT1A1. Mefenamic acid is a potent and selective inhibitor of SULT1A1. Quercetin, a natural flavonoid, is a potent inhibitor of SULT1A1 and SULT1A3. The interindividual variability in drug metabolism is an important factor in drug therapy. The rate of (−)-salbutamol sulfation varied three-fold in human liver (Pacifici et al., 1997a). Environmental factors, such as drug use and eating habits, may contribute to such a variation. The inhibition of sulfotransferases was studied in parallel in human liver and duodenum. On several occasions, the human duodenum sulfotransferase activities were less susceptible than the liver sulfotransferase activities to the inhibition by drugs and quercetin. For example, the IC50 of salbutamol sulfation by mefenamic acid was 23,000-fold greater in the duodenum than in the liver. This suggests that forms of sulfotransferase that catalyze sulfation of salbutamol and are resistant to inhibition by mefenamic acid are present in the duodenum.

REFERENCES

Anderson, R.J., Yoon, J.K., Sinsheimer, E.G., and Jackson, B.L. (1986) Human pituitary phenol sulfotransferase: biochemical properties and activities of the thermostable and thermolabile form. *Neuroendocrinology*, 44: 117–124.

Bamforth, K.J., Dalggliesh, K., and Coughtrie, M.W.H. (1992) Inhibition of human liver steroid sulfotransferase activities by drugs: a novel mechanism of drug toxicity. *Eur. J. Pharmacol.* 228: 15–21.

Bamforth, K.J., Jones, A.L., Roberts, R.C., and Coughtrie, M.W.H. (1993) Common food additives are potent inhibitors of human liver 17α-ethinyloestradiol and dopamine sulfotransferases. *Biochem. Pharmacol.* 46: 1713–1720.

Boberg, E.W., Miller, E.C., Miller, J.A., Poland, A., and Liem, A. (1983) Strong evidence from studies with brachymorphic mice and pentachlorophenol that 1'-sulfooxysafrole is the major ultimate electrophilic and carcinogenic metabolite of 1'-hydroxysafrole in mouse liver. *Cancer Res.* 43: 5163–5173.

Borgstrom, L., Chang-Xiao, L., and Walhagen, A. (1989a) Pharmacokinetics of the enantiomers of terbutaline after repeated oral dosing with racemic terbutaline. *Chirality* 1: 174–177.

Borgstrom, L., Nyberg, L., Jonsson, S., Lindberg, C., and Paulson, J. (1989b) Pharmacokinetic evaluation in man of terbutaline given as separate enantiomers and as the racemate. *Br. J. Clin. Pharmacol.* 27: 49–56.

Brashear, W.T., Kuhnert, B.R., and Wei, R. (1988) Maternal and neonatal urinary excretion of sulfate and glucuronide ritodrine conjugates. *Clin. Pharmacol. Ther.* 44: 634–641.

Campbell, N.R.C., Van Loon, J.A., and Weinsilboum R.M. (1987) Human liver phenol sulfotransferase: assay conditions, biochemical properties and partial purification of isoenzymes of the thermostable form. *Biochem. Pharmacol.* 36: 1435–1446.

Cappiello, M., Franchi, M., Rane, A., and Pacifici, G.M. (1990) Sulphotransferase and its substrate: adenosine-3'-phosphate-5'-phosphosulphate in human fetal liver and placenta. *Dev. Pharmacol. Ther.* 14: 62–65.

Cappiello, M., Giuliani, L., Rane, A., and Pacifici, G.M. (1991) Dopamine sulfotransferase is better developed than p-nitrophenol sulfotransferase in human fetus. *Dev. Pharmacol. Thear.* 16: 83–88.

Chen, X., Zhong, D., and Blume, H. (2000) Stereoselective pharmacokinetics of propafenone and its major metabolite in healthy Chinese volunteers. *Eur. J. Pharm. Sci.* 10: 11–16.

Cheng, A.L., Hsu, C.H., Lin, J.K., Hsu, M.M., Ho, Y.F., Shen, T.S., Ko, J.Y., Lin, J.T., Lin, B.R., Ming-Shiang, W., Yu, H.S., Jee, S.H., Chen, G.S., Chen, T.M., Chen, C.A., Lai, M.K., Pu, Y.S., Pan, M.H., Wang, Y.J., Tsai, C.C., and Hsieh, C.Y. (2001) Phase I clinical trial of curcumin, a chemopreventive agent, in patients with high-risk or premalignant lesions. *Anticancer Res.* 21: 2895–9200.

Coughtrie, M.W.H. and Johnston, L.E. (2001) Interactions between dietary chemicals and human sulfotransferases — molecular mechanism and clinical significance. *Drug Metab. Dispos.* 29: 522–528.

De Santi, C., Pierabissa, A., Mosca, F., Rane, A., and Pacifici, G.M. (2002) Inhibition of phenolsulfotransferase (SULT1A1) by quercetin in human adult and fetal livers. *Xenobiotica* 32: 363–368.

De Santi, C., Pietrabissa, A., Spisni, R., Mosca, F., and Pacifici, G.M. (2000a) Sulfation of resveratrol, a natural product present in grapes and wine, in the liver and duodenum. *Xenobiotica* 30: 609–617.

De Santi, C., Pietrabissa, A., Spisni, R., Mosca, F., and Pacifici, G.M. (2000b) The sulfation of resveratrol, a natural compound present in wine, is inhibited by natural flavonoids. *Xenobiotica* 30: 857–866.

Eaton, E.A., Walle, U.K., Lewis, A.J., Hudson, T., Wilson, A.A., and Walle, T. (1996) Flavonoids, potent inhibitors of the human P-form phenolsulfotransferase. *Drug Metab. Dispos.* 24: 232–237.

Evans, M.E., Walker, S.R., Brittain, R.T., and Paterson, J.W. (1973) The metabolism of salbutamol in man. *Xenobiotica* 3: 113–120.

Falany, C.N. (1991) Molecular enzymology of human liver cytosolic sulphotransferase. *Trends Pharmacol. Sci.* 12: 255–259.

Funkhouser, E.M. and Sharp, G.B. (1995) Aspirin and reduced risk of esophageal carcinoma. *Cancer* 76: 1116–1119.

Ghazali, R.A. and Waring, R.H. (1999) The effects of flavonoids on human phenolsulphotrans-ferases: potential in drug metabolism and chemoprevention. *Life Sci.* 65: 1625–1632.

Gibb, C., Glover, V., and Sandler, M. (1987) *In vitro* inhibition of phenolsulphotransferase by food and drink constituents. *Biochem. Pharmacol.* 36: 2325–2330.

Greenberg, E.R., Baron, J.A., Freeman, D.H., Mandel, J.S., and Haile, R. (1993) Reduced risk of large-bowel adenomas among aspirin users. *J. Nat. Cancer Inst.* 85: 912–916.

Harris, R.M., Hawker R.J., Langman, M.J.S., Singh, S., and Waring, R.H. (1998) Inhibition of phenolsulphotransferase by salicylic acid: a possible mechanism by which aspirin may reduce carcinogenesis. *Gut* 42: 272–275.

Harris, R.M. and Waring, R.H. (1996) Dietary modulation of human platelet phenolsul-photransferase activity. *Xenobiotica* 26: 1241–1247.

Heath, C.W., Thun, M.J., Greenberg, E.R., Levin, B., and Marnett, L.J. (1994) Nonsteroidal antiinflammatory drugs and human cancer. *Cancer* 74: 2885–2888.

Kester, M.H., Bulduk, S., Tibboel, D., Meinl, W., Glatt, H., Falany, C.N., Coughtrie, M.W.H., Bergman, A., Safe, A., Kuiper, G.G., Schuur, A.G., Brow, Brouwer, A., and Visser, T.J. (2000) Potent inhibition of estrogen sulfotransferase by hydroxy-lated PCB metabolites: a novel pathway explaining the estrogenic activity of PCBs. *Endocrinology* 141: 1897–1900.

Klecker, R.W. and Collins, J.M. (1997) Stereoselective metabolism of fenoldopam and its metabolites in human liver microsomes, cytosol, and slices. *J. Cardiovasc. Pharma-col.* 30: 69–74.

Littlewood, J.T., Glover, V., and Sandler, M. (1985) Red wine contains a potent inhibitor of phenolsulphotransferase. *Br. J. Clin. Pharmacol.* 19: 275–278.

Marchetti, F., De Santi, C., Vietri, M., Pietrabissa, A., Spisni, F., Mosca, F., and Pacifici, G.M. (2001) Differential inhibition of human liver and duodenum sulfotransferase activities by quercetin, a flavonoid present in vegetables, fruit and wine. *Xenobiotica* 31: 841–847.

Meisheri, K.D., Cipkus, L.A., and Taylor, C.J. (1988) Mechanism of action of minoxidil sulfate-induced vasodilation a role for increased K^+ permeability. *J. Pharmacol. Exp. Ther.* 245: 751–760.

Morgan, D.J., Paull, J.D., Richmond, B.H., Wilson-Evered, E., and Ziccone, S.P. (1986) Pharmacokinetics of intravenous and oral salbutamol and its sulfate conjugate. *Br. J. Clin. Pharmacol.* 122: 587–593.

Mulder, G.J. and Jakoby, W.B. (1990) Sulphation, in *Conjugation Reactions in Drug Metab-olism*, Mulder, G.J., Ed., Taylor & Francis, London, pp. 107–161.

Mulder, G.J. and Scholtens, E. (1977) Phenol sulphotransferase and uridine diphosphate glucuronyltransferase from rat liver *in vivo* and *in vitro*: 2,6-Dichloro-4-nitrophenol as selective inhibitor sulphation. *Biochem. J.* 165: 553–559.

Otake, Y., Nolan, A.L., Walle, U.K., and Walle, T. (2000) Quercetin and resveratrol potently reduce estrogen sulfotransferase activity in normal human mammary epithelial cells. *J. Steroid Biochem. Mol. Biol.* 73: 265–270.

Pacifici, G.M., Bigotti, R., Marchi, G., and Giuliani, L. (1993b) Minoxidil sulfation in human liver and platelets. *Eur. J. Clin. Pharmacol.* 45: 337–341.

Pacifici, G.M., D'Alessandro, C., Gucci, A., and Giuliani, L. (1997c) Sulfation of the eterocyclic amine 1,2,3,4-tetrahydroisoquinoline in the human liver and intestinal mucosa. *Arch. Toxicol.* 71: 477–481.

Pacifici, G.M. and De Santi, C. (1995) Human sulphotransferase. Classification and metabolic profile of the major isoforms: the point of view of the clinical pharmacologist, in *Advances in Drug Metabolism in Man*, Pacifici, G.M. and Fracchia, G.N., Eds., European Commission, Luxembourg, pp. 312–349.

Pacifici, G.M., De Santi, C., Mussi, A., and Angeletti, C.A. (1996) Interindividual variability in the rate of salbutamol sulfation in the human lung. *Eur. J. Clin. Pharmacol.* 49: 299–303.

Pacifici, G.M., Eligi, M., and Giuliani, L. (1993c) (+) and (–) Terbutaline are sulfated at a higher rate in human intestine than in liver. *Eur. J. Clin. Pharmacol.* 45: 483–487.

Pacifici, G.M., Ferroni, M.A., Temellini, A., Gucci, A., Morelli, M.C., and Giuliani, L. (1994a) Human liver budesonide sulfotransferase is inhibited by testosterone and correlates with testosterone sulfotransferase. *Eur. J. Clin. Pharmacol.* 46: 49–54.

Pacifici, G.M., Giulianetti, B., Quilici, M.C., Spisni, R., Nervi, M., Giuliani, L., and Gomeni, R. (1997a) (–)-Salbutamol sulfation in the human liver and duodenal mucosa: interindividual variability. *Xenobiotica* 27: 279–286.

Pacifici, G.M., Gucci, A., and Giuliani, L. (1997b) Testosterone sulfation and glucuronidation in the human liver: interindividual variability. *Eur. J. Drug Metab. Pharmacokin.* 22: 253–258.

Pacifici, G.M., Kubrich, M., Giuliani, L., de Vries, M., and Rane, A. (1993a) Sulfation and glucuronidation of ritodrine in human foetal and adult tissues. *Eur. J. Clin. Pharmacol.* 44: 259–264.

Pacifici, G.M. and Marchi, G. (1993) Phenol and catechol sulphotransferases in platelets from adult subjects and newborns: a study on the interindividual variability. *Br. J. Clin. Pharmacol.* 36: 593–597.

Pacifici, G.M. and Nottoli, R. (1995) Placental transfer of drugs administered to the mother. *Clin. Pharmacokin.* 28: 235–269.

Pacifici, G.M., Quilici, M.C., Giulianetti, B., Spisni, R., Nervi, M., Giuliani, L., and Gomeni, R. (1998) Ritodrine sulfation in the human liver and duodenal mucosa: interindividual variability. *Eur. J. Drug Metab. Pharmacokin.* 23: 67–74.

Pacifici, G.M. and Rossi, A.M. (2001) Interindividual variability of sulfotransferases in *Interindividual Variability in Drug Metabolism in Humans*, Pacifici, G.M. and Pelkonen, O., Eds., Taylor & Francis, London, pp. 434–459.

Pacifici, G.M., Temellini, A., Castiglioni, M., D'Alessandro, C., Ducci, A., and Giuliani, L. (1994b) Interindividual variability of the human hepatic sulfotransferases. *Chem. Biol. Interact.* 92: 219–231.

Pesola, G.R. and Walle, T. (1993) Stereoselective sulfate conjugation of isoproterenol in humans: comparison of hepatic intestinal and platelet activity. *Chirality* 5: 602–609.

Rao, S.I. and Duffel, M.W. (1991) Inhibition of aryl sulfotransferase by carboxylic acids. *Drug Metab. Dispos.* 19: 543–545.

Rein, G., Glover, V., and Sandler, M. (1982) Multiple forms of phenolsulphotransferase in human tissues: selective inhibition by dichloronitrophenol. *Biochem. Pharmacol.* 31: 1893–1897.

Rein, G., Glover, V., and Sandler, M. (1984) Characterization of human brain phenolsul-photransferase. *J. Neurochem.* 42: 80–85.

Reither, C. and Weinshilboum, R. (1982) Platelet sulfotransferase activity: correlation with conjugation of acetaminophen. *Clin. Pharmacol. Ther.* 32: 612–621.

Richard, K., Hume, R., Kaptein, E., Stanley, E.L., Visser, T.J., and Coughtrie, M.W.H. (2001) Sulfation of thyroid hormone and dopamine during human development: ontogeny of phenol sulfotransferases and arylsulfatase in liver, lung, and brain. *J. Clin. Endocrinol. Metab.* 86: 2734–2742.

Roos, R.A.C., Steenvoorden, J.M.C., Mulder, G.J., and Van Kempen, G.M.J. (1993) Acetaminophen sulfation in patients with Parkinson's disease or Huntington's disease in not impaired. *Neurology* 43: 1373–1376.

Rosenberg, L., Palmer, J.R., Zauber, A.G., Warshauer, M.E., Stolley, P.D., and Shapiro, S. (1991) A hypothesis: nonsteroidal anti-inflammatory drugs reduce the incidence of large-bowel cancer. *J. Nat. Cancer Inst.* 83: 355–358.

Schuur, A.G., Legger, F.F., van Meeteren, M.E., Moonen, M.J., van Leeuwen-Bol, I., Bergman, A., Visser, T.J., and Brouwer, A. (1998a) *In vitro* inhibition of thyroid hormones sulfation by hydroxylated metabolites of halogenated aromatic hydrocarbons. *Chem. Res. Toxicol.* 11: 1075–1081.

Schuur, A.G., Van Leeuwen-bol, I., Jong, W.M., Bergman, A., Coughtrie, M.W.H., Brouwer, A., and Visser, T.J. (1998b) *In vitro* inhibition of thyroid hormone sulfation by polychlorobiphenols: isozyme specificity and inhibition kinetics. *Toxicological. Sci.* 45: 188–194.

Seah, V.M.Y. and Wong, K.P. (1994) 2,6-Dichloro-4-nitrophenol (DCNP), an alternate-substrate inhibitor of phenolsulfotransferase. *Biochem. Pharmacol.* 47: 1743–1749.

Sundaram, R.S., Szumlanski, C., Otterness, D., van Loon, J.A., and Weinshilboum, R.M. (1989) Human intestinal phenol sulfotransferase: assay conditions, activity levels and partial purification of the thermolabile form. *Drug Metab. Dispos.* 17: 255–264.

Tamura, H. and Matsui, M. (2000) Inhibitory effects of green tea and grape juice on the phenol sulfotransferase activity of mouse intestines and human colon carcinoma cell line, Caco-2. *Biol. Pharm. Bull.* 23: 695–699.

Temellini, A., Giuliani, L., and Pacifici, G.M. (1991) Interindividual variability in the glucuronidation and sulphation of ethinyloestradiol in human liver. *Br. J. Clin. Pharmacol.* 31: 661–664.

Thun, M.J., Namboodiri, M.M., and Heath, C.W. (1991) Aspirin use and reduced risk of fatal colon cancer. *N. Engl. J. Med.* 325: 1593–1596.

Tsutsumi, M., Noguchi, O., and Okita, S. (1995) Inhibitory effects of sulfation inhibitors on initiation of pancreatic ductal carcinogenesis by N-nitrosobis(2-oxopropyl)amine in hamsters. *Carcinogenesis* 16: 457–459.

Verbeeck, R.K., Loewen, G.R., Macdonald, J.I., and Herman, R.J. (1990) The effect of multiple dosage on the kinetics of glucuronidation and sulfation of diflunisal in man. *Br. J. Clin. Pharmacol.* 29: 381–389.

Vietri, M., De Santi, C., Pietrabissa, A., Mosca, F., and Pacifici, G.M. (2000a) Inhibition of human liver phenol sulfotransferase by nonsteroidal anti-inflammatory drugs. *Eur. J. Clin. Pharmacol.* 56: 81–87.

Vietri, M., De Santi, C., Pietrabissa, A., Mosca, F., and Pacifici, G.M. (2000b) Fenamates and the potent inhibition of human liver phenol sulphotransferase. *Xenobiotica* 30: 111–116.

Vietri, M., Pietrabissa, A., Mosca, F., Rane, A., and Pacifici, G.M. (2001) Human adult and foetal liver sulphotransferase: inhibition by mefenamic acid and salicylic acid. *Xenobiotica* 31: 153–161.

Vietri, M., Pietrabissa, A., Mosca, F., Spisni, R., and Pacifici, G.M. (2003) Curcumin is a potent inhibitor of phenol sulfotransferase (SULT1A1) in human liver and extrahepatic tissues. *Xenobiotica* 33: 357–363.

Vietri, M., Pietrabissa, R., Spisni, R., Mosca, F., and Pacifici, G.M. (2000c) Differential inhibition of hepatic and duodenal sulfation of (−)-salbutamol and minoxidil by mefenamic acid. *Eur. J. Clin. Pharmacol.* 56: 477–479.

Vietri, M., Pietrabissa, A., Spisni, R., Mosca, F., Pacifici, G.M. (2002c) 7-OH-Flavone is sulfated in human liver and duodenum whereas 5-OH-flavone and 3-OH-flavone are potent inhibitors of SULT1A1 activity and 7-OH-flavone sulfation rate. *Xenobiotica* 32: 563–571.

Vietri, M., Vaglini, F., Cantini, R., Pacifici, G.M. (2002b) Quercetin inhibits the sulfation of R(−)-apomorphine in human brain. *Int. J. Clin. Pharmacol. Ther.* 41: 30–35.

Vietri, M., Vaglini, F., Pietrabissa, A., Spisni, R., Pacifici, G.M. (2002a) Sulfation of R(−)-apomorphine in human liver and duodenum and its inhibition by mefenamic acid, salicylic acid and quercetin. *Xenobiotica* 32: 587–594.

Walle, T., Eaton, E.A., and Walle, U.K. (1995) Quercetin, a potent and specific inhibitor of the human P form phenolsulfotransferase. *Biochem. Pharmacol.* 50: 731–734.

Walle, T. and Walle, U.K. (1990) Stereoselective sulfate conjugation of racemic terbutaline by human liver cytosol. *Br. J. Clin. Pharmacol.* 30: 127–133.

Walle, T. and Walle, U.K. (1991) Stereoselective sulfate conjugation of racemic 4-hydroxypropranolol by human and rat liver cytosol. *Drug. Metab. Dispos.* 19: 448–452.

Walle, T., Walle, U.K., Thornburg, K.R., and Schey, H.L. (1993b) Stereoselective sulfation of albuterol in humans: biosynthesis of the sulfate conjugate by HEP G2 cells. *Drug. Metab. Dispos.* 21: 76–80.

Walle, U.K., Pesola, G.R., and Walle, T. (1993a) Stereoselective sulfate conjugation of salbutamol in humans: comparison of hepatic, intestinal and platelet activity. *Br. J. Clin. Pharmacol.* 35: 413–418.

Weinshilboum, R.M. (1986) Phenol sulphotransferase in humans: properties, regulation, and function. *Federation Proc.* 45: 2223–2228.

Young, W.F., Okazaki, H., Laws, E.R., and Weinshilboum, R.M. (1984) Human brain phenol sulfotransferase: biochemical properties and regional localization. *J. Neurochem.* 43: 706–715.

Ziemniak, J.A., Allison, N., Boppana, V.K., Dubb, J., and Stote, R. (1987) The effect of acetaminophen on the disposition of fenoldopam: competition of sulfation. *Clin. Pharmacol. Ther.* 41: 275–281.

10 Human SULT1A SULTs

Nadine Hempel, Amanda Barnett,
Niranjali Gamage, Ronald G. Duggleby,
Kelly F. Windmill, Jennifer L. Martin, and
Michael E. McManus

CONTENTS

INTRODUCTION

Conjugation with sulfonate (SO_3^{1-}) is an important pathway in the biotransformation of numerous xeno- and endobiotics such as drugs, chemical carcinogens, hormones, bile acids, and neurotransmitters. This reaction is often incorrectly referred to in the literature as sulfation (Banoglu, 2000; Klaassen et al., 1998; Li and Anderson, 1999; Zhu et al., 1993a, 1993b). Xenobiotics and endobiotics may undergo sulfonation directly (phase II metabolism: paracetamol, minoxidil) or following phase I metabolism after gaining a functional hydroxyl group (e.g., N-hydroxy arylamines). The sulfonate donor for these reactions is 3'-phosphoadenosine 5'-phosphosulfate (PAPS), and the transfer of this functional group is catalyzed by a supergene family of enzymes called sulfotransferases (SULTS). In the case of the above substrates, the SULTs that catalyze this reaction are localized in the cytosolic fraction of the cell. Another class of membrane-bound SULTs, identified in the Golgi apparatus, is involved in the sulfonation of proteins, peptides, lipids, and glycosaminoglycans that affect both their structural and functional characteristics (Habuchi, 2000; Negishi et al., 2001; Niehrs et al., 1994). The focus of this review is on the emerging cytosolic forms of SULTs, for which at least 56 cDNAs and 18 gene sequences have been cloned and characterized in organisms ranging from microbes to humans (Nagata and Yamazoe, 2000; Raftogianis, 2003; Rikke and Roy, 1996). To date, five distinct gene families of SULTs have been identified in mammals: SULT1, SULT2, SULT3, SULT4, and SULT5 (Raftogianis, 2003). The SULT3 family has only been found in mouse and rabbit and has been shown to primarily sulfonate amino groups (Raftogianis, 2003). SULT5 has only been isolated from mice, and limited information is available on this family (Nagata and Yamazoe, 2000). In relation to humans, three families have been identified, consisting of ten distinct members: SULT1 — A1, A2, A3, B1, C1, C2, E1; SULT2 — A1 and B1 (SULT2B1a andSULT2B1b); and SULT4 — A1 (Raftogianis, 2003). In this review, we will concentrate on the SULT1A subfamily of enzymes that have been extensively studied and shown to be capable of metabolizing a broad range of drug, xenobiotic, and endobiotic substrates (Table 10.1). The SULT1A members have mainly been described based on their metabolic preferences as aryl, phenol, and monoamine SULTs. Although the term *aryl* is frequently used to collectively describe these enzymes (Bernier et al., 1994a; Duanmu et al., 2000; Nagata et al., 1997; Zhu et al., 1993a), it is technically incorrect, as it refers to the group obtained by dropping a hydrogen atom from the nucleus of an aromatic ring (e.g., phenyl). Thus in line with the new nomenclature system enunciated by Raftogianis et al. (2003), we will refer to these as members of the SULT1A subfamily.

Sulfonation is generally considered to be a detoxification pathway for both xeno- and endobiotics, as the addition of a sulfonate moiety usually results in a more hydrophilic molecule and thus facilitates excretion. In the case of thyroid hormones, sulfonation plays an important role in their irreversible inactivation (Li and Anderson, 1999; Stanley et al., 2001; Visser, 1994). Sulfonation is also a significant pathway involved in the metabolism of numerous drugs (e.g., paracetamol, 17α-ethinylestradiol, minoxidil, and cicletanine) and small endogenous substrates (e.g., estrone, cholesterol, dopamine, bile salts, and testosterone), and SULT1A subfamily members have been shown to play a major role in metabolizing many of these compounds (Table 10.1). In man, β-adrenergic agonists (e.g., isoprenaline, terbutaline, salbutamol, and fenoterol) are metabolized essentially by sulfonation (Hartman et al., 1998), and the sulfonated metabolite of paracetamol forms roughly 50% of the excreted drug (Mitchell et al., 1974). In the case of isoprenaline, metabolism of this drug by SULTs in the small intestine accounts for its low and variable bioavailability (George et al., 1974). In general, drugs and xenobiotics plus their metabolites containing a functional hydroxyl group are likely substrates for sulfonation and glucuronidation. Sulfonation of xenobiotics appears to be a complementary pathway to glucuronidation, with the former being a high affinity, low capacity pathway and the latter being a low affinity, high capacity pathway (Burchell and Coughtrie, 1997; Mulder, 1981). The enzymes responsible for these reactions, SULTs and UDP-glucuronosyltransferases, occupy different cellular compartments, with the former being cytosolic enzymes and the latter being located in the endoplasmic reticulum. Besides performing similar detoxification roles in different cellular compartments, these two phase II enzyme systems also exhibit differences in tissue localization. For example, two major portals of entry of foreign chemicals into the body, the lung and intestine, exhibit high SULT activity but relatively low UDP-glucuronosyltransferase activity (Pacifici et al., 1998).

While metabolism and excretion is undoubtedly the ultimate fate for a broad range of compounds undergoing sulfonation, evidence has emerged in recent years that challenge this simplistic view. For example, the sulfonation of hydroxymethyl polycyclic aromatic hydrocarbons, allylic alcohols, N-hydroxy derivatives of arylamines, and heterocyclic amines can lead to the formation of highly reactive molecules capable of binding to DNA and resulting in mutagenicity and carcinogenicity (Banoglu, 2000; Chou et al., 1995; Turteltaub et al., 1995; Windmill et al., 1997; Yamazoe et al., 1999). Further, sulfonation activates the antihypertensive and hair growth stimulant, minoxidil (Meisheri et al., 1993), while the sulfonated forms of the endogenous compounds pregnenolone and dehydroepiandrosterone (DHEA) are able to regulate neuronal activity via interaction with $GABA_A$, NMDA, and sigma receptors in the mouse and rat brain (Baulieu, 1991; Buu et al., 1984; Park-Chung et al., 1997, 1999). It has been shown that 99% of dopamine, 78% of noradrenaline, and 67% of adrenaline in normal human serum are present in the sulfonated form (Eisenhofer et al., 1999). Similarly, human plasma levels of the sulfonated form of the important hydroxysteroid hormone, DHEA, are 100-fold higher than that of the free form, and metabolic clearance of DHEA-sulfate has been shown to be around 100-fold less than that of free DHEA (Falany, 1997). In fact, both DHEA-sulfate and estrone-sulfate have been shown to have long half-lives in human blood (Hobkirk et al., 1990). As sulfonated catecholamines and steroids are thought not to be active at their appropriate receptors, it was assumed in the past that sulfonation of these molecules constituted an inactivation pathway (Runge-Morris, 1997). However, with the potential for sulfatase enzymes to cleave the sulfonate moiety and reactivate the molecules, it has been proposed that sulfonation of catacholamines and steroid hormones acts as a storage and transport mechanism (Falany and Falany, 1997; Strott, 2002). Thus, tissue-specific SULTs and sulfatases may act in concert to form a regulatory pathway for the action of these important endogenous compounds. Therefore, just as the transfer and removal of phosphate groups through the action of kinases and phosphatases can regulate the function of biological molecules, the addition and removal of a sulfonate group may modify the biological function of xeno- and endobiotics through the action of SULT and sulfatase enzymes.

Although Baumann first reported sulfonate conjugation of xenobiotics in 1876 (Baumann, 1876), an understanding of the enzymes responsible for catalyzing this reaction has lagged behind other enzymes systems involved in xenobiotic metabolism, such as cytochromes P450, flavin-containing monooxygenases, UDP-glucuronosyltransferases, N-acetyltransferases, and glutathione S-transferases. Initially this was in part due to the lack of a reliable source of the cofactor PAPS and the difficulty experienced in purifying these enzymes. The advent of the Foldes and Meek (1973) radiometric assay and its subsequent modifications (Brix et al., 1999a) proved invaluable as a means of measuring SULT activity in subcellular fractions

TABLE 10.1
Examples of (*Hsa*) SULT1A Substrates

SULT1A1

Simple phenolic compounds: *p*-nitrophenol, *m*-nitrophenol, *p*-ethylphenol, *p*-cresol (Brix et al., 1999b; Wilborn et al., 1993); short chain 4-n-alkylphenols (C< 8; Harris et al., 2000); 2-naphthol (Nowell et al., 2000)

Dietary phenolic compounds: catechin hydrate, epicatechin, epigallocatechin gallate, quercetin, myricetin, kaempferol, caffeic acid, chlorogenic acid, n-propyl gallate, and resveratrol (Coughtrie and Johnston, 2001; Pai et al., 2001)

Iodothyronines: 3,3'-diiodothyronine 3,3'-T2, rT3, T4 (Anderson et al., 1995; Kester et al., 1999; Li et al., 2001; Stanley et al., 2001)

Estrogens: estrone, β-estradiol (Falany, 1997); 2-hydroxyestrone, 2-hydroxyestradiol, 4-hydroxyestrone; 4-hydroxyestradiol (Adjei and Weinshilboum, 2002)

Serotonin derivatives: 5-hydroxyindole, 6-hydroxymelatonin, Harmol (Honma et al., 2001)

Catecholamines and derivatives: dopamine, tyramine (Brix et al., 1999b)

Industrial solvent: 2-nitropropane (Kreis et al., 2000)

Synthetic estrogenic compounds: trans-4-hydroxytamoxifen (Nishiyama et al., 2002); diethylstilbestrol (Falany, 1997); 2-methoxy estradiol (Spink et al., 2000)

Drugs: paracetamol (Bonham Carter et al., 1983; Lewis et al., 1996); minoxidol (Eaton et al., 1996; Falany and Kerl, 1990); troglitazone (Honma et al., 2002)

N-hydroxy aryl amines and heterocyclic amines: N-hydroxyacetylaminofluorene, N-hydroxy-2-aminoflourine, N-hydroxy-4,4'-methylene-bis(2-chloroaniline), N-hydroxy-2-amino-1-methyl-6-phenylimidazo[4,5-b]pyridine, and N-hydroxy-2-amino-6-methyldipyrido[1,2-a:3',2'-d]imidazole (Chou et al., 1995; Nowell et al., 2000; Ozawa et al., 1994); N-hydroxy-2-acetyl-amino-5-phenylpyridine (N-OH-2AAPP), N-hydroxy-4-acetylaminobiphenyl (N-OH-4AABP), N-hydroxy-4'fluoro-4 acetylaminobiphenyl (N-OH-4FAABP), N-hydroxy-2-acetylaminonaphthalene (N-OH-2AAN), N-hydroxy-2-acetylaminophenanthrene (N-OH-2AAP), N-hydroxy-4-acetylaminostilbene (N-OH-4AAS; Gilissen et al., 1994), 2-amino-alpha-carboline (A alpha C; King et al., 2000); hydroxymethyl PAHs (Teubner et al., 2002)

SULT1A2

Simple phenolic compounds: *p*-nitrophenol (Zhu et al., 1996); short chain 4-n alkylphenols (C < 8; Harris et al., 2000)

Carcinogens: 2-hydroxyamino-1-methyl-phenylimidazo[4,5-*b*]pyridine (OH-PhIP; Ozawa et al., 1994); N-hydroxy-2-acetylaminofluorene, 2-hydroxylamino-5-phenylpyridine, 1-hydroxymethylpyrine (Meinl et al., 2002)

Estrogens: β-estradiol and catecholestrogens (Adjei and Weinshilboum, 2002)

Drugs: minoxidil (Baker et al., 1994)

Dietary phenolic compounds: catechin, epicatechin, resveratrol (Coughtrie, 1996; Coughtrie and Johnston, 2001)

(continued)

TABLE 10.1 (CONTINUED)
Examples of (*Hsa*) SULT1A Substrates

SULT1A3

Catecholamines and derivatives: dopamine, tyramine (Brix et al., 1999b; Dajani et al., 1999b);
  norepinephrine, epinephrine (Ganguly et al., 1995; Reiter et al., 1983); normetanephrine
  (Honma et al., 2001) 5-hydroxytryptamine (Coughtrie and Johnston, 2001)

Serotonin derivatives: harmol, 6-hydroxy-melatonin (Honma et al., 2001)

Iodothyronines: T3, 3,3′-T2, T4 (Kester et al., 1999)

Simple phenols: *p*-nitrophenol, *p*-ethylphenol (Brix et al., 1999b)

Dietary phenolic compounds: catechin hydrate, epicatechin, epigallocatechin gallate, quercetin,
  myricetin, kaempferol, caffeic acid (Pai et al., 2001); Resveratrol (Coughtrie and Johnston, 2001)

Industrial solvent: 2-nitropropane (Kreis et al., 2000)

Drugs: minoxidil (Ganguly et al., 1995; Kudlacek et al., 1995), isoproterenol, metaproterenol,
  terbutaline, albuterol, salmeterol, formoterol (Coughtrie et al., 1996; Hartman et al., 1998;
  Lewis et al., 1996) salbutamol, isoprenaline, dobutamine (Coughtrie and Johnston, 2001)

---

and of purified proteins. Initial attempts to classify SULTs focused mainly on the metabolic activities of subcellular tissue fractions and their sensitivity to the inhibitor, 2,6-dichloro-4-nitrophenol (DCNP; Bonham Carter et al., 1983; Butler et al., 1983; Hernandez et al., 1991). However, like other xenobiotic metabolizing enzymes such as cytochromes P450, this approach while initially helpful, proved unsatisfactory due to the promiscuous nature of these enzymes: they do not abide by a classification scheme that limits them to a particular functional group (Duffel et al., 2001). While across the broad range of SULTs this has certainly been shown to be true, it nonetheless is appropriate to point out that the early metabolic work on tissue subcellular fractions, especially on SULT1A subfamily members, proved invaluable in highlighting the multiplicity of these enzymes and their particular substrate preferences (Anderson et al., 1981; Duffel and Jakoby, 1981; Weinshilboum, 1992). From early work on human liver (Campbell et al., 1987), platelets (Reiter et al., 1983), and brain (Whittemore et al., 1986), at least two forms of phenol SULTs were identified based on their thermostability, behavior on ion-exchange chromatography, preference for either *p*-nitrophenol or dopamine as a substrate, and sensitivity to the inhibitor DCNP (Van Loon and Weinshilboum, 1984; Weinshilboum, 1992). These two forms were subsequently labeled by Weinshilboum's group (Reiter et al., 1983) as the TS or P (thermostable or phenol, TS-PST) and TL or M (thermolabile or monoamine, TL-PST) sulfonating forms and shown to be responsible for the metabolism of most phenolic compounds. As with all areas of biology, the use of recombinant DNA technology has now enabled us to explore and differentiate between closely related forms of SULT. These studies also led to the characterization of another SULT1A subfamily member, SULT1A2 (ST1A2, HAST4; Ozawa et al., 1995; Zhu et al., 1996).

The TS-PST and TL-PST sulfonating forms referred to above have subsequently been shown to correspond to SULT1A1 and SULT1A3, respectively (Veronese et al., 1994). Indeed, the encoded proteins of their recombinant cDNAs expressed in COS cells exhibit markedly different preferences for both p-nitrophenol and dopamine as substrates. For example, the K_m values for the sulfonation of p-nitrophenol by SULT1A1 and SULT1A3 are approximately 1 and 2000 μM, respectively (Veronese et al., 1994). In contrast, SULT1A3 has a high affinity for dopamine as a substrate ($K_m = 10$ μM) and SULT1A1 ($K_m = 345$ μM) low affinity. In relation to their thermostabilities, SULT1A3 shows no activity toward dopamine as a substrate after treatment at 45°C for 15 minutes. In contrast, SULT1A1 retained approximately 90% of its activity toward p-nitrophenol as a substrate after similar treatment. Unlike SULT1A1 or SULT1A3, SULT1A2 exhibits no activity toward dopamine as a substrate and possesses a K_m for p-nitrophenol sulfonation (~70 μM) between SULT1A1 and SULT1A3 (Zhu et al., 1996). These substrate preferences are interesting from a structural viewpoint, as SULT1A subfamily members share greater than 90% amino acid sequence identity (see Tissue Distribution of the Human SULT1A Subfamily).

As alluded to above, large interindividual variations in phenol SULT activity (5- to 36-fold) in the human population have been reported (Coughtrie et al., 1999; Pacifici and De Santi, 1995; Pacifici and Rossi, 2001; Pacifici et al., 1994, 1997, 1998; Van Loon and Weinshilboum, 1984; Weinshilboum, 1992). However, these activities do not appear to be influenced by age, and there is conflicting evidence about whether gender differences in phenol SULT activity exist. Some studies show differences in human platelet phenol SULT activity between male and female populations (Marazziti et al., 1998; Nowell et al., 2000), and Brittelli and colleagues (1999) observed this difference only in a Finnish population and not in subjects of Italian origin. In contrast, studies using hepatic cytosols have shown no gender difference in either p-nitrophenol or dopamine activities (Pacifici et al., 1994). Inheritance has been shown to play an important part in the regulation of both TS-PST (SULT1A1) and TL-PST (SULT1A3) activities (Price et al., 1988; Weinshilboum, 1988). Further, it was concluded from early work that genetic polymorphisms regulate levels of both the TS-PST and TL-PST forms of phenol SULT (Price et al., 1988; Weinshilboum, 1992). Additional studies utilizing human platelets (Nakamura et al., 1990; Sundaram et al., 1989; Young et al., 1985) showed a correlation between platelet TS-PST activity and the same activity in brain, liver, and small intestine. The same has not been evident for the TL-PST enzyme (Sundaram et al., 1989). Further, genetically determined platelet TS-PST activity has been shown to correlate with individual differences in the sulfonation of orally administered drugs such as paracetamol and methyldopa (Campbell et al., 1985). However, given the complexity of the SULT system and the fact that these enzymes exhibit overlapping substrate specificities, the use of platelet activities as a predictive tool of similar activities in other tissues is questionable. In the past 10 years, the use of molecular biology has broadened our understanding of the tissue distribution and kinetic and structural properties of the SULT1A subfamily, which will be discussed below.

MOLECULAR BIOLOGY OF THE HUMAN
SULT1A SUBFAMILY

Figure 10.1 shows a dendogram of SULT1A forms isolated from seven species including *Homo sapiens* (*Hsa*). It is interesting to note that humans are quite different from other species in having multiple members of the SULT1A subfamily. It is thought that the three human SULT1A members originated as a result of gene duplication or gene duplication plus recombination events (Dooley, 1998a; Rikke and Roy, 1996). Genomic mapping has placed all three SULT1A genes on the short arm of chromosome 16 (position 16p12.1-11.2), which has been confirmed by recent data from the human genome project (NCBI; Aksoy et al., 1994; Dooley and Huang, 1996; Dooley et al., 1993, 1994; Gaedigk et al., 1997; Her et al., 1996; Weinshilboum et al., 1997). Besides the greater than 96% amino acid sequence identity shared between SULT1A1 and SULT1A2, these two enzymes share greater than 93% homology in their gene sequence and are thought to have arisen quite recently in

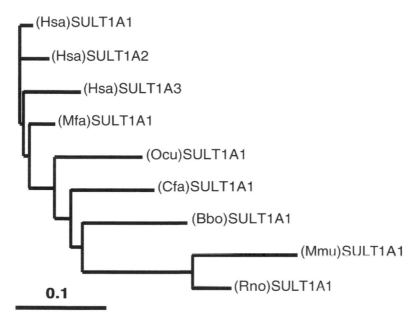

FIGURE 10.1 Fitch Phylogenetic Tree showing the genetic distances between SULT1A family members. The tree was generated by Protdist Software version 3.5 using amino acid sequences. Distance scale represents the number of differences between the sequences. (0.1 means 10% differences between two sequences). The species abbreviations follow the Rafto-gianis et al. (2003) nomenclature such as Hsa: human, Mfa: monkey, Ocu: rabbit, Cfa: dog, Bbo: cow, Mmu: mouse, and Rno: rat. The GenBank Accession numbers are as follows: (Hsa)SULT1A1 (U26309), (Hsa)SULT1A2 (U28169), (Hsa)SULT1A3 (L19956), (Mfa)SULT1A1 (D85514), (Mmu)SULT1A1 (P52840), (Ocu)SULT1A1 (AF360872), (Bbo)SULT1A1 (U35253), (Cfa)SULT1A1 (D29807), and (Rno)SULT1A1 (L19998).

evolution due to gene duplication (Aksoy and Weinshilboum, 1995; Bernier et al., 1996; Dooley, 1998a, 1998b; Gaedigk et al., 1997; Her et al., 1996; Raftogianis et al., 1996). The *SULT1A3* gene differs mainly in its 5′ promoter and intron sequences and shares about 60% sequence identity with the other two *SULT1A* genes, when aligning 8.5 kilobases of gene sequence including 4 kilobases upstream of the ATG start codon (Aksoy and Weinshilboum, 1995; Bernier et al., 1994a; Dooley et al., 1994). It appears that *SULT1A1* is the most likely ortholog of the equivalent animal counterparts based on the enzymes' functional similarities (Riley et al., 2002; Tsoi et al., 2001). No equivalent forms of *SULT1A2* and *SULT1A3* have been identified in other species to date, but based on sequence similarity these two genes could also be considered orthologs of the equivalent animal *SULT1A1* genes. From the available metabolic data across species, it is apparent that through evolutionary pressures, humans have acquired a *SULT1A3* gene whose expressed protein fulfills a specific role in sulfonating catecholamines such as dopamine (Dooley, 1998a).

HUMAN SULT1A1

The SULT1A1 cDNA (originally termed P-PST-1) was the first of three members of this human phenol SULT subfamily to be published and was isolated from a lambda Uni-Zap XR human liver cDNA library (Wilborn et al., 1993). The full-length SULT1A1 cDNA is 1206 base pairs in length and encodes a protein of 295 amino acids that was shown to metabolize both *p*-nitrophenol and minoxidil. In the same year, our laboratory cloned a similar sequence from a liver cDNA library that shared 98.2% identity to the Wilborn et al. (1993) sequence (Zhu et al., 1993b) and termed it human aryl SULT 1 (HAST1). Several other groups have also cloned cDNAs that encode allelic variants of SULT1A1 (Table 10.2). Each of the proteins encoded by the above sequences share >98% identity and therefore fulfill the require-ment of being designated allelic variants.

At least 7 different SULT1A1 allozymes have been identified thus far (Table 10.3), and in a recent study by Raftogianis et al. (1999), the frequencies of 15 allelic variants resulting in 4 amino acid changes were identified in a study of Caucasian liver and blood samples (Table 10.3). Except for a significant difference in the K_m for PAPS exhibited by SULT1A1*3, these amino acid changes had no effect on the ability of recombinant COS cell expressed protein of the allozymes tested to metab-olize *p*-nitrophenol, on their thermostability, or on the IC_{50} of DCNP (Raftogianis et al., 1999). However, phenol SULT activity of platelets from subjects with the SULT1A1*2 allozyme showed lower activity toward *p*-nitrophenol and lower ther-mal stability than subjects expressing the wildtype SULT1A1*1 allele (Raftogianis et al., 1997). In Caucasians, the frequencies of the allozymes SULT1A1*1, 1A1*2, and 1A1*3 were reported as 0.66, 0.33, and 0.012, respectively, similar to the frequencies reported by Carlini et al. (2001). However in the Chinese population, the ratio of SULT1A1*1 is much higher at 0.91 and lower in the African-American population, at 0.48 (Table 10.4). The highest frequency (0.23) of the SULT1A1*3 allele was observed in the African-American population (Carlini et al., 2001).

The high affinity of SULT1A1 for *p*-nitrophenol (K_m ~ 1 μM) compared to other forms of SULT (SULT1A2 ~ 10 μM; SULT1A3 ~ 2000 μM; Raftogianis et al.,

TABLE 10.2
cDNAs and Genes Comprising the (*Hs*a) SULT1A Subfamily

SULT1A		Name by Author	Accession No.	Reference
SULT1A1	cDNA	P-PST-1	L19999	Wilborn et al., 1993
		HAST1	L10819	Zhu et al., 1993b
		ST1A3	X78283	Ozawa et al., 1995
		P-PST	X84654	Jones et al., 1995
		H-PST	U26309	Hwang et al., 1995
		HAST2	L19955	Zhu et al., 1996
			U09031	
		SULT1A1	NM_001055	Her et al., 1996
		SULT1A1	AJ007418	Raftogianis et al., 1996
	Gene	STP STP1	U71086	Dooley et al., 1994
			U54701	Bernier et al., 1996
		TS-PST1	U52852	Raftogianis et al., 1996
SULT1A2	cDNA	ST1A2	X78282	Ozawa et al., 1995
		HAST4v	U28169	Zhu et al., 1996
		HAST4	U28170	Zhu et al., 1996
	Gene	STP2	U76619	Dooley and Huang, 1996
			U34804	Her et al., 1996
			U33886	Gaedigk et al., 1997
SULT1A3	cDNA	HAST3	L19956	Zhu et al., 1993a
		HAST3 INT	L19957	
		TL-PST	U08032	Wood et al., 1994
		hEST	L25275	Bernier et al., 1994a
		m-PST	X84653	Jones et al., 1995
		hm-PST	—	Ganguly et al., 1995
	Gene	STM	U20499	Aksoy et al., 1994
				Aksoy and Weinshilboum, 1995
			U37686	Dooley et al., 1994
			L34160	Bernier et al., 1994a

1999; Veronese et al., 1994; Zhu et al., 1996) has made this substrate an effective metabolic probe for studying the expression of this enzyme in human tissues. Figure 10.2 shows that SULT1A1 is the predominant SULT1A protein expressed in human tissue cytosols. Because of the high sequence homology between SULT1A subfamily members (~90%), it is possible to use an antibody raised against SULT1A1 to simultaneously recognize SULT1A1, SULT1A2, and SULT1A3. On SDS-PAGE we have previously reported estimated molecular weight masses of approximately 32 and 34 kDa for SULT1A1 and SULT1A3, respectively (Veronese et al., 1994). These differences in electrophoretic mobilities would not be expected from similarities in predicted molecular masses for these proteins based on their amino acid sequences (i.e., SULT1A1 34,178 Da, SULT1A2, 34,309 Da; and, SULT1A3 34,196 Da). From

TABLE 10.3
***(Hsa)* SULT1A1 Allozymes**

SULT1A1 Allozyme	Amino Acid									Frequency (Caucasian Liver Samples*)	Reference
	37	90	146	181	213	223	243	282	290		
1A1*1 (wildtype)	Arg	Pro	Ala	Glu	Arg	Met	Val	Glu	Ser	0.67	Ozawa et al., 1995 Raftogianis et al., 1999
1A1*2	Arg	Pro	Ala	Glu	His	Met	Val	Glu	Ser	0.31	Zhu et al., 1993a Raftogianis et al., 1999
1A1*3	Arg	Pro	Ala	Glu	Arg	Val	Val	Glu	Ser	0.016	Raftogianis et al., 1999
1A1*4	Gly	Pro	Ala	Glu	Arg	Met	Val	Glu	Ser	ND[a]	Raftogianis et al., 1999
1A1*5	Arg	Pro	Thr	Gly	His	Met	Val	Glu	Ser	cDNA only[b]	Jones et al., 1995
1A1*6	Arg	Leu	Ala	Glu	Arg	Met	Ala	Glu	Ser	cDNA only[b]	Hwang et al., 1995
1A1*7	Arg	Pro	Ala	Glu	Arg	Met	Val	Lys	Thr	cDNA only[b]	Wilborn et al., 1993

[a] ND = Not Detectable.
[b] cDNA only = no frequency in population investigated, represents different cDNA isolated.

Figure 10.2, it is apparent that the COS cell expressed SULT1A2 protein runs between the 32 kDa SULT1A1 and the 34 kDa SULT1A3. Neither SULT1A2 nor SULT1A3 was detectable in liver cytosols from ten different human subjects under the conditions used (Figure 10.2). However, a recent study by Honma et al. (2001) using liver cytosols from 21 patients, showed the content of SULT1A1 (ST1A3) was approximately 19 times higher (120 ± 38 pmol/mg cytosolic protein) than SULT1A3 (ST1A5; 6.4 ± 2.6 pmol/mg). Falany (1997) has also reported a large difference in the levels of these enzymes in human liver; SULT1A1 (P-PST) activity was 20- to 40-fold higher than SULT1A3 (M-PST) activity. When the SULT1A1 cDNA was first isolated, it was observed that two different mRNA species of SULT1A1 existed and that the mRNA species were dependent on the tissue these were isolated from (Zhu et al., 1993a, 1993b). Zhu and colleagues observed that these sequence differences occur outside the coding region and identified two different untranslated first exons of SULT1A1 depending on the origin from either a liver or a brain cDNA library (Honma et al., 2001; Zhu et al., 1993a, 1993b). Other studies have confirmed the presence of the distal first exon found on the SULT1A1 gene to be present in the major mRNA species found in the liver (Raftogianis et al.,

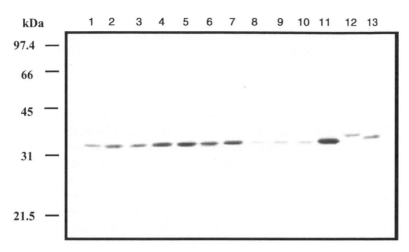

FIGURE 10.2 Immunoblot analysis of human SULT1A proteins expressed in liver cytosols. Lanes 1–10: 100 μg of human liver cytosol from 10 different individuals; Lane 11: COS expressed recombinant SULT1A1 protein (40 μg); Lane 12: COS expressed recombinant SULT1A3 protein (40 μg); Lane 13: COS expressed recombinant SULT1A2 protein (40 μg). The 32 kDa band corresponding to SULT1A1 protein is seen in lanes 1–10. The protein molecular weight bands indicated in extreme left are, from top to bottom: 97.4, 66, 45, 31, and 21.5 kDa.

1996; Wilborn et al., 1993). Similarly, Bernier et al. (1994a) isolated an alternate mRNA species of SULT1A3 from a human placental cDNA library to that isolated by Zhu et al. (1993a) from a brain library, again differing only in the 5′-untranslated region (Bernier et al., 1994a). This suggests that either alternate transcriptional start sites are utilized or alternate splicing occurs posttranscriptionally in different tissues. The answer to this question is currently unknown; however, when the *SULT1A1* gene was cloned, promoter activities of sequences in front of the furthest 5′ exon showed highest promoter activity in human embryonic kidney cells (Bernier et al., 1996). This has been confirmed by our laboratory in several cell lines, including human primary hepatocytes, HepG2, Hep3B, Caco2, MCF-7, and the glioblastoma cell line T98G (Hempel, N., Wang, H., LeCluyse, E. L., McManus, M.E., Negishi, M., unpublished data). However, it remains to be investigated if, for example, a neuronal environment or differential cellular conditions, such as developmental or stress-related environments, favor the utilization of a different promoter or if these mRNA differences are due to alternate splicing. Our ability to understand the substrate specificity of SULT1A1 and its allozymes has recently been enhanced by the publication of the crystal structure of this enzyme (Gamage et al., 2003).

HUMAN SULT1A2

A human SULT1A2 cDNA (ST1A2) was first cloned from a human liver library by Ozawa et al. (1995). Our laboratory subsequently cloned two cDNA forms of this enzyme, HAST4 and HAST4v, that differed by two amino acids (Thr7 to Ile and Thr235 to Asn). The coding region of HAST4 and HAST4v are 97% and 94%

homologous to SULT1A1 and SULT1A3, respectively (Zhu et al., 1996). Interestingly, when both forms are expressed in COS cells, they exhibit markedly different affinities for *p*-nitrophenol, with K_m values of 74 μM and 8 μM for HAST4 and HAST4v, respectively. For the same reaction, SULT1A1 and SULT1A3 exhibit K_m values of 0.7 and 2200 μM, respectively (Zhu et al., 1996). The effect of a change from Asn235 to Thr in SULT1A2 was investigated using the crystal structure of SULT1A1 (Gamage et al., 2003). Asn235 is a highly conserved residue in the SULT1A family. This residue is located at the end of helix α12 and at the beginning of the large flexible loop that undergoes disorder–order transition upon substrate binding (Figure 10.3; see color photo insert following p. 210). The side chain oxygen of this residue forms a hydrogen bond with the main chain nitrogen of Met237, and this interaction could be important for maintaining the stability of the flexible loop interactions (Thr235 does not form this interaction). It is noteworthy that the Phe247, which interacts with both *p*-nitrophenol molecules is located in this flexible loop region. Based on this structural observation, we propose that disruption to flexible loop interactions could affect the enzyme's stability.

In a more recent human population study of DNA isolated from human liver tissue obtained from 61 Caucasian patients, Raftogianis et al. (1999) reported 13 different allelic variants of SULT1A2 that encode four different amino acid changes, which resulted in six different SULT1A2 allozymes (Table 10.5). These changes

```
                        β1      β2        α1                  β3   PSB
SULT1A1    1:  MELIQDTSRP PLEYVKGVPL IKYFAEALGP LQSFQARPDD LLISTYPKSG   50
SULT1A2    1:  MELIQDTSRP PLEYVKGVPL IKYFAEALGP LQSFQARPDD LLISTYPKSG   50
SULT1A3    1:  MELIQDTSRP PLEYVKGVPL IKYFAEALGP LQSFQARPDD LLINTYPKSG   50

                  α2          α3                              α4
SULT1A1   51:  TTWVSQILDM IYQGGDLEKC HRAPIFMRVP FLEFKAPGIP SGMETLKDTP  100
SULT1A2   51:  TTWVSQILDM IYQGGDLEKC HRAPIFMRVP FLEFKVPGIP SGMETLKNTP  100
SULT1A3   51:  TTWVSQILDM IYQGGDLEKC NRAPIYVRVP FLEVNDPGEP SGLETLKDTP  100

                  β4         α5        β5  3'PB       3'PB    α6
SULT1A1  101:  APRLLKTHLP LALLPQTLLD QKVKVVYVAR NAKDVAVSLY HFYHMAKVHP  150
SULT1A2  101:  APRLLKTHLP LALLPQTLLD QKVKVVYVAR NAKDVAVSLY HFYHMAKVYP  150
SULT1A3  101:  PPRLIKSHLP LALLPQTLLD QKVKVVYVAR NPKDVAVSLY HFHRMEKAHP  150

                  α7          α8         β6         α9
SULT1A1  151:  EPGTWDSFLE KFMVGEVSYG SWYQHVQEWW ELSRTHPVLY LFYEDMKENP  200
SULT1A2  151:  HPGTWESFLE KFMAGEVSYG SWYQHVQEWW ELSRTHPVLY LFYEDMKENP  200
SULT1A3  151:  EPGTWDSFLE KFMAGEVSYG SWYQHVQEWW ELSRTHPVLY LFYEDMKENP  200

                  α10         α11        α12
SULT1A1  201:  KREIQKILEF VGHSLPEETV DFMVQHTSFK EMKKNPMTNY TTVPQEFMDH  250
SULT1A2  201:  KREIQKILEF VGRSLPEETV DLMVEHTSFK EMKKTPMTNY TTVRREFMDH  250
SULT1A3  201:  KREIQKILEF VGRSLPEETM DFMVQHTSFK EMKKNPMTNY TTVPQELMDH  250

                  3'PB                α13
SULT1A1  251:  SISPFMRKGM AGDWKTTFTV AQNERFDADY AEKMAGCSLS FRSEL        295
SULT1A2  251:  SISPFMRKGM AGDWKTTFTV AQNERFDADY AEKMAGCSLS FRSEL        295
SULT1A3  251:  SISPFMRKGM AGDWKTTFTV AQNERFDADY AEKMAGCSLS FRSEL        295
```

FIGURE 10.3 Sequence alignment of human SULT1A1, 1A2, and 1A3. Secondary structure elements are numbered based on SULT1A1 structure. Residues conserved in PSB loop and 3′PB site are boxed, and those that differ among the isoforms are shown in red. Residues that line the substrate-binding pocket are highlighted in yellow (based on SULT1A1 structure). (See color photo insert following p. 210.)

TABLE 10.4
Frequency of the SULT1A1*1 and SULT1A1*2 (R213H) Polymorphisms in Different Ethnic Groups

Population Ethnicity	Allele Frequency		#	Mean Age (years)	Female:Male Ratio	Reference
	Wildtype	R213H Mutant				
Caucasian	0.69	0.31	150	—	—	Raftogianis et al., 1997
Caucasian	0.68	0.32	293	53.6	0.56	Coughtrie et al.,
African	0.73	0.27	52	—	—	1999
Japanese	0.83	0.17	143	61	0.81	Ozawa et al., 1999
Caucasian	0.66	0.34	150	40–64	Male	Engelke et al., 2000
Caucasian	0.61	0.39	189	—	1.27	Steiner et al.,
Caucasian with prostate cancer	0.65	0.35	134	—	—	2000
Caucasian	0.659	0.341	156	—	Female	
Caucasian with breast cancer	0.584	0.416	332	—	Female	Zheng et al., 2001
Caucasian	0.64	0.36	211	—	0.66	Nowell et al.,
African	0.74	0.26	40	—	0.66	2000
Caucasian	0.66	0.33	242	—	—	Carlini et al., 2001
African-American	0.48	0.29	70	—	—	
Chinese	0.91	0.08	290	—	Female	
Caucasian	0.68	0.32	402	71.27	0.66	Wong et al., 2002
Caucasian with colorectal cancer	0.67	0.33	383	68.36	0.77	

TABLE 10.5
(*Hsa*) SULT1A2 Allozymes

SULT1A2 Allozyme	Amino Acid				Frequency (Caucasian Liver Samples*)	Reference
	7	19	184	235		
1A2*1 (wildtype)	Ile	Pro	Arg	Asn	0.51	Zhu et al., 1996 Raftogianis et al., 1999
1A2*2	Thr	Pro	Arg	Thr	0.29	Zhu et al., 1996 Raftogianis et al., 1999
1A2*3	Ile	Leu	Arg	Asn	0.18	Ozawa et al., 1995 Raftogianis et al., 1999
1A2*4	Thr	Pro	Cys	Thr	0.008	Raftogianis et al., 1999
1A2*5	Thr	Pro	Arg	Asn	0.008	Raftogianis et al., 1999
1A2*6	Ile	Pro	Arg	Thr	0.008	Raftogianis et al., 1999

resulted in >60-fold variation in the K_m for p-nitrophenol, 10-fold variation in the K_m for PAPS, and ~50-fold variation in the IC_{50} for DCNP on p-nitrophenol sulfonation by these allozymes. Population studies have shown that ethnic differences exist in the expression of these enzymes. Carlini et al. (2001) showed that SULT1A2*1, 1A2*2, and 1A2*3 were present at frequencies of 0.51, 0.39, and 0.10, respectively, in the Caucasian population, whereas in the Chinese population the frequencies were 0.92 for SULT1A2*1 and 0.08 for SULT1A2*2 and the 1A2*3 allele was not detectable. In the African-American population the frequencies of SULT1A2*1, 1A2*2, and 1A2*3 were 0.64, 0.25, and 0.11 respectively (Carlini et al., 2001). In a model *Salmonella typhimurium* system, Meinl et al. (2002) have shown that these polymorphic forms of SULT1A2 have different activities toward certain aromatic amines and amides, which may affect individual susceptibility to these carcinogens. The most striking functional difference between this new SULT1A member, SULT1A2, and SULT1A1 and SULT1A3 is that it exhibited no activity toward dopamine as a substrate (Zhu et al., 1996). Ozawa et al. (1995) have shown that COS cell expressed SULT1A2 (ST1A2) to sulfonate p-nitrophenol, minoxidil, and β-naphthol at significantly lower rates than SULT1A1 (ST1A3). The same authors also reported a large difference in K_m values for p-nitrophenol sulfonation between SULT1A2 and SULT1A1: 20 μM for SULT1A2 and 1.4 μM for SULT1A1. Most studies have focused on the SULT1A1 and SULT1A3 enzymes, as the validity of SULT1A2 as a functionally significant enzyme is still questionable. Little data exist on the tissue distribution of this enzyme, and mRNA profiles suggest lower expression levels in most tissues and cell lines tested compared to SULT1A1 and SULT1A3 (Dooley et al., 2000). These reverse transcription PCR (RT-PCR) studies also suggest the presence of a SULT1A2 mRNA species that is incorrectly spliced, to contain an intron. This may result in at least some portion of the SULT1A2 mRNA never being translated into a functional protein (Dooley et al., 2000). While Figure 10.2 shows that SULT1A2 protein is not detectable in the ten human liver cytosols tested, it nonetheless is interesting to note that Ozawa et al. (1995) were able to show an approximate twofold difference in the levels of both SULT1A2 (ST1A2) and SULT1A1 (ST1A3) mRNA levels in livers from six human subjects.

HUMAN SULT1A3

Initial attempts in our laboratory to clone SULT1A3 from a human liver cDNA library were unsuccessful. However, on screening a human brain cDNA library, two cDNAs were isolated that shared high sequence homology with SULT1A1 (HAST2 and HAST3; Zhu et al., 1993a). The HAST3 sequence cloned encoded a 34-kDa protein that was 93% similar to SULT1A1 and has subsequently, under the new nomenclature, been termed SULT1A3 (Raftogianis, 2003; Zhu et al., 1993a). Characterization of the SULT1A3 expressed protein by DCNP inhibition, substrate specificity, and thermal stability studies was consistent with it being the thermolabile or M-form of SULT (Reiter et al., 1983; Veronese et al., 1994). It has been shown to be the major SULT responsible for the sulfonation of dopamine in the human body. The fidelity of the SULT1A3 sequence was initially confirmed by Wood et al. (1994; TL-PST) and then subsequently by Jones et al. (1995; m-PST) and Ganguly et al.

(1995; hM-PST; Table 10.2). Bernier et al. (1994b) also isolated this cDNA but originally reported it as an estrogen SULT (hEST). To date, only one polymorphism has been reported for the SULT1A3 coding region (Lys234Asp; Iida et al., 2001). As indicated above, all of the SULT1A family share high sequence homology at the amino acid level but vary markedly in their substrate preferences.

STRUCTURE AND FUNCTION OF THE HUMAN SULT1A SUBFAMILY

Overall Crystal Structure

The crystal structure of the major catecholamine SULT, SULT1A3, was the first three-dimensional structure of a human cytosolic SULT solved (Bidwell et al., 1999; Dajani et al., 1999a). Prior to this, Negishi's group had published the crystal structures of the mouse cytosolic estrogen SULT (Kakuta et al., 1997) and the SULT domain of the human membrane-bound heparan sulfate *N-deacetylase/N-SULT* 1 (Kakuta et al., 1999). Since then, additional crystal structures of human cytosolic SULTs have been published: the hydroxysteroid SULT, SULT2A3 (Pedersen et al., 2000), the dehydroepiandrosterone sulfotransferase, DHEA-ST (Rehse, et al., 2002), the estrogen SULT, SULT1E1 (Pedersen et al., 2002), and the phenol SULT SULT1A1 (Gamage et al., 2003). The Bidwell et al. (1999) structure of SULT1A3 diffracted to 2.4Å and was solved with a sulfate ion in the active site, whereas the Dajani et al. (1999a) structure diffracted to 2.5Å and was complexed with 3'-phosphoadenosine 5'-phosphate (PAP). Significantly, both structures showed large stretches of disorder that account for approximately 25% of the SULT1A3 structure (Figure 10.4a; residues 64-77, 84-99, 216-261). This is thought to be due to the lack of a bound substrate.

To gain more insight into the structural and functional features of SULT1A subfamily members we have recently solved the crystal structure of SULT1A1 (Gamage et al., 2003). The SULT1A1 cDNA was expressed as an N-terminal hexa-histidine tagged protein in *Escherichia. coli*, purified, and crystallized in the presence of the desulfonated cofactor PAP and the model xenobiotic substrate *p*-nitrophenol. The crystal diffracted to 1.9Å, and the structure was fully solved except for seven residues missing from the N terminus. The crystal structure clearly shows binding of the PAP and two molecules of *p*-nitrophenol in an extended substrate-binding pocket. From both the crystal structures of human SULT1A1 and SULT1A3, we can conclude that they are comprised of a core α/β domain, similar to other SULTs, which forms the backbone of a central five-stranded parallel β sheet surrounded on either side by helices (Figure 10.4; see color photo insert following p. 210).

PAP Binding

From the initial cloning and sequence alignment studies of SULTs from a variety of species, it became apparent that there were at least four highly conserved regions throughout phylogeny (Varin et al., 1992; Weinshilboum et al., 1997). A number of laboratories utilized these data to construct chimeric cDNAs, performed site-directed

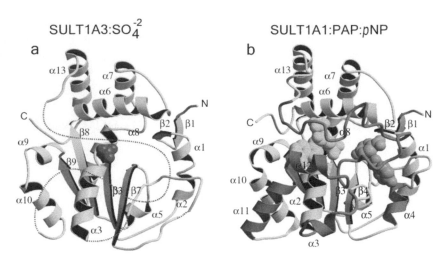

FIGURE 10.4 Crystal structure of human (a) SULT1A3 complexed with sulfate (pink) and (b) SULT1A1 complexed with PAP (green) and *p*-nitrophenol (*p*NP[1] orange and *p*NP[2] blue). Disordered regions of SULT1A3 are shown in blue dotted lines (Gamage, N.U., Duggleby, R.G., Barnett, A.C., Tresillian, M., Latham, C.F., Liyou, N.E., McManus, M.E., and Martin, J.L. (2003) *J Biol Chem* 278:7655–7662. With permission). (See color photo insert following p. 210.)

mutagenesis on these regions, and carried out affinity labeling to gain insight into the substrate and PAPS-binding sites of guinea pig (Komatsu et al., 1994), rat (Tamura et al., 1997), plant (Marsolais and Varin, 1995; Marsolais et al., 1999), and human (Radominska et al., 1996) SULTs. Initially, it was suggested that a motif with the sequence GxxGxxK (GMAGDWK in both SULT1A1 and SULT1A3), which is common to all SULTs, was critical for the interaction between SULT and the cofactor PAPS and considered as the P-loop (Chiba et al., 1995; Driscoll et al., 1995; Komatsu et al., 1994). However, the residues that are in the P-loop, a nucleotide-binding motif found in ATP- and GTP-binding proteins, actually correspond to residues TYPKSGT (45-51) in mouse SULT1E1 (Kakuta et al., 1997). However, results from Zheng et al. (1994), employing chemical affinity labeling of a rat SULT1A SULT IV by an ATP analog, showed covalent attachment to Lys65 and Cys66 in the N-terminal portion of the protein. Following mutagenesis studies on the plant flavonol 3-SULT, Marsolais and Varin (1995) concluded that both Lys59 and Arg276 are involved in PAPS binding through ionic interactions. These results were in part confirmed by Kakuta et al. (1997) when the crystal structure of the mouse estrogen SULT was published (Kakuta et al., 1997). This structure showed that PAP binding involves residues 257-259 (Arg, Lys, Gly) at the beginning of the GXXGXXK region, a second region before the α-helix 6 loop (Arg130) and on α-helix 6 (Ser138). It also showed that the position of the adenine ring of PAPS is determined by Trp53, Thr227, and Phe229.

In studies on SULT1A1 and SULT1A3, we have shown that the inactive cofactor PAP binds in a similar manner to that found by Kakuta et al. (1997) for mouse SULT1E1. The PSB loop (residues 45-51) interacts with the 5′-phosphate of the PAP molecule (Figure 10.5). The amino acid sequence (45-TYPKSGT-51), which

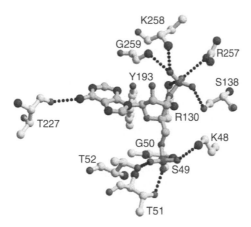

FIGURE 10.5 Human SULT1A1 interactions with PAP. Residues from the PSB loop (K48, S49, G50, T51, T52) form H-bonds to the 5′-phosphate of PAP. Residues R130, S138, R257, K258, and G259 form interactions with the 3′PB site. Residues that are H-bonded to the N6 of the adenine ring are T227 and Y193.

forms the classical strand-loop-helix motif found in all other SULT structures (Kakuta et al., 1997; Yoshinari et al., 2001), is also conserved in the human SULT1A subfamily (Figure 10.3). The backbone amides as well as the side chains of residues 48 to 51 are hydrogen bonded to the 5′-phosphate of PAP (Figure 10.5). In particular Lys48, which hydrogen bonds with the leaving oxygen of the 5′-phosphate group, has been shown by Yoshinari et al. (2001) to act as a general acid in catalysis. This residue is conserved in both SULT1A1 and SULT1A3 and nearly all other SULTs (Kakuta et al., 1998). Therefore, this binding site is considered to play a role in the recognition of PAPS and orientates the sulfate group for sulfonate transfer.

The 3′-phosphate-binding site (3′PB; residues 130, 138, and 257-259) of SULT1A1 is well ordered, showing a conserved strand and helix motif, a structural feature found in all other SULT structures (Kakuta et al., 1997; Pedersen et al., 2000). However, in the human SULT1A3 structure this site is disordered (Bidwell et al., 1999; Dajani et al., 1999a). In the SULT1A1 crystal structure, the two residues of the conserved strand-helix motif (Arg130 at the end of strand 5 and Ser138 from helix 8) interact directly with the 3′-phosphate group of PAP (Figure 10.5). The side-chain interaction of Ser138 with 3′-phosphate of PAP is observed in all known SULT structures. The human SULT1E1 structure, which was solved recently by Pedersen et al. (2002) in the presence of PAPS, showed that Ser138 prevents PAPS undergoing hydrolysis in the absence of substrate. Negishi et al. (2001) have postulated that the introduction of the 3′-phosphate group on the sulfonate donor could be an important force that has evolved in all SULTs. In the SULT1A1 structure, residues 257-RKG-259 at the beginning of the conserved GxxGxxK motif (Figure 10.3) are also found within hydrogen-bonding distance of the oxygen atoms of 3′-phosphate (Figure 10.5). Similar to the mouse SULT1E1 structure, the conserved residues Trp53 and Phe229 form a parallel ring stacking arrangement with the adenine ring of PAP molecule and are stabilized by hydrogen-bond interactions of

Thr227 (strand 12) and Thr193 (strand 6). Thus, SULT1A1 binds PAP in a proper orientation for catalysis similar to other known SULT structures. This conserved catalytic core of the SULT1A1 structure provides further evidence that both cytosolic and membrane-bound SULTs (the SULT domain of human *N-deacetylase/N-SULT* 1) have similar structural motifs that are responsible for PAP binding. Based on this structural similarity, Yoshinari et al. (2001) have concluded that both cytosolic and membrane-bound SULTs probably evolved from a common ancestral gene.

Substrate binding

The substrate-binding region of the SULT1A subfamily will mainly be reviewed based on the recently solved SULT1A1 structure. Figure 10.6, a-c (see color photo insert following p. 210), shows that a deep hydrophobic pocket comprises the substrate-binding site for cytosolic SULTs, whereas the presumed substrate-binding site of the human membrane-bound NST1 (SULT domain of human *N-deacetylase/N-SULT1*), which is responsible for the sulfonation of large molecules such as carbohydrates, glucosaminylglycans, and proteins, has a large open cleft (Figure 10.6d; see color photo insert following p. 210). However, as mentioned above, the resolved structure of SULT1A1 revealed an L-shaped substrate-binding pocket that is larger than that anticipated from the disordered SULT1A3 structure. The residues lining the substrate-binding pocket of SULT1A1 were found to be well ordered and predominantly hydrophobic (Phe81, 142, 24, 84, 76, 247, 255, Ile89, His149, Tyr169, Tyr240, Ile21, Ala146, 86, Met248, Met77, Val243, Pro90, and Val148) and are contributed by helices 1 and 6, strands 2 and 4, and several loop regions (Figure 10.3 and Figure 10.7). Figure 10.3 demonstrates that residues located at positions 76, 77, 84, 86, 89, 146, 148, 149, 247 in the binding pocket are not conserved among SULT1A family members, which is consistent with the varying substrate specificity of these enzymes.

Figure 10.7 (see color photo insert following p. 210) shows that two *p*-nitrophenol molecules are bound to the substrate-binding pocket of SULT1A1. The more tightly bound and deeply buried *p*-nitrophenol molecule (*p*NP[1]) is positioned near the active site. The two highly conserved phenylalanines at positions 81 and 142 appear to form a substrate access gate for *p*NP[1] (Figure 10.7). Petrotchenko et al. (1999) showed that these two residues play a critical role in maintaining the proper structure of the binding pocket. The highly conserved residues His108 and Lys106 of SULT1A1 form hydrogen bonds with the SO_3^{1-} accepting the hydroxyl group of *p*NP[1]. The nitro group of *p*NP[1] interacts with a water molecule and forms van der Waals (vdW) interactions with Phe247, Met248, and Val148, whereas the weakly bound *p*NP[2] molecule does not interact with the catalytic residues but forms vdW interactions with Ile89 and Phe247. In the SULT1A3 structure, similar structural elements are utilized to form the substrate-binding pocket although the residues in loop region 216-261 are disordered (Figure 10.4a). Another unusual feature of the SULT1A3 structure is that part of the presumed substrate-binding pocket is occupied by residues 86-90 from a symmetry related molecule. In the SULT1A1 structure, these residues line the binding pocket of the *p*NP[2] molecule. The disorder that is observed in the SULT1A3 structure is apparently due to lack of the substrate, which may be necessary for disorder–order transition in the substrate-binding pocket in these enzymes.

a   b

c   d

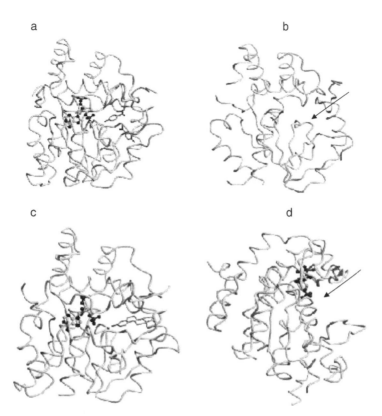

FIGURE 10.6 Crystal structures of four SULTs showing the substrate-binding pocket. (a) Human SULT1A1 with two p-nitrophenol molecules and PAP bound (Gamage, N.U., Duggleby, R.G., Barnett, A.C., Tresillian, M., Latham, C.F., Liyou, N.E., McManus, M.E., and Martin, J.L. (2003) *J Biol Chem* 278:7655–7662. With permission). (b) Human SULT1A3 with SO_4^{2-} bound (Bidwell, L.M., McManus, M.E., Gaedigk, A., Kakuta, Y., Negishi, M., Pedersen, L., and Martin, J.L. (1999) *J Mol Biol* 293:521–530. With permission). (c) Mouse SULT1E1 with E_2 and PAP bound (Kakuta, Y., Pedersen, L.G., Carter, C.W., Negishi, M., and Pedersen, L.C. (1997) *Nat Struct Biol* 4:904–908. With permision). (d) SULT domain of human N-deacetylase/N-SULT 1 (NST1; Kakuta, Y., Sueyoshi, T., Negishi, M., and Pedersen, L.C. (1999) *J Biol Chem* 274:10673–10676. With permission). Substrate molecules are shown in orange and PAP molecule is shown in the ball-and-stick model. Arrow indicates the substrate-binding pocket. (See color photo insert following p. 210.)

REACTION MECHANISM

The precise mechanism of how a sulfonate group is transferred from the cofactor PAPS to the substrate has to a certain degree been in dispute. Initially, Duffel and Jakoby (1981) reported the sulfonation of p-nitrophenol by rat liver aryl transferase as a random Bi Bi mechanism in which substrate and PAPS bind in an independent manner. However, recently Duffel et al. (2001) explained substrate inhibition by p-nitrophenol in terms of a sequential mechanism with PAPS binding first followed by p-nitrophenol. Studies on the catalytic mechanism of recombinant human

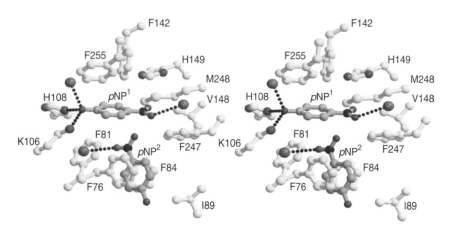

FIGURE 10.7 Stereo view of the active site of human SULT1A1 showing the hydrophobic nature of residues surrounding the ligands. *p*-nitrophenol[1] (*p*NP[1]) and *p*-nitrophenol[2] (*p*NP[2]) are shown in orange and blue, respectively. F142 and F81 form the substrate access gate. H-bonds are shown in black dotted lines (Gamage, N.U., Duggleby, R.G., Barnett, A.C., Tresillian, M., Latham, C.F., Liyou, N.E., McManus, M.E., and Martin, J.L. (2003) *J Biol Chem* 278:7655–7662. With permission). (See color photo insert following p. 210.)

SULT1E1 suggested that sulfonate transfer follows a random Bi Bi mechanism with two dead-end complexes (Zhang, 1998). These same authors also reported that the binding of β-estradiol (E$_2$) to human SULT1E1 resulted in two E$_2$-binding sites/catalytic subunit, suggesting that the enzyme contains an allosteric E$_2$-binding site. In contrast to the random Bi Bi mechanism discussed above, two studies using purified SULTs from human brain (Whittemore et al., 1986) and rhesus monkey liver (Barnes et al., 1986) concluded that the sulfonation reaction proceeds via a sequentially ordered Bi Bi reaction. Varin and Ibrahim (1992) have also reported a similar mechanism for a plant flavonol SULT. When the crystal structure of mouse SULT1E1 complexed with PAP and E$_2$ was solved, it became clear that the core structure resembles the uridine kinase enzyme, indicating a similar mechanism to phosphoryl transfer (i.e., S$_N$2 in-line displacement; Kakuta et al., 1997, 1998). The active site and the transition state mimicked by the mouse SULT1E1-PAP-vanadate ion structure complex provided further evidence for this in-line displacement mechanism for the sulfonate transfer reaction catalyzed by SULTs.

Figure 10.8 shows the sulfonate transfer mechanism based on data from the mouse SULT1E1 (Kakuta et al., 1997; Pedersen et al., 2002) and the human SULT1A1 crystal structure. His108, which is common to all SULTs, acts as a catalytic base and deprotonates the acceptor 3′ OH group of *p*-nitrophenol, which in turn converts it to a strong nucleophile that attacks the sulfur atom of PAPS. This leads to building of negative charge at the bridging oxygen and forces Lys48 to donate its proton to the bridging oxygen, and sulfonate dissociation occurs. The conserved Ser138 appears to prevent PAPS hydrolysis in the absence of substrate (Pedersen et al., 2002). Therefore, based on this structural data, the donor substrate

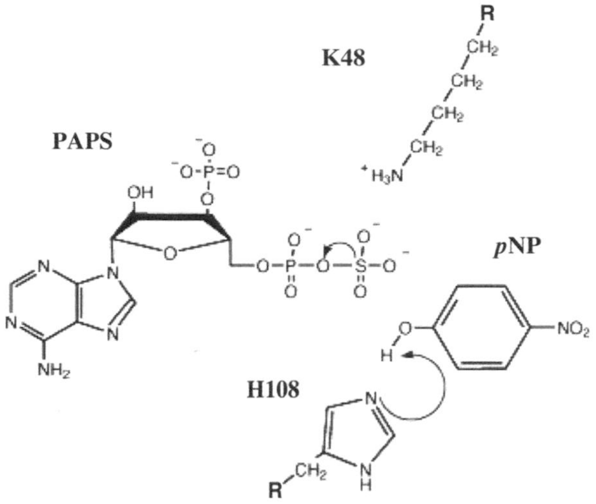

FIGURE 10.8 Schematic representation of sulfuryl transfer mechanism for human SULT1A1 with the substrate *p*-nitrophenol (*p*NP). The reaction takes place via S_N2 in-line displacement as described by Kakuta, Y., Pedersen, L.G., Carter, C.W., Negishi, M., and Pedersen, L.C. (1997) *Nat Struct Biol* 4:904–908. With permision.

PAPS binds first, followed by the binding of sulfonate acceptor substrate, favoring the in-line displacement mechanism (Negishi et al., 2001).

SUBSTRATE SPECIFICITY

While considerable progress has been made in recent times with the publication of five crystal structures of cytosolic SULTs (Bidwell et al., 1999; Dajani et al., 1999a; Gamage et al., 2003; Kakuta et al., 1997; Pedersen et al., 2000, 2002), we are still in the process of fully understanding the principles that underpin the substrate specificity of the individual enzymes. The key to understanding the overlapping but distinct substrate specificities displayed within the SULT1A subfamily most probably lies in the substrate-binding sites of these enzymes. In contrast to the PAPS-binding site, which consists of conserved amino acids, the differing substrate specificities of the closely related SULT1A subfamily members, SULT1A1, SULT1A2, and SULT1A3, suggest that the structure of their substrate-binding pockets has undergone modification during evolution. Therefore, the amino acid residues that contribute to differing substrate specificities are likely to reside in regions that display the most variability across members of this subfamily. The ability to investigate the role of critical amino acids in these variable regions has been aided by the substrate preferences of SULT1A subfamily members: SULT1A1 — high affinity for *p*-nitrophenol and low affinity for dopamine; SULT1A2 — lower affinity for *p*-nitrophenol and no activity toward dopamine as a substrate; and SULT1A3 — high affinity for dopamine and low affinity for *p*-nitrophenol (Dajani et al., 1998; Sakakibara et al., 1998; Veronese et al., 1994; Zhu et al., 1996).

Initial studies aimed at understanding the substrate specificity of SULT were carried out using chimeric constructs of flavonol (Varin et al., 1995) and rat hydroxysteroid SULTs (Tamura et al., 1997). These data suggest that the central regions spanning amino acids 92-194 and 102-164 in the plant and rat enzymes, respectively, determine their substrate specificities. Similarly, Sakakibara et al. (1998) used chimeric constructs to investigate the amino acids determining the substrate specificity of SULT1A1 and SULT1A3 isoforms. A sequence encompassing amino acid residues 84-148 was found to be the substrate-specific domain of both SULT1A1 and SULT1A3 and highlighted the importance of the variable Regions I (residues 84-89) and II (residues 143-148) in determining their distinct enzymatic properties. A related study from our laboratory showed that substrate affinities are mainly determined within the N-terminal end of SULT1A1 (HAST1), SULT1A2 (HAST4), and SULT1A3 (HAST3) and include two regions of high sequence divergence that were termed Region A (residues 44-107) and B (residues 132-164), respectively (Brix et al., 1999a). In parallel with the work of Sakakibara et al. (1998), it was shown by Dajani et al. (1998) that the change of a single amino acid, Glu146Ala, was sufficient to change the catalytic properties and substrate specificity of SULT1A3 so that it mimicked those of SULT1A1. In two related studies, we used a variety of phenolic substrates to functionally characterize the role of the amino acids at position 146 in both SULT1A1 and SULT1A3 (Brix et al., 1999a, 1999b). First, the mutant Ala146Glu in SULT1A1 yielded a SULT1A3-like protein with respect to the K_m for simple phenols such as p-nitrophenol. The mutation Glu146Ala in SULT1A3 yielded a SULT1A1-like protein with respect to the K_m for both phenols and monoamine compounds (e.g., dopamine). These data provided strong evidence that residue 146 is crucial in determining the substrate specificity of both SULT1A1 and SULT1A3. Further, a negatively charged glutamic acid (Glu) at position 146 is crucial for the recognition of dopamine (Brix et al., 1999b).

The importance of specific amino acid interactions in determining the high activity of SULT1A3 toward dopamine as a substrate has been further highlighted by molecular modeling studies (Dajani et al., 1999a; Yoshinari et al., 2001). Unfortunately, these studies have relied on the crystal structure of the mouse SULT1E1 because the structure of SULT1A3 with bound substrate is not currently available (Bidwell et al., 1999; Dajani et al., 1999a). Therefore, we constructed a computer model of SULT1A3 using the coordinates of SULT1A1, which has 93% sequence identity to SULT1A3 (Barnett et al., 2004). Our SULT1A3 model superimposes on the SULT1A3 crystal structure with a root mean square deviation of 1.1Å. PAPS was modeled into SULT1A3, based on the SULT1E1-PAPS structure (PDB code 1HY3; Pederson et al., 2002), and dopamine was docked. With a small shift in orientation, either the 3-OH or 4-OH of dopamine could be placed for sulfonation. However, 3-O sulfonated dopamine is reported to be the predominant metabolite in human (Dajani et al., 1999a); therefore, dopamine was orientated for sulfonation at the 3-O position. As in the SULT1A1 structure, Phe81 and Phe142 form a substrate access gate. In contrast to SULT1A1, the substrate-binding pocket is comprised of charged residues such as Glu and Asp. However, as in SULT1A1, the substrate-binding pocket is large enough to accommodate two molecules of dopamine

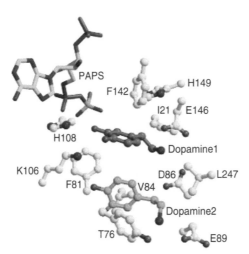

FIGURE 10.9 The substrate-binding pocket of the modeled SULT1A3 structure showing the binding mode of PAPS and two molecules of dopamine. Ligands and residues are represented as ball-and-stick models. Dopamine 1 is shown in dark green and dopamine 2 in light green. The cofactor PAPS is shown as stick model. Residues not shown for clarity are Tyr240, Ile21, Phe24, Pro90, Val243, Met248 (Barnett, A.C., Tsvetanov, S., Gamage, N., Martin, J.L., Duggleby, R.G., and McManus, M.E. (2004) Active site mutations and substrate inhibition in human SULT 1A1 and 1A3. *J Biol Chem* 279:18799–18805. With permission). (See color photo insert following p. 210.)

(Figure 10.9; see color photo insert following p. 210). In this model, the side chain of Glu146 is placed within hydrogen-bonding distance of the amino group of the first dopamine molecule to form a charge interaction, further highlighting the importance of this interaction in dopamine sulfonation (Dajani et al., 1999a; Yoshinari et al., 2001). This interaction has been confirmed by functional studies by Brix et al. (1999b) who have shown that by changing the Glu at position 146 of SULT1A3 to a glutamine (Gln), thereby neutralizing the negative charge at this position, a 360-fold decrease occurred in the specificity constant for dopamine. Further, Dajani et al. (1999a) have reported that mutating Glu146 to an alanine (Ala) in SULT1A3 significantly reduces the K_m values for a range of 4-substituted phenols, suggesting that this residue plays a central role in limiting substrate binding and access to the active site. In a more recent paper, Liu et al. (2000) have shown that the concerted action of three mutations (Asp86Ala, Glu89Ile, and Glu146Ala) is sufficient for the conversion of the substrate phenotype of SULT1A3 (M-PST) to that of SULT1A1 (P-PST). In our SULT1A3 model, Glu89 interacts with the amino group of the second dopamine molecule and Asp86 is positioned close to the first dopamine molecule, in agreement with the above observations. Recent studies have shown that Asp86 is involved in Mn^{2+} stimulation of the Dopa/tyrosine activity of SULT1A3 (Pai et al., 2003). However, Mn^{2+} exerts a smaller stimulatory effect on dopamine sulfonation by binding directly to Asp86 without making a complex with dopamine. This again highlights the importance of these residues, and it is interesting that in the active site of SULT1A1, we have also noted the importance of Ala86 and Ile89. In the crystal structure of SULT1A1 (Figure 10.7), Ile89 has been shown to form a critical interaction with the pNP^2 molecule. While it has been possible to convert the SULT1A3 substrate phenotype to a SULT1A1-like functional protein, achieving the reverse has not been possible to date (Brix et al., 1999b; Liu et al., 2000).

KINETICS OF HUMAN SULT1A1

Substrate inhibition has been reported previously for SULT1A enzymes (Raftogianis et al., 1999; Reiter et al., 1983) though the published studies have generally assumed a Michaelis–Menten model to explain the kinetics of these enzymes (Brix et al., 1999a; Lewis et al., 1996). The kinetic implications of the presence of two p-nitrophenol molecules in the crystal structure of SULT1A1 were investigated (Figure 10.10, a and b). A slight deviation from Michaelis–Menten kinetics (broken line) is observed at low substrate concentrations (Gamage et al., 2003). This could suggest that some positive cooperativity is present for p-nitrophenol binding. The most pronounced feature is the substrate inhibition occurring at higher substrate concentrations (above 2 μM). According to the general kinetic model constructed (Figure 10.10c), p-nitrophenol can bind to the enzyme at site 1 or 2, and occupancy of site 1 does not prevent subsequent binding at site 2. There are two catalytically competent species, ES_1 and ES_1S_2, that form the EP and EPS_2 enzyme-product complexes with rate constants k_1 and k_2, respectively. EP releases the product directly (dissociation constant K_p), while release from EPS_2 requires prior release of p-nitrophenol from site 2 (dissociation constant K_{ps2}). The model fits well to the experimental data as shown by the lines in Figure 10.10, a and b. Therefore, the presence of two molecules in the active site revealed by the structure is a real property of the SULT1A1 enzyme, and substrate inhibition at high concentrations of p-nitrophenol is due to the impeded catalysis when both binding sites are occupied.

The structural model of SULT1A3 showed that the substrate-binding pocket is large enough to accommodate a second molecule of dopamine (Figure 10.9). Similar to other SULT1A enzymes, SULT1A3 exhibits substrate inhibition at high concentrations of dopamine (Ganguly et al., 1995). Based on our kinetic analysis of sulfonation of dopamine by SULT1A3, we suggest that the inhibition we observed with dopamine is also caused by binding of a second substrate molecule in the substrate-binding pocket.

Table 10.1 shows that in addition to sulfonating a range of small molecular weight compounds, SULT1A isoforms are capable of metabolizing larger substrates such as 17β-estradiol (E_2), iodothyronines, 1-hydroxymethylpyrene, 7-hydroxy-7,8,9,10-tet-rahydrobenzo(a)pyrene, 2-naphthylamine, N-hydroxy-2-acetylaminoflorene, and 2-hydroxylamino-5-phenylpyridine (Glatt et al., 2001; Hernandez et al., 1991; Llerena et al., 2001; Meinl et al., 2002). Further, Chou et al. (1995) have reported that the mutagens/carcinogens N-hydroxy-2-acetylaminofluore, N-hydroxy-2-aminoflourene, N-hydroxy-4, 4′-methylene-bis (2-chloroaniline), N-hydroxy-2-amino-1-methyl-6-imidazo[4,5-b]pyridine, and N-hydroxy-2-amino-6-methyldipyrido[1,2-a:3′,2′-d]imidazole are preferentially sulfonated by human SULT1A1. The binding of two molecules of p-nitrophenol in the active site of SULT1A1 highlights the large and very hydrophobic nature of the substrate-binding region of this enzyme. This binding

a  b

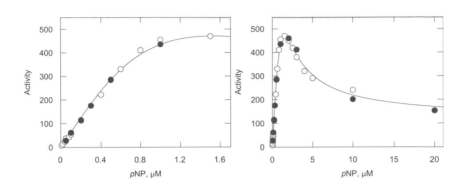

c

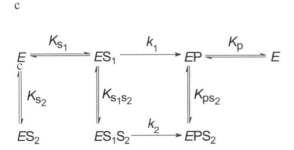

FIGURE 10.10 Kinetics of *p*NP sulfonation by human SULT1A1. (a) *p*NP concentrations up to 1.5 μM. (b) *p*NP concentrations up to 20 μM. Open and closed symbols show the results from two independent experiments. Each data point is the mean of duplicate or triplicate assays; the standard deviation is contained within the dimensions of the symbols. (c) Kinetic model of human SULT1A1. The enzyme (E) can bind *p*NP at site 1 to give ES_1 with a dissociation constant of K_{S1} or at site 2 to give ES_2 with a dissociation constant of K_{S2}. Occupancy of site 2 prevents *p*NP from binding to site 1, while occupancy of site 1 does not prevent *p*NP from binding to site 2 to give ES_1S_2 (Gamage, N.U., Duggleby, R.G., Barnett, A.C., Tresillian, M., Latham, C.F., Liyou, N.E., McManus, M.E., and Martin, J.L. (2003) *J Biol Chem* 278:7655–7662. With permission). The lines in panels (a) and (b) represent the best fit of this model to these data.

site can accept small flat aromatic compounds, larger L-shaped aromatics, and extended planar aromatic or aliphatic ring systems.

Recent studies have suggested that SULT1A1 is primarily responsible for the sulfonation of 3,3′-diiodothyronine (3,3′-T2) in the human placenta and developing liver (Li et al., 2001; Stanley et al., 2001). While other SULTs such as human SULT1B1 (Fujita et al., 1997; Wang et al., 1998), human SULT1C1 (Li et al., 2001), and human SULT1E1 (Kester et al., 1999) are also capable of sulfonating thyroid hormones, it is apparent from their expression levels that they play a minor role compared to SULT1A1 during the ontogeny of the liver (Richard et al., 2001). 3,3'-T2 has been shown to stimulate mitochondrial respiration in different tissues, which is not mediated by the T3 receptor (Li et al., 2001; Moreno et al., 1997). Therefore, it is possible that T2 and its sulfonated product could play a physiological function in the human fetus. Based on the geometry of the two bound p-nitrophenol molecules in the substrate-binding pocket of the SULT1A1 crystal structure, we docked 3,3'-T2 into the SULT1A1 structure (Figure 10.11a; see color photo insert following p. 210). The outcome was a catalytically competent binding model for 3,3'-T2 when Phe247 was changed to an alternate conformation to prevent steric clash with one of the iodine atoms of the substrate. This structural model supports the metabolic results of Richard et al. (2001) and suggests that iodothyronines are endogenous substrates of SULT1A1 (Gamage et al., 2003).

It is reported that SULT1A1 sulfonates the endogenous substrate 17β-estradiol (E_2) with a lower affinity than other substrates (Adjei and Weinshilboum, 2002; Falany and Falany, 1997). Therefore, E_2 was modeled into the SULT1A1 active site using the binding mode identified in the SULT1E1:PAP:E_2 structure (Figure 10.11b). However, E_2 cannot be accommodated in the SULT1A1 structure even with the altered conformation for Phe247, as it makes unfavorable interactions with residues involved in two loops (residues 146-154 between α6 and α7 and 84-90 immediately preceding α4; Figure 10.11c). These two loops close over the active site of SULT1A1 more tightly than they do in the SULT1E1 structure, thus restricting the space available for ligands that bind in an extended conformation (Figure 10.11d). Therefore, we propose that binding of extended fused ring systems such as E_2 to SULT1A1 results in an energy cost due to conformational rearrangement. Sulfonation could occur only if there was a conformational change in the SULT1A1-binding site, and thus the SULT1A1 substrate-binding site must be sufficiently plastic to accept flexible and rigid ligands, as shown in Table 10.1.

Through the elucidation of the crystal structures of SULT1A1 and SULT1A3, plus the kinetic analysis of chimeric constructs and mutated enzymes, a degree of insight has been obtained into the critical amino acids that control the substrate specificity of these SULTs. However, considerably more work is required before we fully understand the distinct but overlapping substrate specificities of SULT1A subfamily members. For example, the reasons SULT1A2 has a low affinity for p-nitrophenol and no dopamine activity, although it shares 97% and 94% homology to SULT1A1 and SULT1A3 respectively, is yet to be elucidated. Further, the fact that SULT1A3 is capable of sulfonating a range of substrates from smaller catechols to larger endogenous substrates, such as iodothyronines (Kester et al., 1999), suggests that its substrate-binding pocket may be flexible, similar to that of SULT1A1. In general, the substrate specificity in SULT1A subfamily members appears to be determined by both binding affinity and proper positioning of a substrate in the active site. For example, SULT1A1 has a predominant hydrophobic-binding pocket, which aids its interaction with substrates such as p-nitrophenol. On the other hand, the substrate-binding pocket of SULT1A3 appears to be less hydrophobic, and specific charge interactions between functional groups govern the basis of its substrate specificity. Overall, the above data suggest some commonality between the

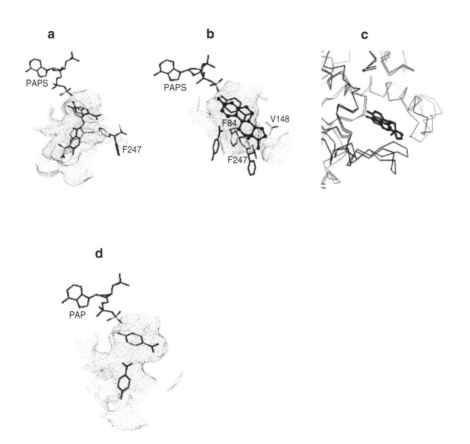

FIGURE 10.11 Human SULT1A1 active site plasticity. (a) T2 (yellow) docked into the active site of human SULT1A1, solvent accessible surface was calculated with Phe247 in an alternate rotamer conformation (pink). p-nitrophenol[1] (orange) and p-nitrophenol[2] (blue) are shown to indicate the orientation. PAPS is modeled at the cofactor-binding site. (b) Solvent accessible surface was calculated for E_2 with Phe247 having an alternate conformation (pink). E_2 makes clashes with Val148, Phe84, and the alternate conformation of Phe247. (c) Superimposition of human SULT1A1 and mouse SULT1E1 showing the striking variation in loop region. (d) The solvent accessible surface for the two pNP molecules as in the human SULT1A1 structure (Gamage, N.U., Duggleby, R.G., Barnett, A.C., Tresillian, M., Latham, C.F., Liyou, N.E., McManus, M.E., and Martin, J.L. (2003) *J Biol Chem* 278:7655–7662. With permission). (See color photo insert following p. 210.)

SULT1A1 and SULT1A3 active sites, which is probably not unexpected since they metabolize many of the same substrates, albeit with different affinities and activities.

TISSUE DISTRIBUTION OF THE HUMAN SULT1A SUBFAMILY

A number of studies have employed metabolic probes and immunohistochemistry techniques to show that SULT1A SULTs exhibit a wide tissue distribution, including liver, lung, brain, skin, platelets, gastrointestinal tissues, and kidney (Cappiello et al., 1990; Heroux et al., 1989; Hume and Coughtrie, 1994; Kudlacek et al., 1995; Zou et al., 1990). However, it was not until 1995, when SULT1A2 was first cloned (Ozawa et al., 1995), that we definitively knew the SULT1A subfamily contained three members, which share >90% sequence identity at the amino acid level. This high sequence identity together with the overlapping substrate specificity of SULT enzymes has compromised the use of immunohistochemistry and metabolic probes to study their tissue and cellular localization. While the early studies on SULT1A localization relied heavily on the predictive nature of *p*-nitrophenol and dopamine-specific SULT activity assays in tissues, they nonetheless gave us significant insight into the tissue-specific expression patterns of these enzymes (Weinshilboum, 1986). For example, it was shown that blood platelets exhibited high SULT1A1 activity (TS-PST; phenol SULT activity) and that it could be correlated with the sulfonation profile observed in other tissues such as brain, liver, kidney, and small intestine (Abenhaim et al., 1981; Anderson et al., 1981; Glatt et al., 2001; Hart et al., 1979; Sundaram et al., 1989; Young et al., 1985). In contrast, no correlation was found between SULT1A3 activity (TL-PST; monoamine SULT activity) in platelets and that observed in brain, liver, or small intestine tissue samples (Campbell et al., 1987; Sundaram et al., 1989; Young et al., 1985). Such studies were also the first to demonstrate the large interindividual differences in phenol SULT activity within the human population and the importance of genetic polymorphisms in controlling SULT1A activity in human tissues (Price et al., 1988, 1989; Weinshilboum, 1992).

It has been possible to show, using an array of approaches such as hybridization histochemistry, immunoblotting (Figure 10.2), and RT-PCR analysis, that SULT1A1, SULT1A2, and SULT1A3 vary in their tissue localization. For example, in the adult liver, SULT1A1 is expressed at high levels, but SULT1A2 and SULT1A3 are almost undetectable (Eisenhofer et al., 1999; Heroux et al., 1989; Richard et al., 2001; Windmill et al., 1997, 1998). In our laboratory, we have used hybridization histochemistry, immunohistochemistry, and immunoblot analysis to study the localization of SULT1A1 and SULT1A3. For localization studies using hybridization histochemistry, we have employed a full-length (1155 bp) SULT1A1 cDNA and a 198 bp PCR fragment (1175-1372) specific for the 3′ end of the untranslated region of SULT1A3. Hybridization studies showed that the SULT1A3 probe was specific, but the full-length SULT1A1 probe detected all SULT1A subfamily members (Windmill et al., 1998). For the identification of SULT1A-specific protein, a polyclonal antibody was raized in the rabbit against the *E. coli*-expressed recombinant human SULT1A1 protein (amino acids 31-286). This antibody was not specific for the SULT1A1 enzyme but was able to immuno-react with all three proteins of the SULT1A subfamily. However, on polyacrylamide gel electrophoresis and subsequent immunoblotting, it is possible to distinguish between the three isozymes based on differential migration on the gel. The presence of both SULT1A mRNA and protein was observed in liver, gastrointestinal tract (stomach, small intestine, and colon), and lung. From immunoblotting analysis, SULT1A1 was predominant in the liver, and high levels of mRNA and protein were observed across the acinus. SULT1A1 and SULT1A3 protein could be detected on immunoblots in cytosolic fractions of stomach, small intestine, and colon (Windmill et al., 1998). From the histological studies, SULT1A mRNA and protein were detected in epithelial cells lining the lumen of the stomach and the gastric pits and in the epithelial cells lining the lumen surface and the crypts of Lieberkuhn of the small intestine and colon. Similarly, SULT1A1 and SULT1A3 were detected in lung cytosols, and histological studies using hybridization histochemistry and immunohistochemistry localized SULT1A mRNA and protein to the epithelial cells of the respiratory bronchioles. The widespread localization of SULT1A mRNA and protein throughout the human gastrointestinal tract and lung suggests they may play a significant role in the extrahepatic detoxification and activation of drugs and xenobiotics (Windmill et al., 1998; Windmill, K.F., Hall, P.M., McManus, M.E., unpublished data).

Enzyme kinetic data obtained from purified recombinant SULT protein have enabled the modeling of catalytic function in tissue cytosols by choosing substrate concentrations that are close to K_m values of the individual SULT1A isozymes. For example, Richard et al. (2001) used 3.3 µM p-nitrophenol and 4.7 µM dopamine, values close to the K_m values of SULT1A1 and SULT1A3 for the respective substrates, to assess the sulfonation of these compounds in human fetal and adult tissues. The authors showed a correlation between the sulfonation of 3,3′-T2 and that of p-nitrophenol, but not to dopamine sulfonation, and concluded that the enzyme responsible for 3,3′-T2 in the tissues tested was SULT1A1 (Richard et al., 2001). Although immunohistochemical detection of tissue sections was not specific for either isoform, immunoblotting of tissue cytosols was able to distinguish the enzymes based on their differential migration on SDS-PAGE. Together with kinetic data, it was revealed that in the fetal brain, sulfonation of dopamine and SULT1A3 protein levels were very low in all areas tested. The sulfonation of 0.1 µM 3,3′-T2 in the brain was shown to be highest in the choroid plexus of the lateral ventricle, and the investigators suggested that the high expression of SULT1A1 in this region could be a mechanism of defense against portally transported toxins, as this region is the most highly vascularized in the developing brain (Richard et al., 2001). Earlier studies also showed SULT1A activity in the adult and fetal brain (Richard et al., 2001; Whittemore et al., 1985, 1986; Young et al., 1985). Young et al. (1985) reported highest dopamine and phenol SULT activity in the cerebral cortex. Phenol SULT activity was also apparent in the anterior pituitary, and activity was 6.5 times higher than in the parts of the brain that are of neuronal origin.

One interesting observation made by Richard et al. (2001) was the developmental expression difference apparent between SULT1A1 and SULT1A3 in lung and liver tissues. Assessing the dopamine and thyroid hormone sulfonation ability of fetal tissues, it was found that high liver SULT1A1 activity was generally retained into adulthood; however, lung activity reduced approximately 10-fold toward 3,3′-T2 (Gilissen et al., 1994; Richard et al., 2001). Sulfonation activity of dopamine was high in the fetal liver and lung but reduced significantly in the postnatal tissues (Pacifici et al., 1993; Richard et al., 2001). Protein levels confirmed these results, suggesting that SULT1A1 and 1A3 are abundantly expressed in the fetal liver and that the SULT1A3 enzyme almost disappears in the adult tissue (Richard et al., 2001). This same pattern has also been observed for SULT1A3 in the kidney (Cappiello et al., 1991). These data suggest an important role for these two SULTs in the protection of the fetus from exogenous toxins and in the homeostasis of hormones such as dopamine and iodothyronines. Immunoblotting of placenta cytosols revealed positive staining for both SULT1A1 and SULT1A3 enzymes, mainly from the cotyledon region where they may have an important role in the xenobiotic metabolism of potentially harmful chemicals entering the fetal circulation from the maternal side (Heroux et al., 1989; Stanley et al., 2001). No immunoreactive band was observed for SULT1A2 in this tissue. Stanley et al. (2001) also showed activities for both dopamine and p-nitrophenol in the placenta, and again p-nitrophenol sulfonation was correlated to that of 3,3'-T2. The available data suggest that SULTs may play a significant role in the phase II metabolism in the placenta, as other conjugating enzymes such as UDP-glucuronosyltransferases seem to be expressed at minimal levels in this tissue.

Another reproductive tissue that shows SULT1A1 activity is the endometrium, where levels do not seem to change according to the menstrual cycle; this is different from the pattern observed with the estrogen SULT (SULT1E1), whose expression appears to be regulated by progesterone (Falany and Falany, 1996). Although SULT1A1 and SULT1A3 have been found only at low levels in mammary gland tissue in immunohistochemical studies, it has been shown with activity assays and immunoreactivity studies that most breast cancer cell lines, including estrogen receptor positive and negative lines, express high levels of both SULT enzymes (Falany and Falany, 1996; Windmill et al., 1998). SULT1A RT-PCR and activity studies have also confirmed the presence of SULT1A1 and 1A3 in carcinoma cell lines, including Caco-2, HepG2, keratinocarcinomas, melanomas, fibrosarcomas, osteosarcoma, and osteoblast cells (Baranczyk-Kuzma et al., 1991; Dooley et al., 2000; Dubin et al., 2001; Satoh et al., 2000). It seems that SULT1A3 is the most abundant form found in the human colon carcinoma cell line Caco2, reflecting its dominance in the normal gastrointestinal tract, whereas SULT1A1 represents the most abundantly expressed SULT in human MCF-7 breast carcinoma cells (Falany et al., 1993; Satoh et al., 2000). RT-PCR studies show that SULT1A1 and SULT1A3 are transcribed ubiquitously throughout many epithelial tissues and cell lines (Dooley et al., 2000). One role of the SULT1A isoforms in the skin is the activation of the topically applied hair growth stimulant minoxidil by sulfo-conjugation, and their distribution in epithelial tissues may serve as both a defense and bioactivation mechanism for xenobiotics entering the body via this route (Dooley et al., 2000). Dooley and colleagues have pointed out that the pattern of SULT1A mRNA expression does not always translate to the formation of protein and have suggested that posttranscriptional and posttranslational modification events may take place. As mentioned above, little is known of the tissue distribution of SULT1A2; however, lower mRNA levels than the other SULT1As are found in liver, kidney, brain, lung, ovary, and some sections of the gastrointestinal tract, and recently SULT1A2 levels have been observed in some bladder tumors (Dooley et al., 2000; Glatt et al., 2001).

GENE STRUCTURE AND REGULATION

The three genes encoding the human SULT1A subfamily are all located in proximity to each other on the short arm of chromosome 16 (16p12.1-p11.2; Aksoy and

Weinshilboum, 1995; Her et al., 1996; Raftogianis et al., 1996). The *SULT1A1* and *SULT1A2* genes are arranged head to tail, approximately 10 kbp apart. *SULT1A1* is located most centromeric and *SULT1A3* most telomeric of the three *SULT1A* genes, the latter being separated from the *SULT1A2* gene by approximately 1.7 Mbp. All three genes contain seven coding exons and alternate untranslated first exons. Two alternatively transcribed exons have been identified for SULT1A1 and SULT1A2 mRNA species and three for SULT1A3 (Aksoy and Weinshilboum, 1995; Bernier et al., 1994a; Raftogianis et al., 1996; Zhu et al., 1993a, 1993b). Our laboratory was the first to identify this phenomenon, which does not influence the coding exons but is limited to the 5′-untranslated region and could potentially be a tissue-specific mechanism. We found that the SULT1A1 cDNA isolated from a human brain library contains 41 bp of untranslated region found immediately upstream of the ATG start codon on the *SULT1A1* gene, whereas the SULT1A1 cDNA isolated from a liver library contained an untranslated exon found 1075 bp from the ATG start codon (Zhu et al., 1993a, 1993b). Raftogianis et al. (1996) found an additional 270 bp untranslated region located 200 bp from the ATG start codon from a liver cDNA library. The untranslated exons of SULT1A2 are homologous to those identified for SULT1A1. SULT1A3 cDNA species found to date have been shown to contain an untranslated region found 1400 bp from the ATG start codon, which displays high sequence identity to the more distal untranslated exon identified for SULT1A1 and SULT1A2 (Aksoy and Weinshilboum, 1995). Another two untranslated exons found approximately 2000 bp from this first untranslated region on the *SULT1A3* gene were also identified in SULT1A3 cDNA species; however, all three untranslated exons have been identified in both brain and liver cDNA libraries, and we cannot conclude that these different SULT1A3 mRNA species are tissue specific (Aksoy and Weinshilboum, 1995; Bernier et al., 1994a; Wood et al., 1994; Zhu et al., 1993b). To date, it remains unclear whether the presence of the different SULT1A 5′-untranslated regions is a result of the utilization of different promoters located upstream of either untranslated exon or differential alternate splicing of mRNA species. We have shown in our laboratory that the distal untranslated region of the *SULT1A1* gene is part of the primary transcript found in the hepatocarcinoma cell lines HepG2 and Hep3B, as well as in primary human hepatocytes. Additionally, the sequence upstream of this untranslated region displayed very high promoter activity when cloned in front of a luciferase reporter vector and transfected into these cell lines. Very little promoter activity was observed for the sequences upstream of the other 5′ untranslated exons identified (Hempel, N., Wang, H., LeCluyse, E.L., McManus, M.E., Negushi, M., unpublished data). This was also shown by Bernier et al. (1996), where the sequence immediately upstream of the most distal 5′-untranslated region of the *SULT1A1* gene displayed higher promoter activity when cloned in front of a CAT reporter gene vector and transfected into kidney 293 cells than the sequence found immediately 5-′ of the ATG start codon. Also the sequence upstream of the homologous 5′untranslated region identified for *SULT1A3* displayed higher promoter activity than the sequences upstream of the other more distal untranslated exon (Bernier et al., 1994a).

On cloning, the human *SULT1A* genes were shown to lack canonical TATA and CCAAT boxes in proximity to their potential transcriptional start sites (Aksoy and

Weinshilboum, 1995; Her et al., 1996; Raftogianis et al., 1996). A lack of TATA boxes is observed in several SULT genes including the rat *SULT1A1* gene (Duanmu et al., 2001). Interestingly, all human *SULT1A* genes contain GC-rich regions in proximity to their transcriptional start sites, which is a common characteristic of TATA-less genes (Smale, 1997). It is therefore possible to speculate that this area within the *SULT1A* gene structure acts as the binding site for the basal transcriptional unit, which recruits RNA polymerase II. The characterization of the human *SULT1A* promoters is currently under investigation in our laboratory.

Very little is known about the regulation of SULT expression in human tissues, although the available data demonstrate that their tissue distribution and transcriptional regulation differs considerably from that observed in animal models. For example, rat SULT1A1 mRNA in primary hepatocytes was significantly increased after treatment with the glucocorticoid dexamethasone (Runge-Morris, 1998). This is contrary to the observation with similar treatment of human primary hepatocytes, which failed to cause a change in human SULT1A1 mRNA levels (Duanmu et al., 2002). This makes the extrapolation of animal data to humans, where many rodent isoforms exhibit dramatic sexual dimorphism (Dunn and Klaassen, 2000), of questionable value. The matter is further complicated by the fact that humans have three members of this subfamily, whereas all other species studied to date have a solitary *SULT1A1* gene.

From the data outlined above, we can conclude that SULT1A1 has a broad tissue distribution, whereas SULT1A3 is more restricted in its expression to tissues such as the intestine, platelets, and brain. A recent study by Richard et al. (2001) found that differential expression patterns exist between SULT1A1 and SULT1A3 during fetal development. These authors showed, using enzyme activity and immunohistochemical studies, that SULT1A3 is expressed at much higher levels in the fetal than adult liver, whereas SULT1A1 levels remain relatively constant (Richard et al., 2001). The factors determining this developmental difference in SULT1A3 regulation are currently unknown. Data on what controls SULT1A2 gene regulation are even more limited, and the endogenous substrate preferences of the SULT1A subfamily members do not provide significant links as to how these genes are regulated in human tissues.

Early work on the regulation of SULTs tested the effects of classical drug metabolizing enzyme inducers, such as glucocorticoids, polycyclic aromatic hydrocarbons, and phenobarbital, using bovine and rodent models (Runge-Morris, 1998; Runge-Morris et al., 1996, 1998; Schauss et al., 1995). The effects of glucocorticoids in the up-regulation of SULT transcription have been widely studied in the rat, particularly their ability to regulate hydroxysteroid SULTs (SULT2 subfamily). Different glucocorticoids and glucocorticoid-like chemicals have been shown to induce both bovine and rat SULT1A1 mRNA levels two to fivefold (Beckmann et al., 1994; Duanmu et al., 2000; Liu and Klaassen, 1996; Liu et al., 1996; Runge-Morris, 1998; Runge-Morris et al., 1996; Schauss et al., 1995). This is accompanied with increased phenol SULT activity and protein levels (Beckmann et al., 1994; Duanmu et al., 2000; Liu and Klaassen, 1996). The up-regulation of these enzymes by glucorticoids in rodents reflects the patterns observed recently with human SULT2A1 (Duanmu et al., 2001; Runge-Morris, 1998; Runge-Morris et al., 1996).

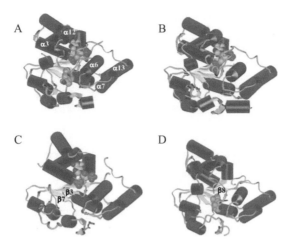

FIGURE 2.2 Folding pattern, secondary structural elements and location of bound ligands revealed by x-ray crystal structures of cytosolic sulfotransferases. (A) mouse Sult1e1 complexed with PAP and estradiol (Kakuta et al., 1997), (B) human SULT1E1 complexed with PAPS (Pedersen et al., 2002), (C) human SULT1A3 complexed with PAP (Dajani et al., 1999a), and (D) human SULT2A1 complexed with dehydroepiandrosterone (Rehse et al., 2002). Red cylinders represent α-helices, yellow arrows represent β-strands, ligands are shown as CPK-models. The figure was created with InsightII. Crystal coordinates for A, B and D are from PDB entry 1aqu.ENT, 1hy3.ENT and 1j99.ENT, respectively.

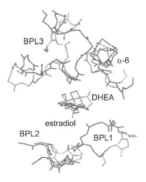

FIGURE 2.5 Peptide backbone structures of the regions lining the substrate binding pocket in mSult1e1 (cyan) with bound estradiol (red) and hSULT2A1 (magenta) with bound DHEA (green).

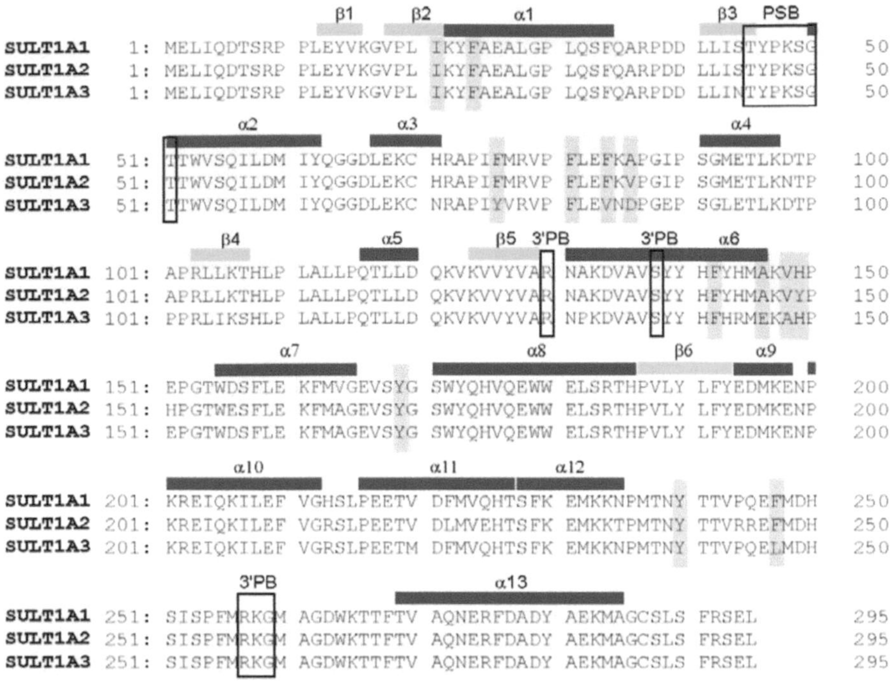

β1 β2 α1 β3 PSB
SULT1A1 1: MELIQDTSRP PLEYVKGVPL IKYFAEALGP LQSFQARPDD LLISTYPKSG 50
SULT1A2 1: MELIQDTSRP PLEYVKGVPL IKYFAEALGP LQSFQARPDD LLISTYPKSG 50
SULT1A3 1: MELIQDTSRP PLEYVKGVPL IKYFAEALGP LQSFQARPDD LLINTYPKSG 50
α2 α3 α4
SULT1A1 51: TTWVSQILDM IYQGGDLEKC HRAPIFMRVP FLEFKAPGIP SGMETLKDTP 100
SULT1A2 51: TTWVSQILDM IYQGGDLEKC HRAPIFMRVP FLEFKVPGIP SGMETLKNTP 100
SULT1A3 51: TTWVSQILDM IYQGGDLEKC NRAPIYVRVP FLEVNDPGEP SGLETLKDTP 100
β4 α5 β5 3'PB 3'PB α6
SULT1A1 101: APRLLKTHLP LALLPQTLLD QKVKVVYVAR NAKDVAVSYY HFYHMAKVHP 150
SULT1A2 101: APRLLKTHLP LALLPQTLLD QKVKVVYVAR NAKDVAVSYY HFYHMAKVYP 150
SULT1A3 101: PPRLIKSHLP LALLPQTLLD QKVKVVYVAR NPKDVAVSYY HFHRMEKAHP 150
α7 α8 β6 α9
SULT1A1 151: EPGTWDSFLE KFMVGEVSYG SWYQHVQEWW ELSRTHPVLY LFYEDMKENP 200
SULT1A2 151: HPGTWESFLE KFMAGEVSYG SWYQHVQEWW ELSRTHPVLY LFYEDMKENP 200
SULT1A3 151: EPGTWDSFLE KFMAGEVSYG SWYQHVQEWW ELSRTHPVLY LFYEDMKENP 200
α10 α11 α12
SULT1A1 201: KREIQKILEF VGHSLPEETV DFMVQHTSFK EMKKNPMTNY TTVPQEFMDH 250
SULT1A2 201: KREIQKILEF VGRSLPEETV DLMVEHTSFK EMKKTPMTNY TTVRREFMDH 250
SULT1A3 201: KREIQKILEF VGRSLPEETM DFMVQHTSFK EMKKNPMTNY TTVPQELMDH 250
3'PB α13
SULT1A1 251: SISPFMRKGM AGDWKTTFTV AQNERFDADY AEKMAGCSLS FRSEL 295
SULT1A2 251: SISPFMRKGM AGDWKTTFTV AQNERFDADY AEKMAGCSLS FRSEL 295
SULT1A3 251: SISPFMRKGM AGDWKTTFTV AQNERFDADY AEKMAGCSLS FRSEL 295

FIGURE 10.3 Sequence alignment of human SULT1A1, 1A2, and 1A3. Secondary structure elements are numbered based on SULT1A1 structure. Residues conserved in PSB loop and 3′PB site are boxed, and those that differ among the isoforms are shown in red. Residues that line the substrate-binding pocket are highlighted in yellow (based on SULT1A1 structure).

SULT1A3:SO$_4^{-2}$
SULT1A1:PAP:pNP
a
b
α13
α7
α6
N
C
β2
β1
β8
α8
α9
α1
β9
α10
β3
β7
α2
α5
α3
α12
β4
α11
α4

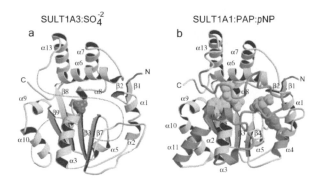

FIGURE 10.4 Crystal structure of human (a) SULT1A3 complexed with sulfate (pink) and (b) SULT1A1 complexed with PAP (green) and *p*-nitrophenol (*p*NP[1] orange and *p*NP[2] blue). Disordered regions of SULT1A3 are shown in blue dotted lines (Gamage et al., 2003; with permission from the *J Biol Chem*).

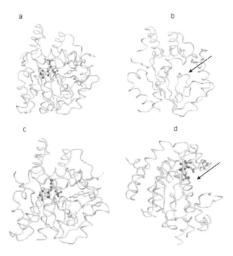

FIGURE 10.6 Crystal structures of four sulfotransferases showing the substrate binding pocket. (a) human SULT1A1 with two *p*-nitrophenol molecules and PAP bound (Gamage et al., 2003) (b) human SULT1A3 with SO_4^{-2} bound (Bidwell et al., 1999) (c) mouse SULT1E1 with E_2 and PAP bound (Kakuta et al., 1997) (d) sulfotransferase domain of human N-deacetylase/N-sulfotransferase 1 (NST1, Kakuta et al., 1999). Substrate molecules are shown in orange and PAP molecule is shown in ball and stick model. Arrow indicates the substrate binding pocket.

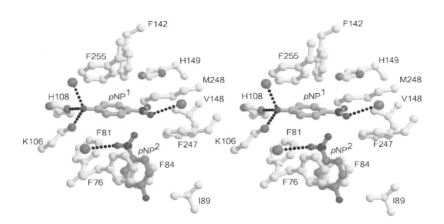

FIGURE 10.7 Stereo view of the active site of human SULT1A1 showing the hydrophobic nature of residues surrounding the ligands. *p*-nitrophenol[1] (*p*NP[1]) and *p*-nitrophenol[2] (*p*NP[2]) are shown in orange and blue, respectively. F142 and F84 form the substrate access gate. H-bonds are shown in black dotted lines (Gamage et al., 2003; with permission from the *J Biol Chem*).

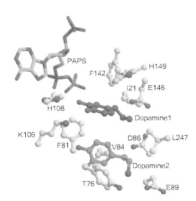

FIGURE 10.9 The substrate binding pocket of the modeled SULT1A3 structure showing the binding mode of PAPS and two molecules of dopamine. Ligands and residues are represented as ball-and-stick models. Dopamine 1 is shown in dark green and dopamine 2 in light green. The cofactor PAPS is shown as stick model. Residues not shown for clarity are Tyr240, Ile21, Phe24, Pro90, Val243, Met248 (Barnett et al., submitted).

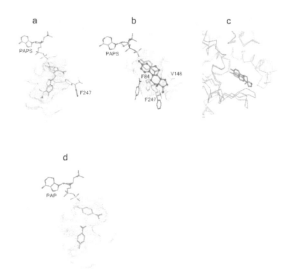

FIGURE 10.11 Human SULT1A1 active site plasticity. (a) T2 (yellow) docked into the active site of human SULT1A1, solvent accessible surface was calculated with Phe247 in an alternate rotamer conformation (pink). *p*-nitrophenol[1] (orange) and *p*-nitrophenol[2] (blue) are shown to indicate the orientation. PAPS is modeled at the cofactor binding site. (b) solvent accessible surface was calculated for E2 with Phe247 having an alternate conformation (pink). E2 makes clashes with Val148, Phe84 and the alternate conformation of Phe247. (c) Superimposition of human SULT1A1 and mouse SULT1E1 showing the striking variation in loop region. (d) the solvent accessible surface for the two *p*NP molecules as in the human SULT1A1 structure (Gamage et al., 2003; with permission from *J Biol Chem*).

Depending on ligand concentration used, this effect is due to either the glucocorticoid receptor or the pregnane X receptor (PXR; Duanmu et al., 2002; Runge-Morris et al., 1999; Sonoda et al., 2002). Interestingly, it has been observed in bovine bronchial epithelial cells that the stimulation of SULT1A1 levels by hydrocortisone can be reduced by administration of retinoic acid (vitamin A), which acts through the retinoic acid receptor (RAR; Beckmann et al., 1994). However, as mentioned above, this induction by glucocorticoids is not observed with human SULT1A1, as dexamtehasone treatment of primary human hepatocytes resulted in no significant change in mRNA levels of this enzyme (Duanmu et al., 2002). When the rat *SULT1a1* promoter was cloned in front of a luciferase reporter vector and transfected into primary rat hepatocytes, promoter activity increased sevenfold upon treatment with 10^{-7} M dexamethasone, due to the presence of a glucocorticoid response element on the rat *Sult1a1* promoter (Duanmu et al., 2001). We have carried out a similar experiment with the human *SULT1A1* promoter in HepG2 cells and have failed to show any change in promoter activity with dexamethasone, even with the glucocorticoid receptor or PXR cotransfected into the cells (Hempel, N., Wang, H., LeCluyse, E.L., McManus, M.E., Negishi, M., unpublished data). Additionally, we were not able to find a homologous glucocorticoid response element in the human *SULT1A1* gene, which differs markedly from the rat *Sult1a1* gene in its promoter region despite their high amino acid sequence homology (>70%).

There are several other examples of SULT regulation by steroids; however, this is not observed for the SULT1A1 enzyme, and generally little data are available on the human SULT1A isoforms. In the rat, DHEA has been shown to down-regulate SULT2A1 expression but to have no effect on SULT1A1 mRNA levels (Runge-Morris et al., 1996). Human Ishikawa endometrial adenocarcinoma cells exhibit no change in SULT1A1 and SULT1A3 activity and protein levels when treated with progesterone; however, estrogen SULT (SULT1E1) activity increased sevenfold following the same progesterone treatment in these cells (Falany and Falany, 1996). This inducibility of SULT1E1 can be observed in the endometrium during the menstrual cycle where its expression follows progesterone surges, with the highest levels being observed in the luteal phase (Rubin et al., 1999) — again emphasizing that progesterone has little effect on SULT1A regulation, as only very minor changes in SULT1A1 and SULT1A3 expression were observed during the menstrual cycle and in early pregnancy (Rubin et al., 1999). SULT1A1 has been shown to have substrate affinity for estrogens, particularly β-estradiol (Falany and Falany, 1997), which suggests that estrogen receptor ligands could be potential modulators of SULT1A1 regulation. An example of this is the significant induction of liver SULT1A1 mRNA after injections of rats with the breast cancer drug and partial estrogen agonist/antagonist tamoxifen (Hellriegel et al., 1996). Similarly, using Serial Analysis of Gene Expression (SAGE), Seth and colleagues found that upon treatment of the human breast cancer cell line ZR75-1 with tamoxifen, SULT1A1 and SULT1A2 mRNA levels were the only transcripts up-regulated from 8000 other genes tested (Seth et al., 2000). Treatment of these cells with β-estradiol also resulted in a significant increase in SULT1A mRNA levels, but this increase was observed much later in the time course of treatment (Seth et al., 2000). This suggests a potential role for estrogens in the regulation of the human *SULT1A1* and *SULT1A2* genes, and the role of the estrogen receptor on the human *SULT1A* promoters is under current investigation in our laboratory. Although human SULT1A1 has been implicated to be a major SULT responsible for thyroid hormone sulfonation (Richard et al., 2001), studies in thyroidectomized rats showed no altered mRNA levels of the rat SULT1A1 isoform (Dunn and Klaassen, 2000). This may correlate with the recent finding suggesting that the rat SULT1A1 enzyme displays no sulfonation activity toward thyroid hormones as substrates, unlike its human ortholog, but rather implicating rat SULT1B1 and rat SULT1C1 as the enzymes responsible for 3,3′-T2 sulfonation (Kester et al., 2003). The role of the barbiturate phenobarbital in the induction of various cytochrome P450 isoforms has been studied extensively. This compound exerts its action on transcription by activating the Constitutive Androstane Receptor (CAR; Sueyoshi and Negishi, 2001). Runge-Morris et al. (1998) found that rat SULT1A1 mRNA levels were decreased by about 40% in livers of rats given phenobarbital, with a similar reduction observed for the hydroxy-steroid SULT, SULT2A1, mRNA levels. Conversely, the mRNA levels of two other rat hydroxysteroid SULTs, SULT2A2 and SULT2A5, were increased by this treatment 4- and 1.6-fold, respectively (Runge-Morris et al., 1998). A decrease in SULT1A1 and SULT2A1 mRNA levels was observed in primary rat hepatocytes that were treated with phenobarbital and various phenobarbital-like inducers (Runge-Morris, 1998). Recently with the discovery of a novel potent human CAR activator Citco, Maglich et al. (2003) investigated the role of this compound on the induction of various phase I and II metabolic enzymes and found an 11-fold induction of SULT1A1 mRNA from human hepatocytes treated with Citco. This effect however was only observed with one of the three hepatocyte donors studied, and our own investigations into the role of CAR on the induction of human SULT1A1 and SULT1A3 failed to show a similar pattern of induction (Hempel, N., Wang, H., LeCluyse, E.L., McManus, M.E., Negishi, M., unpublished data). Other xenobiotic compounds that have been shown to influence rat SULT1A1 levels are the cytochrome P450 inducers β-napthoflavone and 2, 3, 7, 8-tetrachlorodibenzo-p-dioxin (TCDD) and the carcinogen 2-acetylaminofluorene (2AAF). A decrease in liver mRNA levels has been observed following treatment both *in vivo* and in cultured rat hepatocyte with these chemicals (Li et al., 2001; Ringer et al., 1994a, 1994b; Runge-Morris, 1998; Yerokun et al., 1992). Ringer et al. (1994a) observed that aryl SULT activity toward 2-AAF could in part be restored when rats were administered phenobarbital. The precise meaning of such observations in relation to the carcinogenicity of 2-AAF is not entirely clear, but sulfonation of the N-hydroxy-2-acetyl-aminofluorene is a major pathway in its activation to a highly reactive electrophile that can bind to DNA (Thorgeirsson et al., 1983). As can be seen from the above examples, little is known about the regulation of the human *SULT1A* genes, and an obvious species difference in the regulation of these isozymes is apparent.

CONCLUSION

In recent years, the analysis of primary structures of SULTs has enabled us to study the relationship of SULT genes from microbes to humans. However, while attendees at the 3rd International Sulfation Workshop held in Drymen, Scotland, agreed on a common nomenclature, we are still some way from seeing it adopted. For the uninitiated, this makes the ready comparison of work from a variety of laboratories difficult. This is particularly the case when studying SULT1A subfamily members that have been given a multitude of names in species including humans, mice, rats, rabbits, cows, and dogs (Nagata and Yamazoe, 2000). For a more recent analysis of this area, the review by Raftogianis et al. (2003) brings a logical approach to the nomenclature of SULTs. The reason humans possess three members of the SULT1A subfamily whereas rodents, cows, rabbits, and dogs have only a single *SULT1A1* gene is not totally clear. The three human paralogs of this subfamily sit adjacent to each other on chromosome 16, and it is possible to speculate that they arose through a gene conversion event following divergence of humans from the other lineages (Rikke and Roy, 1996). In general, human SULT1A1 and SULT1A3 have distinct but overlapping substrate specificities, and it would be reasonable to conclude that these two paralogs have been maintained in human evolution because they perform different but not necessarily exclusive functions. The physiological role of SULT1A2 is the least understood of the SULT1A subfamily members. It has been cloned from both human liver and colon tissue (Zhu et al., 1996), but outside of this very little is known about its tissue or cellular localization. At the amino acid level, SULT1A2 shares high sequence homology with SULT1A1 and is capable of metabolizing a similar range of phenolic xenobiotics. Further resolution of the role of this orphan enzyme in cellular metabolism will require more specific molecular probes and the elucidation, if any, of its endogenous substrates.

Early work by both Weinshilboum (1992) and Price Evans (1993) demonstrated marked interindividual variation in human SULT activity between patients. In particular, they demonstrated that variation in SULT1A activities in human platelets was controlled by inheritance, thereby providing the first evidence that genetic polymorphisms regulate the levels of SULT activity in human tissues (Weinshilboum, 1988). In a recent catalog of 320 single-nucleotide polymorphisms (SNPs), there were 27 involving SULT1A1, 20 involving SULT1A2, and one involving SULT1A3 (Iida et al., 2001). In relation to SULT1A1, 5 of the 27 SNPs are in coding exons, whereas in the case of SULT1A2, only 3 of the SNPs are in the coding region (Iida et al., 2001). The only SNP identified to date in SULT1A3 is just upstream of the stop codon in exon 8 (Iida et al., 2001). The functional and regulatory significance of these SNPs reported in SULT1A subfamily members is still being assessed (Carlini et al., 2001; Iida et al., 2001; Raftogianis et al., 1999); it is possible that SNPs in regions outside the open reading frame in promoters or introns could affect SULT1A expression patterns.

In light of previous studies on other drug-metabolizing enzymes (cytochromes P450, N-acetyltransferases, glutathione transferase) where inherited differences in enzymatic activities have been linked to disease states (Ingelman-Sundberg, 2001; Meyer and Zanger, 1997; Sheweita and Tilmisany, 2003; Smith et al., 1994), similar associations have also been sought for SULTs. Because SULT1A1 can activate hydroxymethyl polycyclic aromatic hydrocarbons, allylic alcohols, N-hydroxy derivatives of arylamines, and heterocyclic amines to reactive intermediates that can bind to DNA (Banoglu, 2000; Chou et al., 1995; Meinl et al., 2002; Turteltaub et al., 1995; Windmill et al., 1997; Wu et al., 2000; Yamazoe et al., 1999), various studies have looked at its role in human cancer. In particular, the focus has been on the frequency of an arginine (Arg) to histidine (His) change at position 213 (SULT1A1*1 versus SULT1A1*2), resulting from a G to A transition at nucleotide 638 in the coding region of SULT1A1 (Table 10.4). This change alters the recognition site for the restriction enzyme HaeII, which conveniently lends itself to the development of a PCR-restriction fragment polymorphism assay (Coughtrie et al., 1999; Ozawa et al., 1998). Raftogianis et al. (1997) initially showed that patients who were homozygous for the SULT1A1*2 allele had approximately tenfold lower phenol SULT activity than individuals homozygous for the SULT1A*1 allele. While Wang et al. (2002) have demonstrated an association between SULT1A1*2 and lung cancer, no increased risk of colorectal cancer and this polymorphism have been reported (Bamber et al., 2001; Sachse et al., 2002; Wong et al., 2002). A similar negative finding has also been reported in relation to prostate cancer by Steiner et al. (2000) and cancer of the colon, kidney, liver, lung, stomach, and uterine cervix by Peng et al. (2003). An initial study by Seth et al. (2000), employing 444 breast cancer patients and 227 controls, showed no effect of the SULT1A1 genotype on the risk of breast cancer. However, the age of onset was influenced in subjects categorized as early onset breast cancer patients who carried the SULT1A1*1 allele, as patients with this polymorphism were more likely to have other tumors in addition to breast cancer. A more recent study by Zheng et al. (2001) has reported a 1.8-fold increased risk of breast cancer in subjects homozygous for the SULT1A1*2 allele when compared to those homozygous for the SULT1A1*1 allele. Further, Nowell et al. (2002) have reported an approximate threefold increase in breast cancer risk in patients homozygous for SULT1A1*2 allele receiving tamoxifen.

Until recently, SULT1A1 was predominantly seen as an enzyme that was primarily involved in xenobiotic metabolism. This observation was based on it having the widest tissue distribution of any SULT, including liver and intestine, and its broad substrate range (Eisenhofer et al., 1999; Falany, 1997). The recent elucidation of the crystal structure of SULT1A1, which shows the active site of this enzyme to be more versatile than first expected, provides a rational explanation for its substrate diversity. More recent data suggest that SULT1A1 plays a significant role in the metabolism of the iodothyronines, especially 3,3'-T2 (Li et al., 2001; Richard et al., 2001), which we have recently shown binds effectively to the active site of this isozyme (Gamage et al., 2003). 3,3'-T2 is thought to act through nonnuclear pathways to enhance mitochondrial respiration and thus resting metabolism (Li et al., 2001). SULT1A1 is also capable of sulfonating T3 and thyroxine (T4) and may play a more significant role in iodothyronine metabolism than first expected because of its broad tissue distribution. There is however an inbuilt redundancy in relation to thyroid hormone metabolism in humans in addition to SULT1B1, which is considered the major isozyme involved in iodothyronine metabolism (Fujita et al., 1997; Wang et al., 1998), SULT1A1, SULT1A3, SULT1C1, SULT1E1, and SULT2A1 are also involved in this process. Thus the precise physiological reason SULT1A1 exhibits activity toward iodothyronines remains to be elucidated.

At the present time, no solid evidence exists linking a genetic disorder to a defect in a human cytosolic SULT (Strott, 2002). The sulfonation of neurotransmitters such as dopamine, serotonin, adrenaline, and noradrenaline by human SULT1A3 has led to speculation that it plays a role in the pathogenesis of neurological disorders such as Parkinson's disease and migraine (Eisenhofer et al., 1999). A number of studies have used p-nitrophenol (SULT1A1-like activity) and dopamine (SULT1A3-like activity) to measure platelet SULT expression and have compared findings in patients with neurological disorders to control groups (Bongioanni et al., 1996; Marazziti et al., 1992, 1996). These studies have shown increased platelet SULT activity (both SULT1A1 and 1A3) in patients with obsessive–compulsive disorder (Marazziti et al., 1992, 1996), bipolar disorder (Marazziti et al., 1996), and Alzheimer's disease (Bongioanni et al., 1996). In contrast, decreased activity was observed in unipolar depression and migraine patients (Alam et al., 1997; Littlewood et al., 1982). While these studies suggest a role for SULT1As in neurological diseases, it must be pointed out that platelets are somewhat distant from the brain. Further, while a correlation between platelet SULT1A1 activity with that in other tissues has been demonstrated, the same authors showed that platelet SULT1A3 expression was not significantly correlated with levels in other tissues such as the brain (Young et al., 1985). The interpretation of these studies in relation to neurodegenerative disorders will remain speculative until specific data on brain SULT1A expression are obtained.

The available data suggest that SULT1A3 may have evolved to conjugate biogenic amines and thereby plays a significant role in the gut–blood barrier in detoxifying dietary monoamines (Eisenhofer et al., 1999). Indeed, it is becoming increasingly clear that the gastrointestinal tract is the major site of dopamine sulfate production within the body derived from either dietary or endogenous sources of tyrosine, levo-dihydroxyphenylalanine, and dopamine (Goldstein et al., 1999; Strott, 2002). This functional data is also consistent with the observation that SULT1A3 is predominantly expressed in the gastrointestinal tract (Rubin et al., 1996; Windmill et al., 1998). A stronger *prima facie* case for SULT1A3 evolving in humans to handle surges in catecholamine levels is supported by the observations of Richard et al. (2001), who showed that SULT1A3 is expressed in fetal liver but is switched off in the neonate, resulting in its virtual absence in adult livers. They proposed that the expression of SULT1A3 in fetal liver has a protective function against the biological activity of catecholamines, and it is necessary to remove this function prior to birth by switching off SULT1A3 expression. In addition to acting as a barrier to dietary monoamines, gastrointestinal SULT1A3 also plays a major role in the metabolism of orally administered β-adrenergic agonists. Indeed, the β-adrenergic agonists such as salbutamol, isoprenaline, and terbutaline are one of the few groups of drugs in humans that are predominantly metabolized by sulfonation (Coughtrie, 1996). In contrast to dietary catecholamine metabolism, sulfonation of endogenous noradrenaline and adrenaline accounts for <2.5% of the total metabolites excreted by the kidneys, and this amount may be as high as 8% for dopamine (Eisenhofer et al., 1999). This would suggest that SULT1A3 does not play a significant role in endogenous catecholamine metabolism. However, it is difficult to reach this conclusion with total certainty, as the available data are inconclusive. For example, while monoamine oxidase (MAO) is considered the dominant pathway involved in catecholamine metabolism, *in vitro* kinetics have shown SULT1A3 (M-PST) to have greater affinity for dopamine than MAO (Coughtrie, 1996). Further, SULT1A immunoreactive protein has been shown to be present in the human brain, but the inability to distinguish between different SULT1A forms clouds the interpretation of the data, especially in relation to the role of SULT1A3 in catecholamine metabolism (Zou et al., 1990). A more recent study by Richard et al. (2001) using immunoblots and metabolic probes to distinguish between SULT1A1 and SULT1A3 has shown that both isozymes are widely distributed within the developing human fetal brain. These data do not rule out the possibility that in certain discrete cell types, SULT1A3 may be present at higher levels than MAO and thereby more effectively modulate the paracrine action of neurotransmitters. These data highlight the need for a more detailed analysis of this area using modern molecular and functional approaches.

The fact that a number of human cytosolic SULTs are capable of utilizing both catecholamines (SULT1A3 and SULT1A1) and iodothyronines (SULT1B1, SULT1A1, SULT1A3, SULT1C1, SULT1E1, and SULT2A1) as substrates highlights the redundancy that exists within this enzyme system. This is also the case for human cytosolic SULTs involved in xenobiotic and steroid metabolism; this latter area has been expertly covered in reviews by Strott (1996, 2002). Further complexity in unraveling the precise role these enzymes play in cellular defense mechanisms arises when considering that many of the substrates of SULTs are also substrates of UDP-glucuronosyltransferases. However, in analyzing the role different conjugating systems play in such mechanisms, it is important to consider the exposure levels plus the kinetic properties, tissue distribution, and developmental patterns of the isozymes that make up the SULT and UDP-glucuronosyltransferase systems. Indeed, it is generally accepted that sulfonation is a high affinity, low capacity pathway, whereas glucuronidation is a low affinity, high capacity pathway (Mulder, 1981). Based on this latter reasoning, it is possible to speculate that in relation to the low level exposure of dietary carcinogens such as heterocyclic amines, sulfonation may be the critical event in the activation of their N-hydroxy derivatives to form DNA-binding adducts (Turteltaub et al., 1995; Wu et al., 2000).

With the recent publication of the crystal structures of both SULT1A1 and SULT1A3 (Bidwell et al., 1999; Dajani et al., 1999a; Gamage et al., 2003), these data bring the number of human cytosolic SULT structures published to four (Bidwell et al., 1999; Dajani et al., 1999a; Pedersen et al., 2000, 2002; Rehse et al., 2002). This now provides an opportunity to fully understand why different forms of SULT exhibit broad but overlapping substrate specificities and how certain SULTs exhibit strict regio-specificity toward a particular substrate (Song, 2001; Weinshilboum et al., 1997). Understanding the structural basis of such specificity is crucial for elucidating the mechanism and function of these enzymes. It will also aid in predicting the metabolic fate of drugs and chemical carcinogens that are sulfonated and provide a more rational approach to drug design and chemical risk assessment. A more predictive structure/function approach may be the only way to go from a toxicological and drug development viewpoint due to the fact that rodents and other laboratory animals express only a single SULT1A subfamily gene whereas humans have three such genes. The advent of the above structures will also make it possible to explain the effect of specific SNPs on the structure/function relationships of these enzymes.

Very little is known about the transcriptional regulation of SULT gene expression in either human or animal tissues (Song et al., 1998, 2001; Strott, 2002). The available data demonstrate that in rodents, SULT expression appears to be predominantly in hepatic tissue, whereas in humans pronounced extra hepatic patterns of expression are observed (Eisenhofer et al., 1999). These data together with the fact that many rodent forms exhibit dramatic sexual dimorphism, add a degree of complexity to extrapolating gene regulation data from animal models to humans (Coughtrie and Johnston, 2001). A number of studies have demonstrated that the transcription of 5′-untranslated regions of both human SULT1A1 and SULT1A3 is tissue dependent, suggesting alternate splicing and tissue-specific promoter activity (Dooley, 1998a; Her et al., 1998; Strott, 2002; Zhu et al., 1993a). A recent paper by Iida et al. (2001) reported SNPs in the 5′- and 3′-flanking regions, the 5′- and 3′-UTRs, and in a number of introns of both SULT1A1 and SULT1A2, but the effect of these changes on gene expression are as yet to be determined. The genes of the SULT1A subfamily members have all been sequenced (Weinshilboum et al., 1997), and it is anticipated that significant insights will be obtained in the immediate future on how these genes are regulated at the transcriptional level.

Since the initial cloning of a rat SULT2A2 as a senescence marker protein in 1987, at least another 55 eukaryotic SULT isoforms have been identified and functionally characterized (Chatterjee et al., 1987; Raftogianis, 2003). Out of all these SULTs, probably the SULT1A subfamily in humans has received the most attention because of the central role its members play in xenobiotic, drug, catecholamine, and iodothyronine metabolism. The next few years should bring an avalanche of data on the physiological and toxicological significance of these enzymes as we are just acquiring the appropriate molecular tools that will enable us to definitively address these areas. A thorough understanding of the transcriptional regulation and three-dimensional structures of SULT1A subfamily members may also provide new potential drug targets. There is also a need for a more appropriate animal model than rodents to investigate the developmental and physiological importance of the SULT1A subfamily. However, the potential insights that could come from the target disruption of the *SULT1a1* gene in mice are yet to be realized.

REFERENCES

Abenhaim, L., Romain, Y., and Kuchel, O. (1981) Platelet phenolSULT and catecholamines: physiological and pathological variations in humans. *Can J Physiol Pharmacol* 59:300–306.

Adjei, A.A. and Weinshilboum, R.M. (2002) Catecholestrogen sulfation: possible role in carcinogenesis. *Biochem Biophys Res Commun* 292:402–408.

Aksoy, I.A. and Weinshilboum, R.M. (1995) Human thermolabile phenol SULT gene (STM): molecular cloning and structural characterization. *Biochem Biophys Res Commun* 208:786–795.

Aksoy, I.A., Wood, T.C., and Weinshilboum, R. (1994) Human liver estrogen SULT: identification by cDNA cloning and expression. *Biochem Biophys Res Commun* 200:1621–1629.

Alam, Z., Coombes, N., Waring, R.H., Williams, A.C., and Steventon, G.B. (1997) Platelet sulphotransferase activity, plasma sulphate levels and sulphation capacity in patients with migraine and tension headache. *Cephalalgia* 17:761–764.

Anderson, R.J., Babbitt, L.L., and Liebentritt, D.K. (1995) Human liver triiodothyronine SULT: copurification with phenol SULTs. *Thyroid* 5:61–66.

Anderson, R.J., Weinshilboum, R.M., Phillips, S.F., and Broughton, D.D. (1981) Human platelet phenol sulphotransferase: assay procedure, substrate and tissue correlations. *Clin Chim Acta* 110:157–167.

Baker, C.A., Uno, H., and Johnson, G.A. (1994) Minoxidil sulfation in the hair follicle. *Skin Pharmacol* 7:335–339.

Bamber, D.E., Fryer, A.A., Strange, R.C., Elder, J.B., Deakin, M., Rajagopal, R., Fawole, A., Gilissen, R.A., Campbell, F.C., and Coughtrie, M.W (2001) Phenol sulphotransferase SULT1A1*1 genotype is associated with reduced risk of colorectal cancer. *Pharmacogenetics* 11:679–685.

Banoglu, E. (2000) Current status of the cytosolic SULTs in the metabolic activation of promutagens and procarcinogens. *Curr Drug Metab* 1:1–30.

Baranczyk-Kuzma, A., Garren, J.A., Hidalgo, I.J., and Borchardt, R.T. (1991) Substrate specificity and some properties of phenol SULT from human intestinal Caco-2 cells. *Life Sci* 49:1197–1206.

Barnes, S., Waldrop, R., Crenshaw, J., King, R.J., and Taylor, K.B. (1986) Evidence for an ordered reaction mechanism for bile salt: 3′phosphoadenosine-5-phosphosulfate: SULT from rhesus monkey liver that catalyzes the sulfation of the hepatotoxin glycolithocholate. *J Lipid Res* 27:1111–1123.

Barnett, A.C., Tsvetanov, S., Gamage, N., Martin, J.L., Duggleby, R.G., and McManus, M.E. (2004) Active site mutations and substrate inhibition in human SULT 1A1 and 1A3. *J Biol Chem* 279:18799–18805.

Baulieu, E.E. (1991) Neurosteroids: a new function in the brain. *Biol Cell* 71:3–10.

Baumann, E. (1876) Ueber sulfosauren im harn. *Deutsche Chemische Gesellschaft*: 54–58.

Beckmann, J.D., Illig, M., and Bartzatt, R. (1994) Regulation of phenol SULT expression in cultured bovine bronchial epithelial cells by hydrocortisone. *J Cell Physiol* 160:603–610.

Bernier, F., Leblanc, G., Labrie, F., and Luu-The, V. (1994a) Structure of human estrogen and aryl SULT gene: two mRNA species issued from a single gene. *J Biol Chem* 269:28200–28205.

Bernier, F., Lopez Solache, I., Labrie, F., and Luu-The, V. (1994b) Cloning and expression of cDNA encoding human placental estrogen SULT. *Mol Cell Endocrinol* 99:R11–R15.

Bernier, F., Soucy, P., and Luu-The, V. (1996) Human phenol SULT gene contains two alternative promoters: structure and expression of the gene. *DNA Cell Biol* 15:367–375.

Bidwell, L.M., McManus, M.E., Gaedigk, A., Kakuta, Y., Negishi, M., Pedersen, L., and Martin, J.L. (1999) Crystal structure of human catecholamine SULT. *J Mol Biol* 293:521–530.

Bongioanni, P., Donato, M., Castagna, M., and Gemignani, F. (1996) Platelet phenolsulphotransferase activity, monoamine oxidase activity and peripheral-type benzodiazepine binding in demented patients. *J Neural Transm* 103:491–501.

Bonham Carter, S.M., Rein, G., Glover, V., Sandler, M., and Caldwell, J. (1983) Human platelet phenolsulphotransferase M and P: substrate specificities and correlation with *in vivo* sulphoconjugation of paracetamol and salicylamide. *Br J Clin Pharmacol* 15:323–330.

Brittelli, A., De Santi, C., Raunio, H., Pelkonen, O., Rossi, G., and Pacifici, G.M. (1999) Interethnic and interindividual variabilities of platelet SULTs activity in Italians and Finns. *Eur J Clin Pharmacol* 55:691–695.

Brix, L.A., Duggleby, R.G., Gaedigk, A., and McManus, M.E. (1999a) Structural characterization of human aryl sulphotransferases. *Biochem J* 337:337–343.

Brix, L.A., Barnett, A.C., Duggleby, R.G., Leggett, B., and McManus, M.E. (1999b) Analysis of the substrate specificity of human SULTs SULT1A1 and SULT1A3: site-directed mutagenesis and kinetic studies. *Biochemistry* 38:10474–10479.

Burchell, B. and Coughtrie, M.W. (1997) Genetic and environmental factors associated with variation of human xenobiotic glucuronidation and sulfation. *Environ Health Perspect* 105:739–747.

Butler, P.R., Anderson, R.J., and Venton, D.L. (1983) Human platelet phenol SULT: partial purification and detection of two forms of the enzyme. *J Neurochem* 41(3): 630–639.

Buu, N.T., Duhaime, J., and Kuchel, O. (1984) The bicuculline-like properties of dopamine sulfate in rat brain. *Life Sci* 35:1083–1090.

Campbell, N.R., Sundaram, R.S., Werness, P.G., Van Loon, J., and Weinshilboum, R.M. (1985) Sulfate and methyldopa metabolism: metabolite patterns and platelet phenol SULT activity. *Clin Pharmacol Ther* 37:308–315.

Campbell, N.R., Van Loon, J.A., and Weinshilboum, R.M. (1987) Human liver phenol SULT: assay conditions, biochemical properties and partial purification of isozymes of the thermostable form. *Biochem Pharmacol* 36:1435–1446.

Cappiello, M., Giuliani, L., and Pacifici, G.M. (1990) Differential distribution of phenol and catechol sulphotransferases in human liver and intestinal mucosa. *Pharmacology* 40:69–76.

Cappiello, M., Giuliani, L., Rane, A., and Pacifici, G.M. (1991) Dopamine sulphotransferase is better developed than *p*-nitrophenol sulphotransferase in the human fetus. *Dev Pharmacol Ther* 16:83–88.

Carlini, E.J., Raftogianis, R.B., Wood, T.C., Jin, F., Zheng, W., Rebbeck, T.R., and Weinshilboum, R.M. (2001) Sulfation pharmacogenetics: SULT1A1 and SULT1A2 allele frequencies in Caucasian, Chinese and African-American subjects. *Pharmacogenetics* 11:57–68.

Chatterjee, B., Majumdar, D., Ozbilen, O., Murty, C.V.R., and Roy, A.K. (1987) Molecular cloning and characterization of cDNA for androgen-repressible rat liver protein, SMP-2. *J Biol Chem* 262:822–825.

Chiba, H., Komatsu, K., Lee, Y.C., Tomizuka, T., and Strott, C.A. (1995) The 3'-terminal exon of the family of steroid and phenol SULT genes is spliced at the N-terminal glycine of the universally conserved GXXGXXK motif that forms the sulfonate donor binding site. *Proc Natl Acad Sci USA* 92:8176–8179.

Chou, H.C., Lang, N.P., and Kadlubar, F.F. (1995) Metabolic activation of N-hydroxy arylamines and N-hydroxy heterocyclic amines by human SULT(s) *Cancer Res* 55:525–529.

Coughtrie, M.W. (1996) Sulphation catalyzed by the human cytosolic sulphotransferases—chemical defence or molecular terrorism? *Hum Exp Toxicol* 15:547–555.

Coughtrie, M.W., Gilissen, R.A., Shek, B., Strange, R.C., Fryer, A.A., Jones, P.W., and Bamber, D.E. (1999) Phenol sulphotransferase SULT1A1 polymorphism: molecular diagnosis and allele frequencies in Caucasian and African populations. *Biochem J* 337:45–49.

Coughtrie, M.W. and Johnston, L.E. (2001) Interactions between dietary chemicals and human SULTs-molecular mechanisms and clinical significance. *Drug Metab Dispos* 29:522–528.

Dajani, R., Cleasby, A., Neu, M., Wonacott, A.J., Jhoti, H., Hood, A.M., Modi, S., Hersey, A., Taskinen, J., Cooke, R.M., Manchee, G.R., and Coughtrie, M.W. (1999a) X-ray crystal structure of human dopamine SULT, SULT1A3: molecular modeling and quantitative structure-activity relationship analysis demonstrate a molecular basis for SULT substrate specificity. *J Biol Chem* 274:37862–37868.

Dajani, R., Hood, A.M., and Coughtrie, M.W. (1998) A single amino acid, glu146, governs the substrate specificity of a human dopamine SULT, SULT1A3. *Mol Pharmacol* 54:942–948.

Dajani, R., Sharp, S., Graham, S., Bethell, S.S., Cooke, R.M., Jamieson, D.J., and Coughtrie, M.W. (1999b) Kinetic properties of human dopamine SULT (SULT1A3) expressed in prokaryotic and eukaryotic systems: comparison with the recombinant enzyme purified from *Escherichia coli*. *Protein Expr Purif* 16:11–18.

Dooley, T.P. (1998a) Cloning of the human phenol SULT gene family: three genes implicated in the metabolism of catecholamines, thyroid hormones and drugs. *Chem Biol Interact* 109:29–41.

Dooley, T.P. (1998b) Molecular biology of the human phenol SULT gene family. *J Exp Zool* 282:223–230.

Dooley, T.P., Haldeman-Cahill, R., Joiner, J., and Wilborn, T.W. (2000) Expression profiling of human SULT and sulfatase gene superfamilies in epithelial tissues and cultured cells. *Biochem Biophys Res Commun* 277:236–245.

Dooley, T.P. and Huang, Z. (1996) Genomic organization and DNA sequences of two human phenol SULT genes (STP1 and STP2) on the short arm of chromosome 16. *Biochem Biophys Res Commun* 228:134–140.

Dooley, T.P., Mitchison, H.M., Munroe, P.B., Probst, P., Neal, M., Siciliano, M.J., Deng, Z., Doggett, N.A., Callen, D.F., and Gardiner, R.M. (1994) Mapping of two phenol sulphotransferase genes, STP and STM, to 16p: candidate genes for Batten disease. *Biochem Biophys Res Commun* 205:482–489.

Dooley, T.P., Obermoeller, R.D., Leiter, E.H., Chapman, H.D., Falany, C.N., Deng, Z., and Siciliano, M.J. (1993) Mapping of the phenol SULT gene (STP) to human chromosome 16p12.1-p11.2 and to mouse chromosome 7. *Genomics* 18:440–443.

Driscoll, W.J., Komatsu, K., and Strott, C.A. (1995) Proposed active site domain in estrogen SULT as determined by mutational analysis. *Proc Natl Acad Sci USA* 92:12328–12332.

Duanmu, Z., Dunbar, J., Falany, C.N., and Runge-Morris, M. (2000) Induction of rat hepatic aryl SULT (SULT1A1) gene expression by triamcinolone acetonide: impact on minoxidil-mediated hypotension. *Toxicol Appl Pharmacol* 164:312–320.

Duanmu, Z., Kocarek, T.A., and Runge-Morris, M. (2001) Transcriptional regulation of rat hepatic aryl SULT (SULT1A1) gene expression by glucocorticoids. *Drug Metab Dispos* 29:1130–1135.

Duanmu, Z., Locke, D., Smigelski, J., Wu, W., Dahn, M.S., Falany, C.N., Kocarek, T.A., and Runge-Morris, M. (2002) Effects of Dexamethasone on Aryl (SULT1A1)- and Hydroxysteroid (SULT2A1)-SULT Gene Expression in Primary Cultured Human Hepatocytes. *Drug Metab Dispos* 30:997–1004.

Dubin, R.L., Hall, C.M., Pileri, C.L., Kudlacek, P.E., Li, X.Y., Yee, J.A., Johnson, M.L., and Anderson, R.J. (2001) Thermostable (SULT1A1) and thermolabile (SULT1A3) phenol SULTs in human osteosarcoma and osteoblast cells. *Bone* 28:617–624.

Duffel, M.W. and Jakoby, W.B. (1981) On the mechanism of aryl SULT. *J Biol Chem* 256:11123–11127.

Duffel, M.W., Marshal, A.D., McPhie, P., Sharma, V., and Jakoby, W.B. (2001) Enzymatic aspects of the phenol (aryl) SULTs. *Drug Metab Rev* 33:369–395.

Dunn, R.T., Jr. and Klaassen, C.D. (2000) Thyroid hormone modulation of rat sulphotransferase mRNA expression. *Xenobiotica* 30:345–357.

Eaton, E.A., Walle, U.K., Lewis, A.J., Hudson, T., Wilson, A.A., and Walle, T. (1996) Flavonoids, potent inhibitors of the human P-form phenol SULT: potential role in drug metabolism and chemoprevention. *Drug Metab Dispos* 24:232–237.

Eisenhofer, G., Coughtrie, M.W., and Goldstein, D.S. (1999) Dopamine sulphate: an enigma resolved. *Clin Exp Pharmacol Physiol Suppl* 26:S41–S53.

Engelke, C.E., Meinl, W., Boeing, H., and Glatt, H. (2000) Association between functional genetic polymorphisms of human SULTs 1A1 and 1A2. *Pharmacogenetics*, 10:163–169.

Falany, C.N. (1997) Enzymology of human cytosolic SULTs. *Faseb J* 11:206–216.

Falany, C.N. and Kerl, E.A. (1990) Sulfation of minoxidil by human liver phenol SULT. *Biochem Pharmacol* 40:1027–1032.

Falany, J.L. and Falany, C.N. (1996) Regulation of estrogen SULT in human endometrial adenocarcinoma cells by progesterone. *Endocrinology* 137:1395–1401.

Falany, J.L. and Falany, C.N. (1997) Regulation of estrogen activity by sulfation in human MCF-7 breast cancer cells. *Oncol Res* 9:589–596.

Falany, J.L., Lawing, L., and Falany, C.N. (1993) Identification and characterization of cytosolic SULT activities in MCF-7 human breast carcinoma cells. *J Steroid Biochem Mol Biol* 46:481–487.

Foldes, A. and Meek, J.L. (1973) Rat brain phenolSULT — partial purification and some properties. *Biochemica et Biophysica Acta* 327:365–374.

Fujita, K., Nagata, K., Ozawa, S., Sasano, H., and Yamazoe, Y. (1997) Molecular cloning and characterization of rat ST1B1 and human ST1B2 cDNAs, encoding thyroid hormone SULTs. *J Biochem* (Tokyo) 122:1052–1061.

Gaedigk, A., Beatty, B.G., and Grant, D.M. (1997) Cloning, structural organization, and chromosomal mapping of the human phenol SULT STP2 gene. *Genomics* 40:242–246.

Gamage, N.U., Duggleby, R.G., Barnett, A.C., Tresillian, M., Latham, C.F., Liyou, N.E., McManus, M.E., and Martin, J.L. (2003) Structure of a human carcinogen converting enzyme, SULT1A1: structural and kinetic implications of substrate inhibition. *J Biol Chem* 278:7655–7662.

Ganguly, T.C., Krasnykh, V., and Falany, C.N. (1995) Bacterial expression and kinetic characterization of the human monoamine-sulfating form of phenol SULT. *Drug Metab Dispos* 23:945–950.

George, C.F., Blackwell, E.W., and Davies, D.S. (1974) Metabolism of isoprenaline in the intestine. *J Pharm Pharmacol* 26:265–267.

Gilissen, R.A., Bamforth, K.J., Stavenuiter, J.F., Coughtrie, M.W., and Meerman, J.H. (1994) Sulfation of aromatic hydroxamic acids and hydroxylamines by multiple forms of human liver SULTs. *Carcinogenesis* 15:39–45.

Glatt, H., Boeing, H., Engelke, C.E., Ma, L., Kuhlow, A., Pabel, U., Pomplun, D., Teubner, W., and Meinl, W. (2001) Human cytosolic sulphotransferases: genetics, characteristics, toxicological aspects. *Mutat Res* 482:27–40.

Goldstein, D.S., Swoboda, K.J., Miles, J.M., Coppack, S.W., Aneman, A., Holmes, C., Lamensdorf, I., and Eisenhofer, G. (1999) Sources and physiological significance of plasma dopamine sulfate. *J Clin Endocrinol Metab* 84:2523–2531.

Habuchi, O. (2000) Diversity and functions of glycosaminoglycan SULTs. *Biochim Biophys Acta* 1474:115–127.

Harris, R.M., Waring, R.H., Kirk, C.J., and Hughes, P.J. (2000) Sulfation of estrogenic alkylphenols and 17beta-estradiol by human platelet phenol SULTs. *J Biol Chem* 275:159–166.

Hart, R.F., Renskers, K.J., Nelson, E.B., and Roth, J.A. (1979) Localization and characterization of phenol SULT in human platelets. *Life Sci* 24:125–130.

Hartman, A.P., Wilson, A.A., Wilson, H.M., Aberg, G., Falany, C.N., and Walle, T. (1998) Enantioselective sulfation of beta 2-receptor agonists by the human intestine and the recombinant M-form phenolSULT. *Chirality* 10:800–803.

Hellriegel, E.T., Matwyshyn, G.A., Fei, P., Dragnev, K.H., Nims, R.W., R.A. Lubet, R.A., and Kong, A.N. (1996) Regulation of gene expression of various phase I and phase II drug-metabolizing enzymes by tamoxifen in rat liver. *Biochem Pharmacol* 52:1561–1568.

Her, C., Raftogianis, R., and Weinshilboum, R.M. (1996) Human phenol SULT STP2 gene: molecular cloning, structural characterization, and chromosomal localization. *Genomics* 33:409–420.

Her, C., Wood, T.C., Eichler, E.E., Mohrenweiser, H.W., Ramagli, L.S., Siciliano, M.J., and Weinshilboum, R.M. (1998) Human hydroxysteroid SULT SULT2B1: two enzymes encoded by a single chromosome 19 gene. *Genomics* 53:284–295.

Hernandez, J.S., Powers, S.P., and Weinshilboum, R.M. (1991) Human liver arylamine N-SULT activity: thermostable phenol SULT catalyzes the N-sulfation of 2-naphthylamine. *Drug Metab Dispos* 19:1071–1079.

Heroux, J.A., Falany, C.N., and Roth, J.A. (1989) Immunological characterization of human phenol SULT. *Mol Pharmacol* 36:29–33.

Hobkirk, R., Glasier, M.A., and Brown, L.Y. (1990) Purification and some characteristics of an oestrogen sulphotransferase from guinea pig adrenal gland and its non-identity with adrenal pregnenolone sulphotransferase. *Biochem J* 268:759–764.

Honma, W., Kamiyama, Y., Yoshinari, K., Sasano, H., Shimada, M., Nagata, K., and Yamazoe, Y. (2001) Enzymatic characterization and interspecies difference of phenol SULTs, ST1A forms. *Drug Metab Dispos* 29:274–281.

Honma, W., Shimada, M., Sasano, H., Ozawa, S., Miyata, M., Nagata, K., Ikeda, T., and Yamazoe, Y. (2002) Phenol SULT, ST1A3, as the main enzyme catalyzing sulfation of troglitazone in human liver. *Drug Metab Dispos* 30:944–952.

Hume, R. and Coughtrie, M.W. (1994) Phenolsulphotransferase: localization in kidney during human embryonic and fetal development. *Histochem J* 26:850–855.

Hwang, S.R., Kohn, A.B., and Hook, V.Y.H. (1995) Molecular cloning of an isoform of phenol SULT from human brain hippocampus. *Biochem Biophys Res Commun* 207:701–707.

Iida, A., Sekine, A., Saito, S., Kitamura, Y., Kitamoto, T., Osawa, S., Mishima, C., and Nakamura, Y. (2001) Catalog of 320 single nucleotide polymorphisms (SNPs) in 20 quinone oxidoreductase and SULT genes. *J Hum Genet* 46: 225–240.

Ingelman-Sundberg, M. (2001) Genetic susceptibility to adverse effects of drugs and environmental toxicants: the role of the CYP family of enzymes. *Mutation Research* 482:11–19.

Jones, A.L., Hagen, M., Coughtrie, M.W., Roberts, R.C., and Glatt, H. (1995) Human platelet phenolSULTs: cDNA cloning, stable expression in V79 cells and identification of a novel allelic variant of the phenol-sulfating form. *Biochem Biophys Res Commun* 208:855–862.

Kakuta, Y., Pedersen, L.G., Carter, C.W., Negishi, M., and Pedersen, L.C. (1997) Crystal structure of estrogen sulphotransferase. *Nat Struct Biol* 4:904–908.

Kakuta, Y., Petrotchenko, E.V., Pedersen, L.C., and Negishi, M. (1998) The sulfuryl transfer mechanism: crystal structure of a vanadate complex of estrogen SULT and mutational analysis. *J Biol Chem* 273:27325–27330.

Kakuta, Y., Sueyoshi, T., Negishi, M., and Pedersen, L.C. (1999) Crystal structure of the SULT domain of human heparan sulfate N-deacetylase/N-SULT 1. *J Biol Chem* 274:10673–10676.

Kester, M.H., Kaptein, E., Roest, T.J., van Dijk, C.H., Tibboel, D., Meinl, W., Glatt, H., Coughtrie, M.W., and Visser, T.J. (2003) Characterization of rat iodothyronine SULTs. *Am J Physiol Endocrinol Metab* 285:E592–E598.

Kester, M.H., Kaptein, E., Roest, T.J., van Dijk, C.H., Tibboel, D., Meinl, W., Glatt, H., Coughtrie, M.W., and Visser, T.J. (1999) Characterization of human iodothyronine SULTs. *J Clin Endocrinol Metab* 84:1357–1364.

King, R.S., Teitel, C.H., and Kadlubar, F.F. (2000) *In vitro* bioactivation of N-hydroxy-2-amino-alpha-carboline. *Carcinogenesis* 21:1347–1354.

Klaassen, C.D., Liu, L., and Dunn, R.T., Jr. (1998) Regulation of SULT mRNA expression in male and female rats of various ages. *Chem Biol Interact* 109:299–313.

Komatsu, K., Driscoll, W.Y., Koh, Y.C., and Strott, C.A. (1994) A P-loop related motif (GxxGxxK) highly conserved in SULTs is required for binding the activated sulfate donor. *Biochem Biophys Res Commun* 204:1178–1185.

Kreis, P., Brandner, S., Coughtrie, M.W., Pabel, U., Meinl, W., Glatt, H., and Andrae, U. (2000) Human phenol SULTs hP-PST and hM-PST activate propane 2-nitronate to a genotoxicant. *Carcinogenesis* 21:295–299.

Kudlacek, P.E., Anderson, R.J., Liebentritt, D.K., Johnson, G.A., and Huerter, C.J. (1995) Human skin and platelet minoxidil SULT activities: biochemical properties, correlations and contribution of thermolabile phenol SULT. *J Pharmacol Exp Ther* 273:582–590.

Lewis, A.J., Kelly, M.M., Walle, U.K., Eaton, E.A., Falany, C.N., and Walle, T. (1996) Improved bacterial expression of the human P form phenolSULT: applications to drug metabolism. *Drug Metab Dispos* 24:1180–1185.

Li, X. and Anderson, R.J. (1999) Sulfation of iodothyronines by recombinant human liver steroid SULTs. *Biochem Biophys Res Commun* 263:632–639.

Li, X., Clemens, D.L., Cole, J.R., and Anderson, R.J. (2001) Characterization of human liver thermostable phenol SULT (SULT1A1) allozymes with 3,3′,5-triiodothyronine as the substrate. *J Endocrinol* 171:525–532.

Littlewood, J., Glover, V., Sandler, M., Petty, R., Peatfield, R., and Rose, F.C. (1982) Platelet phenolsulphotransferase deficiency in dietary migraine. *Lancet* 1:983–986.

Liu, L. and Klaassen, C.D. (1996) Regulation of hepatic SULTs by steroidal chemicals in rats. *Drug Metab Dispos* 24:854–858.

Liu, L., LeCluyse, E.L., Liu, J., and Klaassen, C.D. (1996) SULT gene expression in primary cultures of rat hepatocytes. *Biochem Pharmacol* 52:1621–1630.

Liu, M.C., Suiko, M., and Sakakibara, Y. (2000) Mutational analysis of the substrate binding/catalytic domains of human M form and P form phenol SULTs. *J Biol Chem* 275:13460–13464.

Llerena, A., Berecz, R., de La Rubia, A., Fernandez-Salguero, P., and Dorado, P. (2001) Effect of thioridazine dosage on the debrisoquine hydroxylation phenotype in psychiatric patients with different CYP2D6 genotypes. *Ther Drug Monit* 23:616–620.

Maglich, J.M., Parks, D.J., Moore, L.B., Collins, J.L., Goodwin, B., Billin, A.N., Stoltz, C.A., Kliewer, S.A., Lambert, M.H., Wilson, T.M., and Moore, J.T. (2003) Identification of a novel human constitutive androstane receptor (CAR) agonist and its use in the identification of CAR target genes *J Biol Chem* 278:17277–17283.

Marazziti, D., Hollander, E., Lensi, P., Ravagli, S., and Cassano, G.B. (1992) Peripheral markers of serotonin and dopamine function in obsessive–compulsive disorder. *Psychiatry Res* 42:41–51.

Marazziti, D., Palego, L., DellOsso, L., Batistini, A., Cassano, G.B., and Akiskal, H.S. (1996) Platelet SULT in different psychiatric disorders. *Psychiatry Res* 65:73–78.

Marazziti, D., Palego, L., Rossi, A., and Cassano, G.B. (1998) Gender-related seasonality of human platelet phenolSULT activity. *Neuropsychobiology* 38:1–5.

Marsolais, F., Laviolette, M., Kakuta, Y., Negishi, M., Pedersen, L.C., Auger, M., and Varin, L. (1999) 3′-Phosphoadenosine 5′-phosphosulfate binding site of flavonol 3-SULT studied by affinity chromatography and 31P NMR. *Biochemistry* 38:4066–4071.

Marsolais, F. and Varin, L. (1995) Identification of amino acid residues critical for catalysis and cosubstrate binding in the flavonol 3-SULT. *J Biol Chem* 270:30458–30463.

Meinl, W., Meerman, J.H., and Glatt, H. (2002) Differential activation of promutagens by alloenzymes of human SULT 1A2 expressed *Salmonella typhimurium*. *Pharmacogenetics* 12:677–689.

Meisheri, K.D., Johnson, G.A., and Puddington, L. (1993) Enzymatic and non-enzymatic sulfation mechanisms in the biological actions of minoxidil. *Biochem Pharmacol* 45:271–279.

Meyer, U.A. and Zanger, U.M. (1997) Molecular mechanisms of genetic polymorphisms of drug metabolism. *Annu Rev Pharmacol Toxicol* 37:269–296.

Mitchell, J.R., Thorgeirsson, S.S., Potter, W.Z., Jollow, D.J., and Keiser, H. (1974) Acetaminophen-induced hepatic injury: protective role of glutathione in man and rationale for therapy. *Clin Pharmacol Ther* 16:676–684.

Moreno, M., Lanni, A., Lombardi, A., and Goglia, F. (1997) How the thyroid controls metabolism in the rat: different roles for triiodothyronine and diiodothyronines. *J Physiol* 505:529–538.

Mulder, G.J. (1981) *Sulfation of Drugs and Other Compounds*, CRC Press, Boca Raton.

Nagata, K. and Yamazoe, Y. (2000) Pharmacogenetics of SULT. *Annu Rev Pharmacol Toxicol* 40:159–176.

Nagata, K., Yoshinari, K., Ozawa, S., and Yamazoe, Y. (1997) Arylamine activating SULT in liver. *Mutat Res* 376:267–272.

Nakamura, J., Mizuma, T., Hayashi, M., and Awazu, S. (1990) Correlation of phenol sulphotransferase activities in the liver and platelets of rat. *J Pharm Pharmacol* 42:207–208.

Negishi, M., Pedersen, L.G., Petrotchenko, E., Shevtsov, S., Gorokhov, A., Kakuta, Y., and Pedersen, L.C. (2001) Structure and function of SULTs. *Arch Biochem Biophys* 390:149–157.

Niehrs, C., Beisswanger, R., and Huttner, W.B. (1994) Protein tyrosine sulfation, 1993—an update. *Chem Biol Interact* 92:257–271.

Nishiyama, T., Ogura, K., Nakano, H., Ohnuma, T., Kaku, T., Hiratsuka, A., Muro, K., and Watabe, T. (2002) Reverse geometrical selectivity in glucuronidation and sulfation of cis- and trans-4-hydroxytamoxifens by human liver UDP- glucuronosyltransferases and SULTs. *Biochem Pharmacol* 63:1817–1830.

Nowell, S., Ambrosone, C.B., Ozawa, S., MacLeod, S.L., Mrackova, G., Williams, S., Plaxco, J., Kadlubar, F.F., and Lang, N.P. (2000) Relationship of phenol SULT activity (SULT1A1) genotype to SULT phenotype in platelet cytosol. *Pharmacogenetics* 10:789–797.

Nowell, S., Sweeney, C., Winters, M., Stone, A., Lang, N.P., Hutchins, L.F., Kadlubar, F.F., and Ambrosone, C.B. (2002) Association between SULT 1A1 genotype and survival of breast cancer patients receiving tamoxifen therapy. *J National Cancer Institute*, 94:1635–1640.

Ozawa, S., Chou, H.C., Kadlubar, F.F., Nagata, K., Yamazoe, Y., and Kato, R. (1994) Activation of 2-hydroxyamino-1-methyl-6-phenylimidazo[4,5-b] pyridine by cDNA-expressed human and rat arylSULTs. *Jpn J Cancer Res* 85:1220–1228.

Ozawa, S., Nagata, K., Shimada, M., Ueda, M., Tsuzuki, T., Yamazoe, Y., and Kato, R. (1995) Primary structures and properties of two related forms of aryl SULTs in human liver. *Pharmacogenetics* 5:S135–S140.

Ozawa, S., Tang, Y.M., Yamazoe, Y., Kato, R., Lang, N.P., and Kadlubar, F.F. (1998) Genetic polymorphisms in human liver phenol SULTs involved in the bioactivation of N-hydroxy derivatives of carcinogenic arylamines and heterocyclic amines. *Chem Biol Interact* 109:237–248.

Ozawa, S., Shimizu, M., Katoh, T., Miyajima, A., Ohno, Y., Matsumoto, Y., Fukuoka, M., Tang, Y.M., Lang, N.P., and Kadlubar, F.F. (1999) Sulfating-activity and stability of cDNA-expressed allozymes of human phenol SULT, ST1A3*1 ((213)Arg) and ST1A3*2 ((213)His), both of which exist in Japanese as well as Caucasians. *J Biochem (Tokyo)* 126:271–277.

Pacifici, G.M., Kubrich, M., Giuliani, L., deVries, M., Rane, A. (1993) Sulphation and glucoronidation of ritodrine in human foetal and adult tissues. *Eur J Clin Pharmacol* 44:259–264.

Pacifici, G.M., Temellini, A., Castiglioni, M., D'Alessandro, C., Ducci, A., and Giuliani, L. (1994) Interindividual variability of the human hepatic SULTs. *Chem Biol Interact* 92:219–231.

Pacifici, G.M. and De Santi, C. (1995) Human sulphotransferase. Classification and metabolic profile of the major isoforms: the point of view of the clinical pharmacologist, in *Advances in Drug Metabolism in Man*, Pacifici, G.M. and Fracchia, G.N., Eds., European Commission, Luxembourg, pp. 312–349.

Pacifici, G.M., Giulianetti, B., Quilici, M.C., Spisni, R., Nervi, M., Giuliani, L., and Gomeni, R. (1997) (–)-Salbutamol sulfation in the human liver and duodenal mucosa: interindividual variability. *Xenobiotica* 27:279–286.

Pacifici, G.M., Quilici, M.C., Giulianetti, B., Spisni, R., Nervi, M., Giuliani, L., and Gomeni, R. (1998) Ritodrine sulfation in the human liver and duodenal mucosa: interindividual variability. *Eur J Drug Metab Pharmacokin* 23:67–74.

Pacifici, G.M. and Rossi, A.M. (2001) Interindividual variability of SULTs, in *Interindividual Variability in Drug Metabolism in Humans*, Pacifici, G.M. and Pelkonen, O., Eds., Taylor & Francis, London, pp. 434–459.

Pai, T.G., Suiko, M., Sakakibara, Y., and Liu, M.C. (2001) Sulfation of flavonoids and other phenolic dietary compounds by the human cytosolic SULTs. *Biochem Biophys Res Commun* 285:1175–1179.

Pai, T.G., Oxendine, I., Sugahara, T., Suiko, M., Sakakibara Y., and Liu, M.C. (2003) Structure function relationships in the stereospecific and manganese-dependent 3, 4 dihydroxy-phenylalanine/tyrosine sulfating activity of human monoamine from phenol sulfotransferse, SULT1A3. *J Biol Chem* 278:1525–1532.

Park-Chung, M., Malayev, A., Purdy, R.H., Gibbs, T.T., and Farb, D.H. (1999) Sulfated and unsulfated steroids modulate gamma-aminobutyric acid A receptor function through distinct sites. *Brain Res* 830:72–87.

Park-Chung, M., Wu, F.S., Purdy, R.H., Malayev, A.A., Gibbs, T.T., and Farb, D.H. (1997) Distinct sites for inverse modulation of N-methyl-D-aspartate receptors by sulfated steroids. *Mol Pharmacol* 52:1113–1123.

Pedersen, L.C., Petrotchenko, E., Shevtsov, S., and Negishi, M. (2002) Crystal structure of the human estrogen SULT-PAPS complex: evidence for catalytic role of Ser137 in the sulfuryl transfer reaction. *J Biol Chem* 277:17928–17932.

Pedersen, L.C., Petrotchenko, E.V., and Negishi, M. (2000) Crystal structure of SULT2A3, human hydroxysteroid SULT. *FEBS Lett* 475:61–64.

Peng, C.T., Chen, J.C., Yeh, K.T., Wang, Y.F., Hou, M.E., Lee, T.P., Shih, M.C., Chang, J.Y., and Chang, J.C. (2003) The relationship among the polymorphisms of SULT1A1,1A2 and different types of cancers in Taiwanese. *Int J Mol Med* 11:85–89.

Petrotchenko, E.V., Doerflein, M.E., Kakuta, Y., Pedersen, L.C., and Negishi, M. (1999) Substrate gating confers steroid specificity to estrogen SULT. *J Biol Chem* 274:30019–30022.

Price, R.A., Cox, N.J., Spielman, R.S., Van Loon, J.A., Maidak, B.L., and Weinshilboum, R.M. (1988) Inheritance of human platelet thermolabile phenol SULT (TL PST) activity. *Genet Epidemiol* 5:1–15.

Price, R.A., Spielman, R.S., Lucena, A.L., van Loon, J.A., Maidak, B.L., and Weinshilboum, R.M. (1989) Genetic polymorphism from human platlet thermostable phenol SULTs (TS PST) activity. *Genetics* 122:905–914.

Price Evans, D.A. (1993) *Genetic Factors in Drug Therapy: Clinical and Molecular Pharmacogenetics*. Cambridge University Press, Cambridge, UK, pp. 364–370.

Radominska, A., Drake, R.R., Zhu, X., Veronese, M.E., Little, J.M., Nowell, S.M., McManus, M.E., Lester, R., and Falany, C.N. (1996) Photoaffinity labelling of human recombinant SULTs with 2-azidoadenosine 3′, 5′- [5′ 32P] biphosphate. *J Biol Chem* 271:3195–3199.

Raftogianis, R.B., Her, C., and Weinshilboum, R.M. (1996) Human phenol SULT pharmacogenetics: STP1 gene cloning and structural characterization. *Pharmacogenetics* 6:473–487.

Raftogianis, R.B., Wood, T.C., Otterness, D.M., Van Loon, J.A., and Weinshilboum, R.M. (1997) Phenol SULT pharmacogenetics in humans: association of common SULT1A1 alleles with TS PST phenotype. *Biochem Biophys Res Commun* 239:298–304.

Raftogianis, R.B., Wood, T.C., and Weinshilboum, R.M. (1999) Human phenol SULTs SULT1A2 and SULT1A1: genetic polymorphisms, allozyme properties, and human liver genotype-phenotype correlations. *Biochem Pharmacol* 58:605–616.

Raftogianis, R.B., Coughtrie, M.W.H., Freimuth, R.R., Buck, J., and Weinshilboum, R.M. (2003) A nomenclature system for the cytosolic SULT (SULT) superfamily, in press.

Rehse, P., Zhou, M., and Lin, S.-X. (2002) Crystal structure of human dehydroepiandrosterone SULT in complex with substrate. *Biochem J* 364:165–171.

Reiter, C., Mwaluko, G., Dunnette, J., Van Loon, J., and Weinshilboum, R. (1983) Thermolabile and thermostable human platelet phenol SULT: substrate specificity and physical separation. *Naunyn Schmiedebergs Arch Pharmacol* 324:140–147.

Richard, K., Hume, R., Kaptein, E., Stanley, E.L., Visser, T.J., and Coughtrie, M.W. (2001) Sulfation of thyroid hormone and dopamine during human development: ontogeny of phenol SULTs and arylsulfatase in liver, lung, and brain. *J Clin Endocrinol Metab* 86:2734–2742.

Rikke, B.A. and Roy, A.K. (1996) Structural relationships among members of the mammalian SULT gene family. *Biochim Biophys Acta* 130:331–338.

Riley, E., Bolton-Grob, R., Liyou, N., Wong, C., Tresillian, M., and McManus, M.E. (2002) Isolation and characterization of a novel rabbit SULT isoform belonging to the SULT1A subfamily. *Int J Biochem Cell Biol* 34:958–969.

Ringer, D.P., Howell, B.A., Norton, T.R., Woulfe, G.W., Duffel, M.W., and Kosanke, S.D. (1994a) Evidence of two separate mechanisms for the decrease in aryl SULT activity in rat liver during early stages of 2-acetylaminofluorene-induced hepatocarcinogenesis. *Mol Carcinog* 9:2–9.

Ringer, D.P., Yerokun, T., and Khan, A.S. (1994b) Molecular mechanisms for the regulation of aryl SULT IV expression during 2-acetylaminofluorene-induced hepatocarcinogenesis in rat. *Chem Biol Interact* 92:343–350.

Rubin, G.L., Harrold, A.J., Mills, J.A., Falany, C.N., and Coughtrie, M.W. (1999) Regulation of sulphotransferase expression in the endometrium during the menstrual cycle, by oral contraceptives and during early pregnancy. *Mol Hum Reprod* 5:995–1002.

Rubin, G.L., Sharp, S., Jones, A.L., Glatt, H., Mills, J.A., and Coughtrie, M.W. (1996) Design, production and characterization of antibodies discriminating between the phenol- and monoamine-sulphating forms of human phenol sulphotransferase. *Xenobiotica* 26:1113–1119.

Runge-Morris, M. (1998) Regulation of SULT gene expression by glucocorticoid hormones and xenobiotics in primary rat hepatocyte culture. *Chem Biol Interact* 109:315–327.

Runge-Morris, M., Rose, K., Falany, C.N., and Kocarek, T.A. (1998) Differential regulation of individual SULT isoforms by phenobarbital in male rat liver. *Drug Metab Dispos* 26:795–801.

Runge-Morris, M., Rose, K., and Kocarek, T.A. (1996) Regulation of rat hepatic SULT gene expression by glucocorticoid hormones. *Drug Metab Dispos* 24:1095–1101.

Runge-Morris, M., Wu, W., and Kocarek, T.A. (1999) Regulation of rat hepatic hydroxysteroid SULT (SULT2-40/41) gene expression by glucocorticoids: evidence for a dual mechanism of transcriptional control. *Mol Pharmacol* 56:1198–1206.

Runge-Morris, M.A. (1997) Regulation of expression of the rodent cytosolic SULTs. *Faseb J* 11:109–117.

Sachse, C., Smith, G., Wilkie, M.J., Barrett, J.H., Waxman, R., Sullivan, F., Forman, D., Bishop, D.T., Wolf, C.R., and Colorectal Cancer Study Group. (2002) A pharmacogenetic study to investigate the role of dietary carcinogens in the etiology of colorectal cancer. *Carcinigenesis* 23:1839–1849.

Sakakibara, Y., Takami, Y., Nakayama, T., Suiko, M., and Liu, M.C. (1998) Localization and functional analysis of the substrate specificity/catalytic domains of human M-form and P-form Phenol. *J Biol Chem* 273:6242–6247.

Satoh, T., Matsui, M., and Tamura, H. (2000) SULTs in a human colon carcinoma cell line, Caco-2. *Biol Pharm Bull* 23:810–814 in process citation.

Schauss, S.J., Henry, T., Palmatier, R., Halvorson, L., Dannenbring, R., and Beckmann, J.D. (1995) Characterization of bovine tracheobronchial phenol sulphotransferase cDNA and detection of mRNA regulation by cortisol. *Biochem J* 311:209–217.

Seth, P., Lunetta, K.L., Bell, D.W., Gray, H., Nasser, S.M., Rhei, E., Kaelin, C.M., Iglehart, D.J., Marks, J.R., Garber, J.E., Haber, D.A., and Polyak, K. (2000) Phenol SULTs: hormonal regulation, polymorphism, and age of onset of breast cancer. *Cancer Res* 60:6859–6863.

Sheweita, S.A. and Tilmisany, A.K. (2003) Cancer ans Phase II drug-metabolizing enzymes. *Curr Drug Metab.* 4:45–58.

Smale, S.T. (1997) Transcription initiation from TATA-less promoters within eukaryotic protein-coding genes. *Biochim Biophys Acta* 1351:73–88.

Smith, C.A., Smith, G., and Wolf, C.R. (1994) Genetic polymorphisms in xenobiotic metabolism. *Eur J Cancer* 30A:1921–1935.

Song, C.S., Echchgadda, I., Baek, B.S., Ahn, S.C., Oh, T., Roy, A.K., and Chatterjee, B. (2001) Dehydroepiandrosterone SULT gene induction by bile acid activated farnesoid X receptor. *J Biol Chem* 276:42549–42556.

Song, C.S., Jung, M.H., Kim, S.C., Hassan, T., Roy, A.K., and Chatterjee, B. (1998) Tissue-specific and androgen-repressible regulation of the rat dehydroepiandrosterone SULT gene promoter. *J Biol Chem* 273:21856–21866.

Song, W.C. (2001) Biochemistry and reproductive endocrinology of estrogen SULT. *Ann NY Acad Sci* 948:43–50.

Sonoda, J., Xie, W., Rosenfeld, J.M., Barwick, J.L., Guzelian, P.S., and Evans, R.M. (2002) Regulation of a xenobiotic sulfonation cascade by nuclear pregnane X receptor (PXR) *Proc Natl Acad Sci USA* 99:13801–13806.

Spink, B.C., Katz, B.H., Hussain, M.M., Pang, S., Connor, S.P., Aldous, K.M., Gierthy, J.F., and Spink, D.C. (2000) SULT1A1 catalyzes 2-methoxyestradiol sulfonation in MCF-7 breast cancer cells. *Carcinogenesis* 21:1947–1957.

Stanley, E.L., Hume, R., Visser, T.J., and Coughtrie, M.W. (2001) Differential expression of SULT enzymes involved in thyroid hormone metabolism during human placental development. *J Clin Endocrinol Metab* 86:5944–5955.

Steiner, M., Bastian, M., Schulz, W.A., Pulte, T., Franke, K.H., Rohring, A., Wolff, J.M., Seiter, H., and Schuff-Werner, P. (2000) Phenol sulphotransferase SULT1A1 polymorphism in prostate cancer: lack of association. *Arch Toxicol* 74:222–225.

Strott, C.A. (1996) Steroid SULTs. *Endocr Rev* 17:670–697.

Strott, C.A. (2002) Sulfonation and molecular action. *Endocr Rev* 23:703–732.

Sueyoshi, T. and Negishi, M. (2001) Phenobarbital response elements of cytochrome P450 genes and nuclear receptors. *Annu Rev Pharmacol Toxicol* 41:123–143.

Sundaram, R.S., Van Loon, J.A., Tucker, R., and Weinshilboum, R.M. (1989) Sulfation pharmacogenetics: correlation of human platelet and small intestinal phenol SULT. *Clin Pharmacol Ther* 46:501–509.

Tamura, H., Morioka, Y., Homma, H., and Matsui, M. (1997) Construction and expression of chimeric rat liver hydroxysteroid SULT isozymes. *Arch Biochem Biophys* 341:309–314.

Temellini, A., Giuliani, L., and Pacifici, G.M. (1991) Interindividual variability in the glucuronidation and sulphation of ethinyloestradiol in human liver. *Br J Clin Pharmacol* 31:661–664.

Teubner, W., Meinl, W., and Glatt, H. (2002) Stable expression of rat SULT 1B1 in V79 cells: activation of benzylic alcohols to mutagens. *Carcinogenesis* 23:1877–1884.

Thorgeirsson, S.S., Glowinski, I.B., and McManus, M.E. (1983) Genotoxicity of N-acetylarylamines in the salmonella/hepatocyte system. *Reviews in Biochemical Toxicology* 5:349–386.

Tsoi, C., Falany, C.N., Morgenstern, R., and Swedmark, S. (2001) Identification of a new subfamily of sulphotransferases: cloning and characterization of canine SULT1D1. *Biochem J* 356:891–897.

Turteltaub, K.W., Vogel, J.S., Frantz, C., Felton, J.S., and McManus, M.E. (1995) Assessment of the DNA adduction and pharmacokinetics of PhIP and MeIOx in rodents at doses approximating human exposure using the technique of accelerator mass spectrometry (AMS) and 32P-postlabeling. *Princess Takamatsu Symp* 23:93–102.

Van Loon, J. and Weinshilboum, R.M. (1984) Human platelet phenol SULT: familial variation in thermal stability of the TS form. *Biochem Genet* 22:997–1014.

Varin, L. and Ibrahim, R. (1992) Novel flavonol 3-SULT. Purification, kinetic properties and partial amino acid sequence. *J Biol Chem* 267:1858–1863.

Varin, L., DeLuca, V., Ibrahim, R.K., and Brisson, N. (1992) Molecular characterization of two plant flavonol SULTs. *Proc Natl Acad Sci USA* 89:1286–1290.

Varin, L., Marsolais, F., and Brisson, N. (1995) Chimeric flavonol SULTs define a domain responsible for substrate and position specificities. *J Biol Chem* 270:12498–12502.

Veronese, M.E., Burgess, W., Zhu, X., and McManus, M.E. (1994) Functional characterization of two human sulphotransferase cDNAs that encode monoamine- and phenol-sulphating forms of phenol sulphotransferase: substrate kinetics, thermal-stability and inhibitor-sensitivity studies. *Biochem J* 302 (Pt 2):497–502.

Visser, T.J. (1994) Role of sulfation in thyroid hormone metabolism. *Chem Biol Interact* 92:293–303.

Wang, J., Falany, J.L., and Falany, C.N. (1998) Expression and characterization of a novel thyroid hormone-sulfating form of cytosolic SULT from human liver. *Mol Pharmacol* 53:274–282.

Wang, Y., Spitz, M.R., Tsou, A.M., Zhang, K., Makan, N., and Wu, X. (2002) SULT (SULT) 1A1 polymorphism as a predisposition factor for lung cancer: a case-control analysis. *Lung Cancer* 35:137–142.

Weinshilboum, R. (1988) Phenol SULT inheritance. *Cell Mol Neurobiol* 8:27–34.

Weinshilboum, R. (1986) Phenol SULT in humans — properties, regulation and function. *Fed Proc* 45:2223–2228.

Weinshilboum, R. (1992) SULT pharmacogenetics, in *Pharmacogenetics of Drug Metabolism*. Kalow, W., Ed., Pergamon Press, New York, pp. 227–242.

Weinshilboum, R.M., Otterness, D.M., Aksoy, I.A., Wood, T.C., Her, C., and Raftogianis, R.B. (1997) Sulfation and SULTs 1: SULT molecular biology: cDNAs and genes. *Faseb J* 11:3–14.

Whittemore, R.M., Pearce, L.B., and Roth, J.A. (1985) Purification and kinetic characterization of a dopamine-sulfating form of phenol SULT from human brain. *Biochemistry* 24:2477–2482.

Whittemore, R.M., Pearce, L.B., and Roth, J.A. (1986) Purification and kinetic characterization of a phenol-sulfating form of phenol SULT from human brain. *Arch Biochem Biophys* 249:464–471.

Wilborn, T.W., Comer, K.A., Dooley, T.P., Reardon, I.M., Heinrikson, R.L., and Falany, C.N. (1993) Sequence analysis and expression of the cDNA for the phenol-sulfating form of human liver phenol SULT. *Mol Pharmacol* 43:70–77.

Windmill, K.F., Christiansen, A., Teusner, J.T., Bhasker, C.R., Birkett, D.J., Zhu, X., and McManus, M.E. (1998) Localization of aryl SULT expression in human tissues using hybridization histochemistry and immunohistochemistry. *Chem Biol Interact* 109:341–346.

Windmill, K.F., McKinnon, R.A., Zhu, X., Gaedigk, A., Grant, D.M., and McManus, M.E. (1997) The role of xenobiotic metabolizing enzymes in arylamine toxicity and carcinogenesis: functional and localization studies. *Mutat Res* 376:153–160.

Wong, C.F., Liyou, N., Leggett, B., Young, J., Johnson, A., and McManus, M.E. (2002) Association of the SULT1A1 R213H polymorphism with colorectal cancer. *Clin Exp Pharmacol Physiol* 29:754–758.

Wood, T.C., Aksoy, I.A., Aksoy, S., and Weinshilboum, R.M. (1994) Human liver thermolabile phenol SULT: cDNA cloning, expression and characterization. *Biochem Biophys Res Commun* 198:1119–1127.

Wu, R.W., Panteleakos, F.N., Kadkhodayan, S., Bolton-Grob, R., McManus, M.E., and Felton, J.S. (2000) Genetically modified Chinese hamster ovary cells for investigating SULT-mediated cytotoxicity and mutation by 2-amino-1-methyl-6-phenylimidazo[4,5-b]pyridine. *Environ Mol Mutagen* 35:57–65.

Yamazoe, Y., Nagata, K., Yoshinari, K., Fujita, K., Shiraga, T., and Iwasaki, K. (1999) SULT catalyzing sulfation of heterocyclic amines. *Cancer Lett* 143:103–107.

Yerokun, T., Etheredge, J.L., Norton, T.R., Carter, H.A., Chung, K.H., Birckbichler, P.J., and Ringer, D.P. (1992) Characterization of a complementary DNA for rat liver aryl SULT IV and use in evaluating the hepatic gene transcript levels of rats at various stages of 2-acetylaminofluorene-induced hepatocarcinogenesis. *Cancer Res* 52:4779–4786.

Yoshinari, K., Petrotchenko, E.V., Pedersen, L.C., and Negishi, M. (2001) Crystal structure-based studies of cytosolic SULT. *J Biochem Mol Toxicol* 15:67–75.

Young, W.F., Laws, E.R., Jr., Sharbrough, F.W., Jr., and Weinshilboum, R.M. (1985) Human phenol SULT: correlation of brain and platelet activities. *J Neurochem* 44:1131–1137.

Zhang, H., Varmalova, O., Falany, F.M., and Lehy, T.S. (1998) Sulfuryl transferase: the catalytic mechanism of human estrogen SULT. *J Biol Chem* 273:10888–10892.

Zheng, W., Xie, D., Cerhan, J.R., Sellers, T.A., Wen, W., and Folsom, A.R. (2001) SULT 1A1 polymorphism, endogenous estrogen exposure, well-done meat intake, and breast cancer risk. *Cancer Epidemiol Biomarkers Prev* 10:89–94.

Zheng, Y., Bergold, A., and Duffel, M.W. (1994) Affinity labeling of aryl SULT IV: identification of a peptide sequence at the binding site for 3'-phosphoadenosine-5'-phosphosulfate. *J Biol Chem* 269:30313–30319.

Zhu, X., Veronese, M.E., Bernard, C.C., Sansom, L.N., and McManus, M.E. (1993a) Identification of two human brain aryl SULT cDNAs. *Biochem Biophys Res Commun* 195:120–127.

Zhu, X., Veronese, M.E., Iocco, P., and McManus, M.E. (1996) cDNA cloning and expression of a new form of human aryl SULT. *Int J Biochem Cell Biol* 28:565–571.

Zhu, X., Veronese, M.E., Sansom, L.N., and McManus, M.E. (1993b) Molecular characterization of a human aryl SULT cDNA. *Biochem Biophys Res Commun* 192:671–676.

Zou, J.Y., Petney, R., and Roth, J.A. (1990) Immunohistochemical detection of phenol SULT-containing neurons in human brain. *J Neurochem* 55:1154–1158.

11 Hydroxysteroid Sulfotransferases SULT2A1, SULT2B1a, and SULT2B1b

Charles A. Strott

CONTENTS

INTRODUCTION

The human hydroxysteroid sulfotransferase family (SULT2) that sulfonates steroids and sterols is divided into two subfamilies, i.e., SULT2A1 and SULT2B1 (Nagata and Yamazoe, 2000). Additionally, the SULT2B1 subfamily consists of two isoforms designated SULT2B1a and SULT2B1b (Her et al., 1998). SULT2A1 is commonly referred to as dehydroepiandrosterone (DHEA) sulfotransferase because DHEA is considered the preferred substrate although this isozyme has a broad substrate predilection (Falany et al., 1989; Hernandez et al., 1992). It would seem that such labeling is arbitrary as SULT2A1, depending on the species, can have an equal or greater propensity for other neutral steroids, e.g., pregnenolone (Fuda et al., 2002). In contrast to SULT2A1, the SULT2B1 isoforms sulfonate DHEA less efficiently and have a much narrower substrate preference (Javitt et al., 2001). Importantly, the SULT2B1 isoforms are structurally distinct from SULT2A1, as well as all other known cognate cytosolic sulfotransferases. Furthermore, the SULT2B1 isoforms, which differ only at their amino-terminal ends, exhibit a more defined substrate selectivity, e.g., SULT2B1a avidly sulfonates pregnenolone but poorly sulfonates cholesterol, whereas SULT2B1b functions as a cholesterol sulfotransferase (Fuda et al., 2002). It is notable that the substrate specificity demonstrated by the SULT2B1 isoforms is in part dependent on the uniqueness of their amino-terminal ends (Fuda et al., 2002). The fact that the SULT2A1 and SULT2B1 isozymes are differentially expressed and display dissimilar substrate propensities strongly suggests that they have explicit biologic roles to play. The challenge is to determine what those biologic roles might actually be.

For many years, the common view of steroid/sterol sulfates was that they simply represented water-soluble metabolic end products suitable for elimination; however, with the discovery that steroid/sterol sulfates can also act as potent bioregulators, that perspective is no longer tenable (Denning et al., 1995; Ikuta et al., 1994; Paul and Purdy, 1992). In addition, it is notable that steroid sulfates circulate at concentrations severalfold higher than their unconjugated forms (Nieschlag et al., 1973; Nishikawa and Strott, 1983), a phenomenon resulting from the avid binding of steroid sulfates to serum proteins, especially albumin (Puche and Nes, 1962). As a consequence, steroid sulfates are cleared slowly (~ 2 orders of magnitude more slowly than their unconjugated counterparts) from the circulation (Wang and Bulbrook, 1967; Wang et al., 1967). In the genomic action of steroid hormones only the unconjugated hormone can interact with its cognate nuclear receptor (Hähnel et al., 1973). Thus, the high concentration of circulating steroid sulfates creates a pool of potential hormones or prehormones following removal of the sulfate moiety by a steroid sulfohydrolase (Haning et al., 1989; Pion et al., 1966).

CLONING, TISSUE EXPRESSION, AND BIOCHEMICAL CHARACTERIZATION

SULT2A1

The gene for human SULT2A1 is approximately 17 kb in size, is composed of 6 exons (Luu-The et al., 1995a; Otterness et al., 1995), and is located on chromosome 19 in the region of 19q13.3 (Otterness et al., 1995b). The cDNA for human SULT2A1 contains an open reading frame of 855 nucleotides that encodes for a 285 amino acid protein (Kong et al., 1992; Otterness et al., 1992). The amino acid sequence of human SULT2A1 is presented in Figure 11.1 where it is aligned with the SULT2B1 isoforms.

Reports on the expression of SULT2A1 in adult human tissues have varied considerably depending on the reporting laboratory and methodology employed. For example, northern analysis involving 16 human tissues revealed expression of SULT2A1 in the adrenal and liver but not the small intestine (Luu-The et al., 1995); however, another study utilizing northern analysis clearly detected SULT2A1 expression in the small intestine (Otterness et al., 1995a). Neither northern analysis demonstrated expression of SULT2A1 in brain, placenta, thymus, heart, testis, ovary, colon, leukocytes, prostate, pancreas, lung, kidney, or skeletal muscle. Immunocytochemistry and western analysis, which have been used to examine a limited number

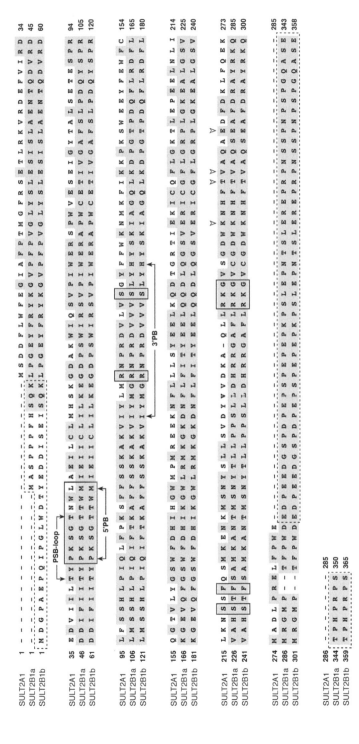

FIGURE 11.1 Amino acid alignment of human hydroxysteroid sulfotransferases SULT2A1, SULT2B1a, and SULT2B1b. Shading indicates identities. Dashed lines outline the extended amino- and carboxy-terminal ends of SULT2B1a and SULT2B1b. Boxed residues delineate conserved amino acids involved in PAPS cofactor interaction (Negishi et al., 2001; Yoshinari et al., 2001). The nucleotide-binding P-loop (PSB) and the 5'- and 3'-phosphate binding motifs (5'PB and 3'PB) are indicated by solid arrows (Negishi et al., 2001; Yoshinari et al., 2001). Open arrowheads specify conserved residues in the KxxxTVxxE dimerization motif (Petrotchenko et al., 2001). Multiple alignment analysis was carried out using the MacVector 7.0 system, which is based on the ClustalW algorithm (Thompson et al., 1994).

of adult human tissues, demonstrated SULT2A1 expression in the liver, adrenal, stomach, and jejunum (Comer and Falany, 1992; Her et al., 1996; Kennerson et al., 1983; Tashiro et al., 2000); however, SULT2A1 was not detected in the kidney (Sharp et al., 1993). Results with RT-PCR have varied substantially. For instance, prominent expression of SULT2A1 was noted in the liver and adrenal with modest expression in the duodenum and ovary, whereas no expression was detected in the stomach, colon, kidney, prostate, or brain (Dooley et al., 2000). In another RT-PCR study, there was strong expression of SULT2A1 in the colon, ovary, prostate, small intestine, and stomach as well as the liver and adrenal, with modest expression in brain, kidney, placenta, spleen, thymus, and thyroid (Javitt et al., 2001). Neither RT-PCR study revealed expression of SULT2A1 in lung or skin. It is notable that three different methods have consistently demonstrated robust expression of human SULT2A1 in the liver and adrenal gland, leaving little doubt about expression of this SULT2 isozyme in these two organ systems. Furthermore, in the human adrenal, SULT2A1 expression is confined to the zona reticularis, whereas there is a lack of expression in the two other cortical zones as well as the medulla (Barker et al., 1994; Falany et al., 1995; Kennerson et al., 1983). Although not totally consistent, the combined studies involving RT-PCR, immunocytochemistry, and northern analysis do strongly argue for the expression of SULT2A1 in the stomach and small intestine. In summary, the expression of SULT2A1 in adult human tissues other than the liver, adrenal, and gastrointestinal tract, if it occurs, will require additional confirmation. For instance, there is evidence by activity measurements suggesting that SULT2A1 expression occurs in the testis (Laatikainen et al., 1969, 1971; Ruokonen et al., 1972; Vihko and Ruokonen, 1975).

Immunocytochemical analysis of human fetal tissues revealed SULT2A1 expression in the adrenal, liver, intestine, testicular Leydig cells, and kidney (Barker et al., 1994; Parker and Schimmer, 1994). SULT2A1 was not detected by immunostaining in the fetal heart, brain, thymus, spleen, pancreas, skeletal muscle, or lung (Parker et al., 1994).

Human SULT2A1 has a broad substrate propensity and will sulfonate a variety of steroids in addition to DHEA, including pregnenolone, estradiol, estrone, testosterone, and androsterone (Falany, 1997; Falany et al., 1994, 1995). Additionally, bile acids are sulfonated by SULT2A1 (Radominska et al., 1990; Song et al., 2001). Interestingly, whereas SULT2A1 is capable of sulfonating cholesterol metabolites, e.g., 27-hydroxycholestserol, it is unable to effectively sulfonate cholesterol itself (Javitt et al., 2001).

SULT2B1

The human *SULT2B1* gene (Figure 11.2), which localizes to chromosome 19q13.3, approximately 500 kb telomeric to the location of the gene for *SULT2A1*, contains an alternative exon I and thus encodes for two mRNAs (Her et al., 1998). The cDNAs for human SULT2B1a and SULT2B1b contain open reading frames that encode for proteins consisting of, respectively, 350 and 365 amino acids (Her et al., 1998). The amino acid sequences of the SULT2B1 isoforms differing only at their amino-terminal ends are depicted in Figure 11.1 where they are aligned with the SULT2A1 isozyme.

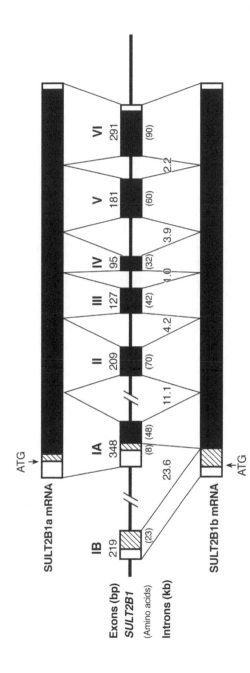

FIGURE 11.2 Schema of the human *SULT2B1* gene demonstrating alternative exons IA and IB. The mRNAs for SULT2B1a and SULT2B1b are shown above and below, respectively, the linearized gene structure. Open boxes represent the 5′- and 3′-untranslated regions. The hatched boxes represent unique amino acid coding regions, whereas the solid boxes represent common amino acid coding regions. The DNA sequence of human *SULT2B1* was obtained from the human genome project and analyzed using cDNA of the SULT2B1 isoforms. A nucleotide alignment analysis, based on the ClustalW algorithm (Thompson et al., 1994), was carried out using the MacVector 7.0 system.

As with SULT2A1, reports on the expression of the SULT2B1 isoforms in human tissues vary considerably depending on the reporting laboratory and the methodology employed. Northern analysis performed using a probe that would detect both SULT2B1 isoforms revealed strong expression in the placenta, prostate, and trachea with faint expression in the small intestine and lung, whereas there was no expression in the heart, brain, spinal cord, liver, skeletal muscle, kidney, pancreas, ovary, testis, stomach, colon, adrenal, thyroid, and bone marrow (Her et al., 1998). RT-PCR revealed that SULT2B1a is expressed in the colon, kidney, ovary, and skin in addition to those tissues detected by northern analysis (Javitt et al., 2001). RT-PCR further revealed that the SULT2B1b isoform is more widely and vigorously expressed than SULT2B1a (Javitt et al., 2001). Another RT-PCR analysis of adult human tissues revealed SULT2B1a and SULT2B1b expression in skin, esophagus, stomach, duodenum, colon, lung, trachea, and prostate, whereas neither isoform was detected in the kidney, liver, ovary, or brain (Dooley et al., 2000). A third RT-PCR study reported expression of SULT2B1a in the colon and ovary but a lack of expression in brain, heart, kidney, spleen, liver, lung, small intestine, muscle, stomach, testis, placenta, pituitary, thyroid, adrenal, pancreas, uterus, prostate, fat, mammary, and leukocytes (Geese and Raftogianis, 2001). On the other hand, SULT2B1b was expressed in the liver, small intestine, placenta, uterus, and prostate as well as the colon and ovary (Geese and Raftogianis, 2001). Interestingly, whereas expression of SULT2B1a but not SULT2B1b was detected in the human fetal brain by RT-PCR (Geese and Raftogianis, 2001), neither isoform has been found to be expressed in the adult brain in any of the studies reported. Real-time PCR demonstrated SULT2B1a expression only in the placenta, prostate, and skin, whereas SULT2B1b expression was noted in the colon, kidney, lung, small intestine, stomach, and thyroid, in addition to the placenta, prostate, and skin (Figure 11.3). Furthermore, expression of SULT2B1b in the placenta, prostate, and skin was markedly higher than its expression in other tissues and was three- to tenfold higher than the expression of SULT2B1a in these tissues (Figure 11.3).

Substrate preferences for the SULT2B1 isoforms, in contradistinction to that for SULT2A1, are more narrowly focused although there is some degree of overlap. The SULT2B1a isoform vigorously sulfonates pregnenolone, modestly sulfonates DHEA, and weakly sulfonates cholesterol (Figure 11.4). Interestingly, however, cholesterol has the lowest K_m for SULT2B1a of the three substrates examined (Table 11.1). In contrast to SULT2B1a, the SULT2B1b isoform avidly sulfonates cholesterol and is thus considered a cholesterol sulfotransferase (Figure 11.4). Although SULT2B1b will also sulfonate pregnenolone (Figure 11.4), its efficiency (k_{cat}/K_m) for doing so is only about one-tenth of its efficiency for sulfonating cholesterol (Table 11.1). SULT2B1b sulfonates DHEA in a fashion similar to the SULT2B1a isoform (Figure 11.4); however, their efficiencies for sulfonating DHEA are quite low in comparison to SULT2A1 (Table 11.1). It is notable that kinetic analyses clearly distinguish the two human SULT2 isozymes (Table 11.1). Importantly, neither SULT2B1 isoform, in contrast to SULT2A1, to all intents and purposes, will sulfonate estradiol, estrone, testosterone, androsterone, or bile acids (Geese and Raftogianis, 2001; Javitt et al., 2001).

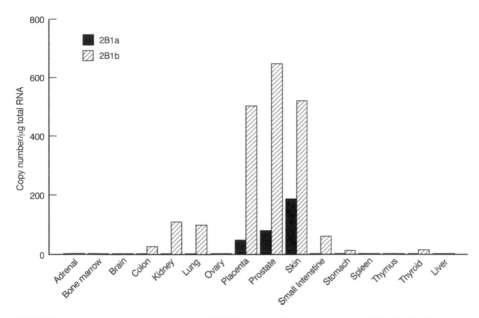

FIGURE 11.3 Expression of human SULT2B1a (solid columns) and SULT2B1b (hatched columns) mRNAs as determined by quantitative PCR. Real-time PCR was performed using a fluorescence temperature cycler and SYBR Green I as a double-stranded DNA-specific binding dye. This technique continuously monitors the cycle-by-cycle accumulation of fluorescently labeled PCR product. The top of each column signifies the average copy number of duplicate determinations (author's unpublished data).

Of further interest is the finding that the SULT2B1b isoform has rather strict structural requirements for optimal reactivity and catalytic efficiency that are related to the configuration of the fused sterol ring as well as the location and spatial orientation of the hydroxyl acceptor group. For instance, 5α-reduction of cholesterol produces the metabolite cholestanol, a planar molecule similar to cholesterol, which SULT2B1b will effectively sulfonate (Figure 11.5). On the other hand, SULT2B1b poorly sulfonates coprostanol (Figure 11.5), the cholesterol metabolite resulting from 5β-reduction, which creates a severe bend in the molecule. These findings indicate the importance of the configuration of the fused-ring structure. SULT2B1b efficiently sulfonates the 3β-hydoxyl moiety of cholesterol, whereas it does not sulfonate the 3α-hydroxyl array of epicholesterol, indicating the importance of the spatial orientation of the acceptor group (Figure 11.5). What's more, although SULT2B1b will sulfonate the 3β-hydroxyl moiety of 27-hydroxycholesterol, it will not sulfonate the 27-hydroxyl group of this oxysterol, indicating selectivity for the acceptor location (Javitt et al., 2001).

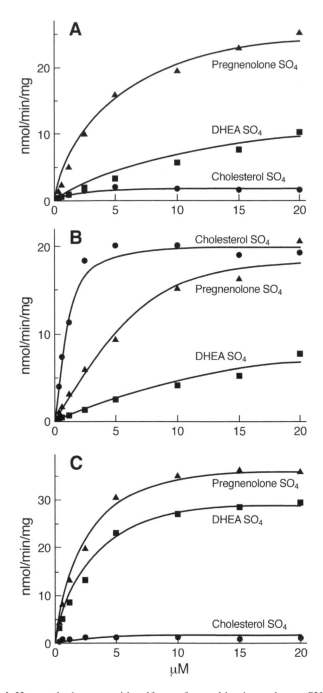

FIGURE 11.4 Human hydroxysteroid sulfotransferase kinetic analyses. SULT2B1a (A), SULT2B1b (B), and SULT2A1 (C) were overexpressed as GST fusion proteins in bacteria, cleaved, and affinity purified. Isolated proteins (>90% pure) were analyzed using pregnenolone, cholesterol, and dehydroepiandrosterone (DHEA) as substrates (Fuda et al., 2002).

TABLE 11.1
Human SULT2 Kinetic Data

Substrate	Cholesterol	Pregnenolone	DHEA
SULT2B1a			
k_{cat} (s^{-1} × 10^{-3})	1.60 ± 0.68	17.0 ± 3.30	17.3 ± 1.22
K_m (M × 10^{-6})	2.30 ± 1.55	4.44 ± 0.71	24.6 ± 0.98
k_{cat}/K_m (M^{-1}S^{-1} × 10^3)	0.87 ± 0.42	3.84 ± 0.49	0.70 ± 0.03
SULT2B1b			
k_{cat} (s^{-1} × 10^{-3})	13.0 ± 4.50	29.0 ± 4.20	8.28 ± 4.05
K_m (M × 10^{-6})	1.20 ± 0.49	17.3 ± 2.73	18.4 ± 5.56
k_{cat}/K_m (M^{-1}S^{-1} × 10^3)	11.1 ± 1.20	1.68 ± 0.40	0.43 ± 0.09
SULT2A1			
k_{cat} (s^{-1} × 10^{-3})	WA	27.3 ± 5.10	20.1 ± 5.22
K_m (M × 10^{-6})	WA	1.93 ± 0.31	3.74 ± 0.18
k_{cat}/K_m (M^{-1}S^{-1} × 10^3)	WA	14.6 ± 4.39	5.40 ± 1.53

Data are presented as the Mean ± SD and are derived from three experiments in duplicate. WA indicates weak activity precluding accurate measurement.

STRUCTURE AND FUNCTION

SULT2A1

The first steroid sulfotransferase structure to be solved was that of mouse estrogen sulfotransferase (SULT1E1; Kakuta et al., 1997). Mouse SULT1E1 was cocrystallized with PAP (desulfonated PAPS) followed by soaking with estradiol. This enabled the structural features important for both cofactor and substrate binding to be determined. Subsequently, the three-dimensional structure of three human steroid sulfotransferases were solved, i.e., dopamine/catecholamine sulfotransferase SULT1A3 (Bidwell et al., 1999; Dajani et al., 1999), SULT2A1 (Pedersen et al., 2000), and SULT1E1 (Pedersen et al., 2002). Importantly, key structural elements important in protein–cofactor interaction of cytosolic sulfotransferases are completely conserved and superimposable in the four enzymes whose structures have been solved (Kakuta et al., 1997; Pedersen et al., 2002; Yoshinari et al., 2001). Furthermore, structural features of the cofactor-binding site, i.e., the PSB loop (P-loop motif found at phosphate binding sites of nucleotide-binding proteins), as well as regions interacting with the 5′ (5′PB) and 3′ (3′PB) phosphate groups of PAPS are also highly conserved (cf. Figure 11.1). It is important to note, however, that only the crystal structure of mouse SULT1E1 has been solved in the presence of substrate (Kakuta et al., 1997), and this achievement has not as yet been attained with any SULT2 enzyme. Interestingly, however, a comprehensive analysis of the three-dimensional structure of soluble sulfotransferases including human SULT2A1 has recently been presented (Negishi et al., 2001). All soluble sulfotransferases are

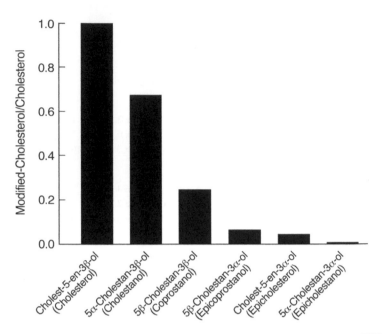

FIGURE 11.5 Sulfonation of cholesterol and cholesterol metabolites by human SULT2B1b. SULT2B1b was prepared by overexpression as a GST fusion protein in bacteria followed by cleavage and affinity purification. Sulfotransferase activity was determined using cholesterol or modified cholesterol at a 20 μM substrate concentration. Results are expressed as the amount of modified cholesterol sulfate formed (nmol/min/mg protein) relative to the amount of cholesterol sulfate formed (nmol/min/mg protein; reproduced by permission of N. B. Javitt, unpublished data).

globular proteins composed of a single α/β domain with the characteristic five-stranded β-sheet, which constitutes the core of the PAPS-binding and catalytic sites (Negishi et al., 2001). The conserved catalytic core has led to a proposal for a common mechanism involved in the sulfonate transfer reaction (Negishi et al., 2001).

Steroid sulfotransferases are generally homodimers in solution, and the structural elements responsible for the dimerization have been identified (Petrotchenko et al., 2001). A ten-residue segment near the carboxy terminus of SULT proteins forms a hydrophobic zipper-like structure further enforced by ion pairs at both ends. This amino acid stretch containing a critical valine, which is conserved as a KxxxTVxxxE motif in nearly all cytosolic sulfotransferases including the human SULT2 isozymes (cf. Figure 11.1), appears to be the common protein–protein interaction motif mediating dimerization phenomena (Petrotchenko et al., 2001). Nevertheless, the physiological significance of dimerization in the function of cytosolic sulfotransferases remains to be defined (Negishi et al., 2001).

SULT2B1

While the human SULT2B1 isoforms are considered to be hydroxysteroid sulfotransferases, they are, nevertheless, structurally distinct when compared to SULT2A1 (cf.

Figure 11.1). In comparing the SULT2A1 and SULT2B1 isozymes, the most out-standing distinction between them is the extended amino- and carboxy-terminal ends of the latter proteins. Overall, the SULT2A1 and SULT2B1 proteins are ~37% identical. However, if the extended amino- and carboxy-terminal ends of the SULT2B1 isoforms are excluded, identities increase to ~48%. All previously cloned members of the mammalian cytosolic sulfotransferase superfamily, i.e., estrogen and phenol sulfotransferases as well as the hydroxysteroid sulfotransferases, have sizes that range from 282 to 295 amino acids (Rikke and Roy, 1996; Weinshilboum et al., 1997), whereas human SULT2B1a and SULT2B1b consist of 350 and 365 amino acids, respectively. The extended amino- and carboxy-terminal ends of the SULT2B1 iso-forms, notwithstanding, there is a significant structural similarity between the SULT2A1 and SULT2B1 isozymes in their core regions. Most notably, key structural elements important in protein–PAPS interaction as well as the dimerization motif, which are described above under SULT2A1, are conserved in the SULT2B1 isoforms (cf. Figure 11.1).

The human SULT2B1 isoforms, which differ only at their amino-terminal ends, are produced by employment of an alternative exon I along with differential splicing (Her et al., 1998). The functional significance of the extended carboxy-terminal end of the SULT2B1 isoforms is not presently appreciated. One speculation is that this region, which is proline-enriched, might play a role in protein–protein interactions (Javitt et al., 2001). Notably, the terminal 53 amino acids of the relatively long carboxyl ends, which are common to both proteins, can be removed without causing a significant change in the catalytic behavior of either isoform (Figure 11.6). On the other hand, removal of the unique amino termini of the two isoforms yields inter-esting results. That is, removal of the 23 residues from the amino terminus of SULTT2B1b, which are unique to this isoform (cf. Figure 11.1), results in an almost complete loss of cholesterol sulfotransferase activity (Figure 11.6), whereas removal of the 8 residues from the amino terminus of SULT2B1a, which are unique to this isoform (cf. Figure 11.1), does not alter pregnenolone sulfotransferase activity (Fig-ure 11.6). It is noteworthy that exon IB of the human *SULT2B1* gene encodes for only the unique amino-terminal 23 amino acids of SULT2B1b, whereas exon IA encodes for the unique amino-terminal 8 amino acids of SULT2B1a plus an additional 48 amino acids that are common to both isoforms. Thus, if the gene for human *SULT2B1* (cf. Figure 11.2) employs exon IB, cholesterol sulfotransferase is synthesized, whereas if the gene employs exon IA, pregnenolone sulfotransferase is produced (Fuda et al., 2002). This realization strongly suggests that differential expression of the SULT2B1 isoforms has significant biologic implications.

BIOLOGY OF STEROID/STEROL SULFONATION

SULT2A1

The lack of substrate specificity obscures the biologic roles of SULT2A1, which in humans is primarily expressed in the liver, gastrointestinal tract, and adrenal cortex. Taken together these characteristics suggest that the evolutionary plan for human SULT2A1 lay in the area of general metabolism and detoxification, whereby

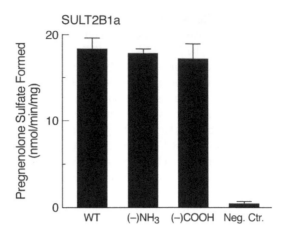

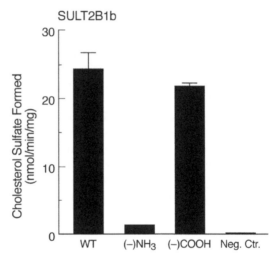

FIGURE 11.6 Sulfotransferase activity after truncation of the amino- and carboxy-terminal ends of SULT2B1a and SULT2B1b. Wild type (WT) and constructs of each isoform lacking either the unique amino-terminal end (-NH₃) or 52 amino acids from the common carboxy-terminal end (-COOH; cf. Figure 11.1) were prepared as GST fusion proteins, cleaved, affinity purified, and assayed for either pregnenolone (SULT2B1a) or cholesterol (SULT2B1b) sulfotransferase activity. Negative control (Neg. Ctr.) indicates the result when a WT preparation was assayed in the absence of the PAPS cofactor. The top of each column represents the mean of five replicates, and error bars indicate the standard deviation (Fuda et al., 2002).

hydrophobic compounds are converted into hydrophilic substances antecedent to elimination. Such a view is supported, in part, by the fact that human SULT2A1 has broad substrate specificity and will sulfonate a variety of steroids including androgens and estrogens (Falany, 1997; Falany et al., 1994, 1995; Falany and Wilborn, 1994). In addition, human SULT2A1 also sulfonates bile acids (Radominska et al., 1990) and is considered the principal bile acid sulfonating enzyme in human liver where it functions to protect the liver from the hepatotoxic effects of secondary bile acids (Falany, 1991). Additionally, sulfonation of secondary bile acids reduces their cytotoxic effect on colon epithelial cells (Halvorsen et al., 1999). Finally, given the fact that bile acids are end products of cholesterol catabolism (Russell, 1999), it is of interest that the sulfonation of bile acids by facilitating their clearance also contributes to cholesterol homeostasis (Yamashita and Setchell, 1994).

SULT2A1 may have more specific physiologic roles in nonhuman species than it does in humans. For instance, in the rat there is evidence that hepatic SULT2A1 plays an important role in regulating androgen responsiveness of the liver during maturation and aging by controlling the degree of testosterone sulfoconjugation (Roy, 1992). Androgen sensitivity, which in the rat liver goes through age-related phases, can be monitored by the androgen-dependent expression of certain target genes. During the androgen-insensitive phase in immature animals, hepatic SULT2A1 expression and concomitant testosterone sulfoconjugation are high, whereas during the androgen-sensitive phase in mature animals expression and sulfoconjugation are low only to rise again in older animals when the liver once more becomes androgen insensitive (Roy, 1992). In other studies involving the rat, expression of SULT2A1 in the brain has been demonstrated by activity, western analysis, and RT-PCR (Aldred and Waring, 1999; Rajkowski et al., 1997; Shimada et al., 2001). Furthermore, the expression of SULT2A1 in rat brain correlates with the presence of DHEA sulfate and pregnenolone sulfate in this tissue (Corpechot et al., 1981, 1983). The significance of this finding is that the sulfates of DHEA and pregnenolone are potent antagonists of the $GABA_A$ receptor that regulates chloride channels and thus act as excitatory neurosteroids (Majewska, 1992; Paul and Purdy, 1992). Although the sulfates of DHEA and pregnenolone have been quantified in multiple regions of the human brain (Lanthier and Patwardhan, 1986), their involvement as functional neurosteroids, as in the case of the rat brain, has not been described. Thus, the relevance to human neurochemistry of these interesting observations involving the rat brain remains problematic (Wolf and Kirschbaum, 1999).

No discussion of the biologic significance of SULT2A1 in humans can occur without the subject of DHEA sulfate coming up. That is because DHEA sulfate is the most abundant steroid in the circulation, principally as a result of adrenal secretion (Baulieu et al., 1965; Vande Wiele et al., 1963). In the adrenal, DHEA sulfate arises solely from the cortical zona reticularis (Endoh et al., 1996), and as noted above, expression of SULT2A1 in the human adrenal is confined to this region of the cortex (Kennerson et al., 1983). Very little plasma DHEA sulfate emanates from the testis or ovary (Migeon et al., 1957). The fascination with DHEA sulfate in humans relates to its unique production and secretory pattern.

The production rate of estriol, the estrogen of pregnancy, can reach >60 mg/day during the last trimester of gestation (Katz and Kappas, 1967) and is derived primarily from DHEA sulfate, which is produced in large amounts by the fetal adrenal (Siiteri and Macdonald, 1963, 1966). Postnatally, as a result of fetal zone involution, adrenal secretion of DHEA sulfate falls rapidly to a low level and remains low during early childhood, a period associated with a poorly developed zona reticularis (Dhom, 1973). Then adrenal secretion of DHEA sulfate starts to rise again before puberty, parallel with the progressive development of the zona reticularis (Gell et al., 1998; Suzuki et al., 2000). Circulating DHEA sulfate reaches a high level once more in young adulthood, after which it progressively declines throughout life (Hopper and Yen, 1975; Hornsby, 1995; Orentreich et al., 1984). From its peak in the 3rd decade, circulating DHEA sulfate falls at a rate of ~2% per year, reaching a residual value of 10 to 20% during the 8th through 9th decades (Baulieu, 1996). Despite an extensive effort covering many years, an explanation for both the youthful increase and age-related decrease in the production of DHEA sulfate by the human adrenal has not been forthcoming. Furthermore, the physiologic significance of this phenomenon continues to be discussed and argued without a clear resolution and thus to this day it remains an arcane phenomenon (Baulieu, 1996; Laughlin and Barrett-Connor, 2000; Yen, 2001).

SULT2B1

In contrast to the broad range of substrate reactivity demonstrated by SULT2A1, the SULT2B1 isoforms display narrower substrate preferences. This fact in association with the finding that the SULT2B1 isoforms are differentially expressed has interesting biologic implications. Thus, in considering the significance of the distinct substrate predilections demonstrated by the human SULT2B1 isoforms, particularly in relation to their differential expression patterns, two organ systems loom as especially interesting, i.e., skin and brain. It is now recognized that cholesterol sulfate plays an essential role in skin development and creation of the epidermal barrier (Hanley et al., 2001; Jetten et al., 1989; Kawabe et al., 1998; Kuroki et al., 2000). Furthermore, as determined by real-time PCR, expression of the human SULT2B1b isoform, which we now recognize as a cholesterol sulfotransferase, is higher in skin than in other organ systems with the possible exception of the placenta and prostate (cf. Figure 11.3). Although expression of the SULT2B1 isoforms in the adult central nervous system remains to be examined in great detail, expression is negative by northern analysis and RT-PCR, the human fetal brain, as determined by RT-PCR, has been shown to express only the SULT2B1a isoform (Geese and Raftogianis, 2001). Interestingly, the mouse ortholog of SULT2B1a appears to be almost exclusively expressed in the central nervous system (Shimizu et al., 2003). The significance of these observations is that in animal models, pregnenolone sulfate, which is produced most efficiently by the action of the SULT2B1a isoform (cf. Table 11.1), is recognized as an essential neurosteroid (Alomary et al., 2001; Baulieu et al., 2001; Engel and Grant, 2001; Plassart-Schiess and Baulieu, 2001).

The importance of cholesterol sulfate and thus the enzyme that produces it cannot be overstated. For instance, cholesterol sulfate is widely distributed in human tissues (Bleau et al., 1972, 1974; Drayer and Lieberman, 1967; Moser et al., 1966; Roberts, 1987); it has a blood production rate that ranges from 35 to 163 mg/day (Gurpide et al., 1966), and it is as abundant in the circulation as DHEA sulfate where their concentrations essentially overlap (Bergner and Shapiro, 1981; Muskiet et al., 1983; Serizawa et al., 1987, 1989). Although a clear physiologic role for cholesterol sulfate is not always apparent, various functions have been considered and explored. For instance, regulation of cholesterol synthesis (Williams et al., 1985), sperm capacitation (Langlais et al., 1981), and thrombin and plasmin activities (Iwamori et al., 1999) have been demonstrated. Furthermore, cholesterol sulfate has been shown to activate specific protein kinase C isozymes (Denning et al., 1995), particularly the η isoform (Ikuta et al., 1994). Cholesterol sulfate can also serve as a substrate for adrenal (Mason and Hemsell, 1982) and ovarian steroidogenesis (Tuckey, 1990). However, the clearest and most investigated physiologic role for cholesterol sulfate is in keratinocyte differentiation and development of the epidermal barrier (Jetten et al., 1989; Kawabe et al., 1998). For example, cholesterol sulfate induces transcription of the gene for transglutaminase I, an essential protein involved in the cross-linking of barrier proteins (Kawabe et al., 1998). Additionally, cholesterol sulfate by functioning as a transcriptional regulator increases mRNA and protein levels of involucrin (Hanley et al., 2001), a major cross-linked protein constituent of the insoluble cornified cell envelope of stratified squamous epithelia (Blumenberg and Tomic-Canic, 1997; Steinert and Marekov, 1997).

REFERENCES

Aldred, S. and Waring, R.H., 1999, Localization of dehydroepiandrosterone sulphotransferase in adult rat brain. *Brain Research Bulletin*, 48, 291–296.

Alomary, A.A., Fitzgerald, R.L., and Purdy, R.H., 2001, Neurosteroid analysis. *International Review of Neurobiology*, 46, 97–115.

Barker, E.V., Hume, R., Hallas, A., and Coughtrie, M.W.H., 1994, Dehydroepiandrosterone sulfotransferase in the developing human fetus: quantitative biochemical and immunological characterization of the hepatic, renal, and adrenal enzymes. *Endocrinology*, 134, 982–989.

Baulieu, E.-E., Corpechot, C., Dray, F., Emiliozzi, R., Lebeau, M.-C., Mauvais-Jarvis, P., and Robel, P., 1965, An andrenal-secreted androgen: dehydroepiandrosterone sulfate. Its metabolism and a tentative generalization on the metabolism of other steroid conjugates in man. *Recent Progress in Hormone Research*, 21, 411–500.

Baulieu, E.E., 1996, Dehydroepiandrosterone (DHEA): a fountain of youth? *Journal of Clinical Endocrinology and Metabolism*, 81, 3147–3151.

Baulieu, E.E., Robel, P., and Schumacher, M., 2001, Neurosteroids: beginning of the story. *International Review of Neurobiology*, 46, 1–31.

Bergner, E.A. and Shapiro, L.J., 1981, Increased cholesterol sulfate in plasma and red blood cell membranes of steroid sulfatase deficient patients. *Journal of Clinical Endocrinology and Metabolism*, 53, 221–223.

Bidwell, L.M., McManus, M.E., Gaedigk, A., Kakuta, Y., Negishi, M., Pedersen, L., and Martin, J.L., 1999, Crystal structure of human catecholamine sulfotransferase. *Journal of Molecular Biology*, 293, 521–530.

Bleau, G., Bodley, F.H., Longpre, J., Chapdelaine, A., and Roberts, K.D., 1974, Cholesterol sulfate I: occurrence and possible function as an amphipathic lipid in the membrane of the human erythrocyte. *Biochimica et Biophysica Acta*, 352, 1–9.

Bleau, G., Chapdelaine, A., and Roberts, K.D., 1972, The assay of cholesterol sulfate in biological material by enzymatic radioisotopic displacement. *Canadian Journal of Biochemistry*, 50, 277–286.

Blumenberg, M. and Tomic-Canic, M., 1997, Human epidermal keratinocyte: keratinization processes. *EXS*, 78, 1–29.

Comer, K.A. and Falany, C.N., 1992, Immunological characterization of dehydroepiandrosterone sulfotransferase from human liver and adrenal. *Molecular Pharmacology*, 41, 645–651.

Corpechot, C., Robel, P., Axelson, M., and Sjovall, J., 1981, Characterization and measurement of dehydroepiandrosterone sulfate in rat brain. *Proceedings of the National Academy of Sciences, USA*, 78, 4704–4707.

Corpechot, C., Synguelakis, M., Talha, S., Axelson, M., Sjovall, J., Vihko, R., Baulieu, E.E., and Robel, P., 1983, Pregnenolone and its sulfate ester in the rat brain. *Brain Research*, 270, 119–125.

Dajani, R., Cleasby, A., Neu, M., Wonacott, A.J., Jhoti, H., Hood, A.M., Modi, S., Hersey, A., Taskinen, J., Cooke, R.M., Manchee, G.R., and Coughtrie, M.W.H., 1999, X-ray crystal structure of human dopamine sulfotransferase, SULT1A3. *Journal of Biological Chemistry*, 274, 37862–37868.

Denning, M.F., Kazanietz, M.G., Blumberg, P.M., and Yuspa, S.H., 1995, Cholesterol sulfate activates multiple protein kinase C isozymes and induces granular cell differentiation in cultured murine keratinocytes. *Cell Growth and Differentiation*, 6, 1619–1626.

Dhom, G., 1973, The prepuberal and puberal growth of the adrenal (adrenarche). *Beitrage zur Pathologie*, 150, 357–377.

Dooley, T.P., Haldeman-Cahill, R., Joiner, J., and Wilborn, T.W., 2000, Expression profiling of human sulfotransferase and sulfatase gene superfamilies in epithelial tissues and cultured cells. *Biochemical and Biophysical Research Communications*, 277, 236–245.

Drayer, N.M. and Lieberman, S., 1967, Isolation of cholesterol sulfate from human aortas and adrenal tumors. *Journal of Clinical Endocrinology and Metabolism*, 27, 136–139.

Endoh, A., Kristiansen, S.B., Casson, P.R., Buster, J.E., and Hornsby, P.J., 1996, The zona reticularis is the site of biosynthesis of dehydroepiandrosterone and dehydroepiandrosterone sulfate in the adult adrenal cortex resulting from its low expression of 3β-hydroxysteroid dehydrogenase. *Journal of Clinical Endocrinology and Metabolism*, 81, 3558–3565.

Engel, S.R. and Grant, K.A., 2001, Neurosteroids and behavior. *International Review of Neurobiology*, 46, 321–348.

Falany, C.N., 1991, Molecular enzymology of human liver cytosolic sulfotransferases. *Trends in Pharmacological Sciences*, 12, 255–259.

Falany, C.N., 1997, Enzymology of human cytosolic sulfotransferases. *FASEB Journal*, 11, 206–216.

Falany, C.N., Comer, K.A., Dooley, T.P., and Glatt, H., 1995, Human dehydroepiandrosterone sulfotransferase: purification, molecular cloning and characterization. *Annals of the New York Academy of Science*, 774, 59–72.

Falany, C.N., Vazquez, M.E., and Kalb, J.M., 1989, Purification and characterization of human dehydroepiandrosterone sulphotransferase. *Biochemical Journal*, 260, 641–646.

Falany, C.N., Wheeler, J., Oh, T.S., and Falany, J.L., 1994, Steroid sulfation by expressed human cytosolic sulfotransferases. *Journal of Steroid Biochemistry and Molecular Biology*, 48, 369–375.

Falany, C.N. and Wilborn, T.W., 1994, Biochemistry of cytosolic sulfotransferases involved in bioactivation. *Advances in Pharmacology*, 27, 301–329.

Fuda, H., Lee, Y.C., Shimizu, C., Javitt, N.B., and Strott, C.A., 2002, Mutational analysis of human hydroxysteroid sulfotransferase SULT2B1 isoforms reveals that exon 1B of the *SULT2B1* gene produces cholesterol sulfotransferase, whereas exon 1A yields pregnenolone sulfotransferase. *Journal of Biological Chemistry*, 277, 36161–36166.

Geese, W.J. and Raftogianis, R.B., 2001, Biochemical characterization and tissue distribution of human SULT2B1. *Biochemical and Biophysical Research Communications*, 288, 280–289.

Gell, J.S., Carr, B.R., Sasano, H., Atkins, B., Margraf, L., Mason, J.I., and Rainey, W.E., 1998, Adrenarche results from development of a 3βa-hydroxysteroid dehydrogenase-deficient adrenal reticularis. *Journal of Clinical Endocrinology and Metabolism*, 83, 3695–3701.

Gurpide, E., Roberts, K.D., Welch, M.T., Bandy, L., and Lieberman, S., 1966, Studies on the metabolism of blood-borne cholesterol sulfate. *Biochemistry*, 5, 3352–3362.

Hähnel, R., Twaddle, E., and Ratajczak, T., 1973, The specificity of the estrogen receptor of human uterus. *Journal of Steroid Biochemistry*, 4, 21–31.

Halvorsen, B., Kase, B.F., Prydz, K., Garagozlian, S., Andresen, M.S., and Kolset, S.O., 1999, Sulphation of lithocholic acid in the colon–carcinoma cell line CaCo-2. *Biochemical Journal*, 343, 533–539.

Haning, J.R., Jr., Chabot, M., Flood, C.A., Hackett, R., and Longcope, C., 1989, Metabolic clearance rate (MCR) of dehydroepiandrosterone sulfate (DS), its metabolism to androstenedione, testosterone and dihydrotestosterone, and the effect of increased plasma DS concentration on DS MCR in normal women. *Journal of Clinical Endocrinology and Metabolism*, 69, 1047–1052.

Hanley, K., Wood, L., Ng, D.C., He, S.S., Lau, P., Moser, A., Elias, P.M., Bikle, D.D., Williams, M.L., and Feingold, K.R., 2001, Cholesterol sulfate stimulates involucrin transcription in keratinocytes by increasing Fra-1, Fra-2, and Jun D. *Journal of Lipid Research*, 42, 390–398.

Her, C., Raftogianis, R., and Weinshilboum, R.M., 1996, Human phenol sulfotransferase (STP)2 gene: molecular cloning, structural characterization and chromosomal localization. *Genomics*, 33, 409–420.

Her, C., Wood, T.C., Eichler, E.E., Mohrenweiser, H.W., Ramagli, L.S., Siciliano, M.J., and Weinshilboum, R.M., 1998, Human hydroxysteroid sulfotransferase SULT2B1: two enzymes encoded by a single chromosome 19 gene. *Genomics*, 53, 284–295.

Hernandez, J.S., Watson, R.W.G., Wood, T.C., and Weinshilboum, R.M., 1992, Sulfation of estrone and 17-β-estradiol in human liver: catalysis by thermostable phenol sulfotransferase and by dehydroepiandrosterone sulfotransferase. *Drug Metabolism and Disposition: The Biological Fate of Chemicals*, 20, 413–422.

Hopper, B.R. and Yen, S.S.C., 1975, Circulating concentrations of dehydroepiandrosterone and dehydroepiandrosterone sulfate during puberty. *Journal of Clinical Endocrinology and Metabolism*, 40, 458–461.

Hornsby, P.J., 1995, Biosynthesis of DHEAS by the human adrenal cortex and its age-related decline. *Annals of the New York Academy of Sciences*, 774, 29–46.

Ikuta, T., Chida, K., Tajima, O., Matsuura, Y., Iwamori, M., Ueda, Y., Mizuno, K., Ohno, S., and Kuroki, T., 1994, Cholesterol sulfate, a novel activator for the eta isoform of protein kinase C. *Cell Growth and Differentiation*, 5, 943–947.

Iwamori, M., Iwamori, Y., and Ito, N., 1999, Regulation of the activities of thrombin and plasmin by cholesterol sulfate as a physiological inhibitor in human plasma. *Journal of Biological Chemistry*, 125, 594–601.

Javitt, N.B., Lee, Y.C., Shimizu, C., Fuda, H., and Strott, C.A., 2001, Cholesterol and hydroxycholesterol sulfotransferases: identification, distinction from dehydroepiandrosterone sulfotransferase, and differential tissue expression. *Endocrinology*, 142, 2978–2984.

Jetten, A.M., George, M.A., Nervi, C., Boone, L.R., and Rearick, J.I., 1989, Increased cholesterolsulfate and cholesterol sulfotransferase activity in relation to the multistep process of differentiation in human epidermal keratinocytes. *Journal of Investigative Dermatology*, 92, 203–209.

Kakuta, Y., Pedersen, L.G., Carter, C.W., Negishi, M., and Pedersen, L.C., 1997, Crystal structure of estrogen sulphotransferase. *Nature Structural Biology*, 4, 904–908.

Katz, F.H. and Kappas, A., 1967, The effects of estradiol and estriol on plasma levels of cortisol and thyroid hormone-binding globulins and on aldosterone and cortisol secretion rates in man. *Journal of Clinical Investigation*, 46, 1768–1777.

Kawabe, S., Ikuta, T., Ohba, M., Chida, K., Ueda, Y., Yamanishi, K., and Kuroki, T., 1998, Cholesterol sulfate activates transcription of transglutaminase 1 gene in normal human keratinocytes. *Journal of Investigative Dermatology*, 111, 1098–1102.

Kennerson, A.R., McDonald, D.A., and Adams, J.B., 1983, Dehydroepiandrosterone sulfotransferase localization in human adrenal glands: a light and electron microscopic study. *Journal of Clinical Endocrinology and Metabolism*, 56, 786–790.

Kong, A.-N.T., Yang, L., Ma, M., Tao, D., and Bjornsson, T.D., 1992, Molecular cloning of the alcohol/hydroxysteroid form (hST$_a$) of sulfotransferase from human liver. *Biochemical and Biophysical Research Communications*, 187, 448–454.

Kuroki, T., Ikuta, T., Kashiwagi, M., Kawabe, S., Ohba, M., Huh, N., Mizuno, K., Ohno, S., Yamada, E., and Chida, K., 2000, Cholesterol sulfate, an activator of protein kinase C mediating squamous cell differentiation: a review. *Mutation Research*, 462, 189–195.

Laatikainen, T., Laitinen, E.A., and Vihko, R., 1969, Secretion of neutral steroid sulfates by the human testis. *Journal of Clinical Endocrinology*, 29, 219–224.

Laatikainen, T., Laitinen, E.A., and Vihko, R., 1971, Secretion of free and sulphate conjugated neutral steroids by the human testis: effect of administration of human chorionic gonadotrophin. *Journal of Clinical Endocrinology*, 32, 59–64.

Langlais, J., Zollinger, M., Plante, L., Chapdelaine, A., Bleau, G., and Roberts, K.D., 1981, Localization of cholesteryl sulfate in human spermatozoa in support of a hypothesis for the mechanism of capacitation. *Proceedings of the National Academy of Sciences USA*, 78, 7266–7270.

Lanthier, A. and Patwardhan, V.V., 1986, Sex steroids and 5-en-3β-hydroxysteroids in specific regions of the human brain and cranial nerves. *Journal of Steroid Biochemistry*, 25, 445–449.

Laughlin, G.A. and Barrett-Connor, E., 2000, Sexual dimorphism in the influence of advanced aging on adrenal hormone levels: the Rancho Bernardo Study. *Journal of Clinical Endocrinology and Metabolism*, 85, 3561–3568.

Luu-The, V., Dufort, I., Paquet, N., Reimnitz, G., and Labrie, F., 1995, Structural characterization and expression of the human dehydroepiandrosterone sulfotransferase gene. *DNA and Cell Biology*, 14, 511–518.

Majewska, M.D., 1992, Neurosteroids: endogenous bimodal modulators of the GABA-A receptor: mechanism of action and physiological significance. *Progress in Neurobiology*, 38, 379–395.

Mason, J.I. and Hemsell, P.G., 1982, Cholesterol sulfate metabolism in human fetal adrenal mitochondria. *Endocrinology*, 111, 208–213.

Migeon, C.J., Keller, A.R., Lawrence, B., and Shepard, T.H., 1957, Dehydroepiandrosterone and androsterone levels in human plasma. Effect of age and sex, day-to-day and diurnal variations. *Journal of Clinical Endocrinology and Metabolism*, 17, 1051–1062.

Moser, H.W., Moser, A.B., and Orr, J.C., 1966, Preliminary observations on the occurrence of cholesterol sulfate in man. *Biochimica et Biophysica Acta*, 116, 146–155.

Muskiet, F.A., Jansen, G., Wolthers, B.G., Marinkovic-Ilsen, A., and van Voorst Vader, P.C., 1983, Gas-chromatographic determination of cholesterol sulfate in plasma and erythrocytes, for the diagnosis of recessive X-linked ichthyosis. *Clinical Chemistry*, 29, 1404–1407.

Nagata, K. and Yamazoe, Y., 2000, Pharmacogenetics of sulfotransferase. *Annual Review of Pharmacology and Toxicology*, 40, 159–176.

Negishi, M., Pedersen, L.G., Petrotchenko, E., Shevtsov, S., Gorokhov, A., Kakuta, Y., and Pedersen, L.C., 2001, Structure and function of sulfotransferases. *Archives of Biochemistry and Biophysics*, 390, 149–157.

Nieschlag, E., Loriaux, D.L., Ruder, H.J., Zucker, I.R., Kirschner, M.A., and Lipsett, M.B., 1973, The secretion of dehydroepiandrosterone and dehydroepiandrosterone sulphate in man. *Journal of Endocrinology*, 57, 123–134.

Nishikawa, T. and Strott, C.A., 1983, Unconjugated and sulfoconjugated steroids in plasma and zones of the adrenal cortex of the guinea pig. *Steroids*, 41, 105–119.

Orentreich, N., Brind, J.L., Rizer, R.L., and Vogelman, J.H., 1984, Age changes and sex differences in serum dehydroepiandrosterone sulfate concentrations throughout adulthood. *Journal of Clinical Endocrinology and Metabolism*, 59, 551–555.

Otterness, D.M., Her, C., Aksoy, S., Kimura, S., Wieben, E.D., and Weinshilboum, R.M., 1995a, Human dehydroepiandrosterone sulfotransferase gene: molecular cloning and structural characterization. *DNA and Cell Biology*, 14, 331–341.

Otterness, D.M., Mohrenweiser, H.W., and Weinshilboum, R.M., 1995b, Dehydroepiandrosterone sulfotransferase gene (STD): localization to human chromosome 19q13.3. *Cytogenetics and Cell Genetics*, 70, 45–47.

Otterness, D.M., Weiben, E.D., Wood, T.C., Watson, W.G., Madden, B.J., McCormick, D.J., and Weinshilboum, R.M., 1992, Human liver dehydroepiandrosterone sulfotransferase: molecular cloning and expression of cDNA. *Molecular Pharmacology*, 41, 865–872.

Parker, C.R., Jr., Falany, C.N., Stockard, C.R., Stankovic, A.K., and Grizzle, W.E., 1994, Immunochemical localization of dehydroepiandrosterone sulfotransferase in human fetal tissues. *Journal of Clinical Endocrinology and Metabolism*, 78, 234–236.

Parker, K.L. and Schimmer, B.P., 1994, The role of nuclear receptors in steroid hormone production. *Seminars in Cancer Biology*, 5, 317–325.

Paul, S.M. and Purdy, R.H., 1992, Neuroactive steroids. *FASEB Journal*, 6, 2311–2322.

Pedersen, L.C., Petrotchenko, E., Shevtsov, S., and Negishi, M., 2002, Crystal structure of the human estrogen sulfotransferase-PAPS complex. *Journal of Biological Chemistry*, 277, 17928–17932.

Pedersen, L.C., Petrotchenko, E.V., and Negishi, M., 2000, Crystal structure of SULT2A3, human hydroxysteroid sulfotransferase. *FEBS Letters*, 475, 61–64.

Petrotchenko, E.V., Pedersen, L.C., Borchers, C.H., Tomer, K.B., and Negishi, M., 2001, The dimerization motif of cytosolic sulfotransferases. *FEBS Letters*, 490, 39–43.

Pion, R.J., Conrad, S.H., and Wolf, B.J., 1966, Pregnenolone sulfate: an efficient precursor for the placental production of progesterone. *Journal of Clinical Endocrinology and Metabolism*, 26, 225–226.

Plassart-Schiess, E. and Baulieu, E.E., 2001, Neurosteroids: recent findings. *Brain Research Reviews*, 37, 133–140.

Puche, R.C. and Nes, W.R., 1962, Binding of dehydroepiandrosterone sulfate to serum albumin. *Endocrinology*, 70, 857–863.

Radominska, A., Comer, K.A., Zimniak, P., Falany, J., Iscan, M., and Falany, C.N., 1990, Human liver steroid sulphotransferase sulphates bile acids. *Biochemical Journal*, 272, 597–604.

Rajkowski, K.M., Robel, P., and Baulieu, E.E., 1997, Hydroxysteroid sulfotransferase activity in the rat brain and liver as a function of aging. *Steroids*, 62, 427–436.

Rikke, B.A. and Roy, A.K., 1996, Structural relationships among members of the mammalian sulfotransferase gene family. *Biochimica et Biophysica Acta*, 1307, 331–338.

Roberts, K.D., 1987, Sterol sulfates in the epididymis; synthesis and possible function in the reproductive process. *Journal of Steroid Biochemistry*, 27, 337–341.

Roy, A.K., 1992, Regulation of steroid hormone action in target cells by specific hormone-inactivating enzymes. *Proceedings of the Society for Experimental Biology and Medicine*, 199, 265–272.

Ruokonen, A., Laatikainen, T., Laitinen, E.A., and Vihko, R., 1972, Free and sulfate-conjugated neutral steroids in human testis tissue. *Biochemistry*, 11, 1411–1416.

Russell, D.W., 1999, Nuclear orphan receptors control cholesterol catabolism. *Cell*, 97, 539–542.

Serizawa, S., Nagai, T., and Sato, Y., 1987, Simplified method of determination of serum cholesterol sulfate by reverse phase thin-layer chromatography. *Journal of Investigative Dermatology*, 89, 580–587.

Serizawa, S., Nagai, T., and Sato, Y., 1989, Simplified determination of serum cholesterol sulfate by gas–liquid chromatography combined with cyclohexylsilane-bonded phase column purification. *Archives for Dermatological Research*, 281, 411–416.

Sharp, S., Barker, E.V., Coughtrie, M.W.H., Hume, R., and Lowenstein, P.R., 1993, Immunochemical characterization of a dehydroepiandrosterone sulfotransferase in rats and humans. *European Journal of Biochemistry*, 211, 539–548.

Shimada, M., Yoshinari, K., Tanabe, E., Shimakawa, E., Kobashi, M., Nagata, K., and Yamazoe, Y., 2001, Identification of ST2A1 as a rat brain neurosteroid sulfotransferase mRNA. *Brain Research*, 920, 222–225.

Shimizu, C., Fuda, H., Yanai, H., and Strott, C.A., 2003, Conservation of the hydroxysteroid sulfotransferase SULT2B1 gene structure in the mouse: pre- and postnatal expression, kinetic analysis of isoforms, and comparison with prototypical SULT2A1. *Endocrinology*, 144, 1186–1193.

Siiteri, P.K. and MacDonald, P.C., 1963, The utilization of circulating dehydroepiandrosterone sulfate for estrogen synthesis during human pregnancy. *Steroids*, 2, 713–730.

Siiteri, P.K. and MacDonald, P.C., 1966, Placental estrogen biosynthesis during human pregnancy. *Journal of Clinical Endocrinology and Metabolism*, 7, 751–761.

Song, C.S., Echchgadda, I., Baek, B.-S., Ahn, S.C., Roy, A.K., and Chatterjee, B., 2001, Dehydroepiandrosterone sulfotransferase gene induction by bile acid activated farnesoid X receptor. *Journal of Biological Chemistry*, 276, 42549–42556.

Steinert, P.M. and Marekov, L.N., 1997, Direct evidence that involucrin is a major early isopeptide cross-linked component of the keratinocyte cornified cell envelope. *Journal of Biological Chemistry*, 272, 2021–2030.

Suzuki, T., Sasano, H., Takeyama, J., Kaneko, C., Freije, W.A., Carr, B.R., and Rainey, W.E., 2000, Developmental changes in steroidogenic enzymes in human postnatal adrenal cortex: immunohistochemical studies. *Clinical Endocrinology*, 53, 739–747.

Tashiro, A., Sasano, H., Nishikawa, T., Yabuki, N., Muramatsu, Y., Coughtrie, M.W., Nagura, H., and Hongo, M., 2000, Expression and activity of dehydroepiandrosterone sulfotransferase in human gastric mucosa. *Journal of Steroid Biochemistry and Molecular Biology*, 72, 149–154.

Thompson, J.D., Higgins, D.G., and Gibson, T.J., 1994, ClustalW: improving the sensitivity of progressive multiple sequence alignment through sequence weighting, position-specific gap penalties and weight matrix choice. *Nucleic Acids Research*, 22, 4673–4680.

Tuckey, R.C., 1990, Side-chain cleavage of cholesterol sulfate by ovarian mitochondria. *Journal of Steroid Biochemistry and Molecular Biology*, 37, 121–127.

Vande Wiele, R.L., MacDonald, P.C., Gurpide, E., and Lieberman, S., 1963, Studies on the secretion and interconversion of the androgens. *Recent Progress in Hormone Research*, 19, 275–290.

Vihko, R. and Ruokonen, A., 1975, Steroid sulphates in human adult testicular steroid synthesis. *Journal of Steroid Biochemistry*, 6, 353–356.

Wang, D.Y. and Bulbrook, R.D., 1967, The metabolic clearance rates of dehydroepiandrosterone, testosterone and their sulphate esters in man, rat and rabbit. *Journal of Endocrinology*, 38, 307–318.

Wang, D.Y., Bulbrook, R.D., and Coombs, M.M., 1967, Metabolic clearance rates of pregnenolone, 17-acetoxypregnenolone and their sulphate esters in man and rabbit. *Journal of Endocrinology*, 39, 395–403.

Weinshilboum, R.M., Otterness, D.M., Aksoy, I.A., Wood, T.C., Her, C., and Raftogianis, R.B., 1997, Sulfotransferase molecular biology: cDNAs and genes. *FASEB Journal*, 11, 3–14.

Williams, M.L., Hughes-Fulford, M., and Elias, P.M., 1985, Inhibition of 3-hydroxy-3-methylglutaryl coenzyme A reductase activity and sterol synthesis by cholesterol sulfate in cultured fibroblasts. *Biochimica et Biophysica Acta*, 845, 349–357.

Wolf, O.T. and Kirschbaum, C., 1999, Actions of dehydroepiandrosterone and its sulfate in the central nervous system: effects on cognition and emotion in animals and humans. *Brain Research Review*, 30, 264–288.

Yamashita, H. and Setchell, K.D.R., 1994, Metabolism and effect of 7-oxo-lithocholic acid 3-sulfate on bile flow and biliary lipid secretion in rats. *Hepatology*, 20, 663–671.

Yen, S.S.C., 2001, Dehydroepiandrosterone sulfate and longevity: new clues for an old friend. *Proceedings of the National Academy of Sciences, USA*, 98, 8167–8169.

Yoshinari, K., Petrotchenko, E.V., Pedersen, L.C., and Negishi, M., 2001, Crystal structure-based studies of cytosolic sulfotransferase. *Journal of Biochemistry and Molecular Toxicology*, 15, 67–75.

12 Species Differences in Cytosolic Sulfotransferases

Kiyoshi Nagata, Miki Shimada, and Yasushi Yamazoe

CONTENTS

INTRODUCTION

Sulfate conjugation mainly occurs on a hydroxyl moiety of chemicals through the transfer of the sulfuryl group from 3′-phosphoadenosine-5′-phosphosulfate (PAPS) to the sulfate accepter. The esterified group may come from phenol, hydroxylamine, hydroxamic acid, or alcohol. Other types of sulfo-conjugations are the formation of sulfamate or thiosulfate from the amino or thiol group, respectively (Mulder, 1984).

The biological sulfate form was found in 1876 by Baumann (De Meio, 1975). He isolated potassium phenyl sulfate from the urine of dogs that had been fed phenol. The formation of sulfated compounds by enzymatic reaction was indicated by Arnolt and De Meio in 1941 by the study of phenol conjugation in mammalian liver (De Meio, 1975). Robbins and Limann in 1957 demonstrated that PAPS was required for this enzyme reaction as the cofactor (Robbins and Lipmann, 1957). Referred to as sulfokinase, sulfotransferase mediated the transfer reaction in the presence of PAPS (De Meio, 1975; Robbins and Lipmann, 1957). Sulfotransferases were largely classified into two distinct groups from their subcellular localization (De Meio, 1975). One is cytosolic sulfotransferase catalyzing sulfation of small hydrophobic chemicals. Another is a membrane-bound sulfotransferase mediating formation of sulfated protein or glycosides such as glycosaminoglycans. In this review, cytosolic sulfotransferase is the focus.

Cytosolic sulfotransferase catalyzes sulfation of steroids, neurotransmitters, thyroid hormones, drugs, and carcinogens (Mulder, 1984; Strott, 1996). Most of the sulfated compounds are inactive, but some of them retain their biological activities. For example, cholesterol sulfate stimulates the differentiation of skin epidermal cells (Elias et al., 1984). Neurosteroids, such as pregnenolone sulfate, however, bind and suppress the GABA receptor (Corpechot et al., 1981).

In the characterization of isolated proteins, cytosolic sulfotransferases had been named based on their distinct catalytic activities. Catalytic activities of typical substrates for these sulfotransferases in experimental animal species are summarized as shown in Table 12.1 (Okuda et al., 1989; Shiraga et al., 1995). Clear species differences of sulfating activity on typical substrates are observed among experimental animals and humans. In particular, liver sulfamation for 4-[(5-chloro-2-oxo-3(2H)-benzothiazolyl)acetyl]-piperazine (DETR) and aniline is greatly varied. The activities are high in rabbit and guinea pig, but low in dog and mouse. Sulfation of 5-hydroxymethy-chrysene (5-HCR) and dehydroepiandrostene (DHEA) is lower in mouse than in rat, guinea pig, and hamster. On the other hand, sulfation of p-nitrophenol (p-NP) and 2-naphthol (2-NT) is very high in livers of all of the animal species examined except for humans. In addition, sex differences in those activities are also observed in rat and mouse. However, it remained unclear how many and which forms of sulfotransferases existed in tissues in the past.

CLASSIFICATION AND NOMENCLATURE OF SULT GENES IDENTIFIED FROM VERTEBRATES

The number of SULT cDNAs has been isolated not only from animals but also from plants and bacteria in past years, and their primary structures have been identified (Marsolais and Varin, 1998; Nagata and Yamazoe, 2000). In vertebrates, more than 50 sulfotransferases have been isolated and classified into seven gene families from their deduced amino acid homologies in animals so far. These families share less than 40% homology with each other. The SULT1 and SULT2 families, which show catalytic activities to phenols and alcoholics, are further divided into five and three subfamilies, respectively, in this review. Each subfamily retains 60% or higher homology in the members. Other families such as SULT3, SULT4, SULT5, SULT6, and SULT7 include only one subfamily at present. It is very interesting how many forms of cytosolic sulfotransferase exist in animals. A dendrogram of vertebrate

TABLE 12.1
Species and Sex Differences in Hepatic Sulfotransferase
Activities toward Various Substrates

Species	Sex	Formation of Sulfate Esters (pmol/mg protein /min)					
		5-HCR	DHEA	p-NP	2-NT	DETR	Aniline
Human	M/F	—	280	—	130	460	—
Monkey	M	—	—	—	1120	150	10
	F	—	—	—	730	170	10
Dog	M	—	—	—	1130	n.d.	10
	F	—	—	—	1040	n.d.	40
Rabbit	M	—	—	—	1570	1910	230
	F	—	—	—	1550	600	90
Rat	M	40	957	197	1940	40	20
	F	168	1885	85	1070	220	40
Mouse	M	3.7	0.46	141	1500	n.d.	10
	F	17	2.27	173	1720	20	50
Guinea Pig	M	76	318	432	740	390	740
	F	85	356	346	590	450	560
Hamster	M	69	497	59	—	—	—
	F	84	432	40	—	—	—

Reactions were carried out at pH 7.4. —: not determined, n.d.: not detected, 5-HCR: 5-hydroxymethyl-chrysene, DHEA: dehydroepiandrostene, p-NP: p-nitrophenol, 2-NT: 2-naphthol, DETR: 4-[(5-chloro-2-oxo-3(2H)-benzothiazolyl)acetyl]-piperazine.

sulfotransferases made from the homology of their amino acid sequences is shown in Figure 12.1. These data were drawn from the DNA database on the homepage of the National Center for Biotechnology Information (NCBI; http://www.ncbi.nlm. nih.gov/; Nagata and Yamazoe, 2000).

The cytosol sulfotransferase homepage (http://www.fccc.edu/research/labs/ raftogianis/sult/index.html) has appeared owing to Rebecca Blanchard. The nomenclature system used on this homepage is well received for the most part. The sulfotransferase nomenclature shown in Figure 12.1 is basically according to ours as previously reported (Nagata and Yamazoe, 2000) and that of the cytosol sulfotransferase (SULT) homepage; novel sulfotransferases, which are not found in the homepage, were arbitrarily named. Names in parentheses in Figure 12.1 are from the nomenclature of the homepage, and accession numbers of cDNA from the DNA database are also shown. Correspondence of our nomenclature to sulfotransferase names used in this review was described previously (Nagata and Yamazoe, 2000). The human gene analysis has been completed, and the results can be searched from NCBI as the genome draft sequence. SULT5, SULT6, and SULT7 genes among seven gene families are not found in the human genome database. The whole SULT1D1 gene is found in the human genome, but it is a putative pseudogene due to a splicing error (Meinl and Glatt, 2001). Three genes sharing high similarity in nucleotide sequence with SULT1C1, SULT2C1, and SULT3A1 cDNA can be

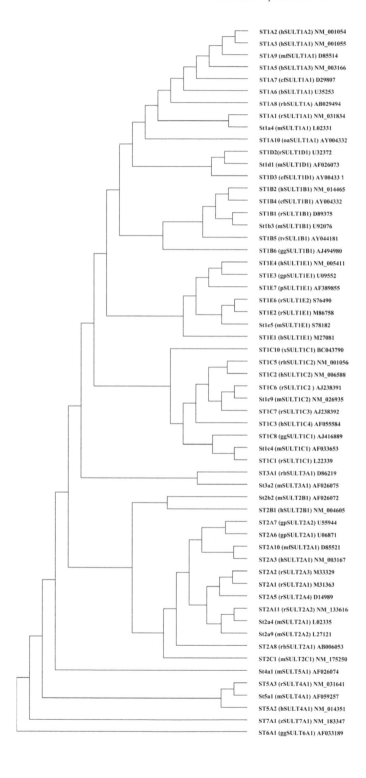

FIGURE 12.1 Dendrogram of cytosolic sulfotransferases previously reported based on their primary amino acid sequences. This dendrogram was made with the GeneWorks program (Intel Genetics). The nomenclature used in this review follows Nagata and Yamazoe (2000) and the SULT homepage (http://www.fccc.edu/research/labs/raftogianis/sult/index.html) with modifications in some forms. Abbreviation used for species names: ck: *Gallus gallus* (chicken); h: *Homo sapiens* (human); r: *Rattus norvegicus* (rat); m: *Mus musculus* (mouse); gp: *Cavia porcellus* (guinea pig); rb: *Oryctolagus cuniculus* (rabbit); cf: *Canis familiaris* (dog); b: *Bos taurus* (bovine); p: *Sus scrofa* (pig); mf: *Macaca fascicularis* (monkey); tv: *Trichosurus vulpecula* (possum); oa: *Ornithorhynchus anatin* (duck-mole); z: *Danio rerio* (zebrafish); x: *Xenopus laevis*.

detected partially, but as predicted some exons are disrupted in humans. In this review, therefore, SULT is used for gene and family names, and ST is used for individual form names.

Sulfotransferases, including the SULT1A subfamily, show more than 70% similarity with each other in their amino acid sequences. A single SULT1A gene has been identified from experimental animal species, while three SULT1A genes have been found in humans. Enzymatic properties are well conserved among these enzymes except for ST1A5 (Honma et al., 2001). Based on the dendrogram, ST1A5 as well as ST1A2 and ST1A3 are an orthologous form to experimental animal ST1A forms. Sulfotransferases including in SULT2A subfamilies show more than 60% similarity with each other except for ST2A8 (Yoshinari et al., 1998a). A single ST2A cDNA and gene have been identified from the human cDNA library and found in the human genome database, respectively. However, multiple ST2A cDNAs have been isolated from rat, mouse, and guinea pig (Nagata and Yamazoe, 2000). Amino acid sequences of ST2A forms are highly conserved between monkey and human and between rat and mouse (Table 12.2). More than 77% similarity is observed within species forms of primate to rodent, whereas the homology between interspecies is less than 70%. This may imply that ST2A forms are changing in their amino acid sequences faster than other families of sulfotransferase. Thus, individual ST2A forms in different species seem to be orthologous with one another. In the DNA database, a novel ST cDNA showing a relatively high similarity with previously identified ST2A forms was found. This new form was identified from mouse aorta and vein but not yet characterized in enzymatic property. This form shows less than 60% similarity in the amino acid sequence to all ST2A forms. Even compared with mouse ST2A forms, this form shared only 58 to 59% similarity. Therefore the authors would like to propose naming it St2c1 to avoid complicating the nomenclature.

In the nomenclature system of Blanchard's SULT homepage, correspondence of the SULT1C genes between species is obscure. The number of sulfotransferase forms classified into the SULT1C subfamily has been identified. There are three groups of SULT1C genes in vertebrates (Nagata and Yamazoe, 2000). One is the SULT1C1 gene group, including ST1C1 and St1c4, which share 94.4% of their similarity at amino acid levels. Recently, a fragment of the SULT1C1 gene was found in humans as mentioned above (Nagata et al., 1997). Furthermore, a novel ST1C cDNA found in the gene database was isolated from chickens. The deduced amino acid sequence shares a higher similarity with ST1C1 and St1c4 (about 68%) than other ST1C forms (about 58 to 62%). These results suggest that this form may belong to the SULT1C1 gene. The SULT1C2 genes were identified from humans, rats, and rabbits. SULT1C3

TABLE 12.2
Amino Acid Similarity of ST2A Forms

	ST2A1	ST2A2	ST2A3	St2a4	ST2A5	ST2A6	ST2A7	ST2A8	St2a9	ST2A10	ST2A11	St2c1
ST2A1	100	89.8	62.1	78.2	86.8	65.2	66.7	60.4	78.2	62.4	78.2	56.7
ST2A2	89.8	100	63.5	77.8	83.5	65.2	65.9	59.3	78.2	63.5	77.8	56.5
ST2A3	62.1	63.5	100	64.9	61.8	62.5	63.9	68.2	63.9	88.8	65.2	59.3
St2a4	78.2	77.8	64.9	100	78.6	65.6	64.4	61.4	96.5	64.2	95.8	58.9
ST2A5	86.8	83.5	61.8	78.6	100	64.9	64.9	59.4	78.2	62.5	78.5	56.8
ST2A6	65.2	65.2	62.5	65.6	64.9	100	86.8	58.4	64.9	64.9	66.7	58.9
ST2A7	66.7	65.9	63.9	64.4	64.9	86.8	100	61.9	63.8	66.3	64.5	56.8
ST2A8	60.4	59.3	68.2	61.4	59.4	58.4	61.9	100	61.4	68.9	61.4	55.9
St2a9	78.2	78.2	63.9	96.5	78.2	64.9	63.8	61.4	100	64.2	96.1	58.2
ST2A10	62.4	63.5	88.8	64.2	62.5	64.9	66.3	68.9	64.2	100	65.5	59.6
ST2A11	78.2	77.8	65.2	95.8	78.5	66.7	64.5	61.4	96.1	65.5	100	59.2
St2c1	56.7	56.5	59.3	58.9	56.8	58.9	56.8	55.9	58.2	59.6	59.2	100

Underlined numbers indicate amino acid similarities between rat and mouse ST2A forms.
Numbers in boxes indicate amino acid similarities between guinea pig ST2A forms (ST2A6 and ST2A7).
Shaded numbers indicate amino acid similarities between human and monkey ST2A forms (ST2A3 and ST2A10).

is only named in rat ST1C7 but shares a high similarity with ST1C6 (92.6%), ST1C2 (84.5%), ST1C9 (83.1%), and ST1C5 (81.8%). These results suggest that ST1C7 as well as ST1C6 is orthologous to other species in the SULT1C2 gene. The ST1C3 form is a third group, termed SULT1C4, on the sulfotransferase home page. There is only one gene characterized at present. The amino acid sequence shares at most 64% similarity (Table 12.3). The above data on the relationship between SULT1C1, SULT1C2/3, and SULT1C4 genes are paralogous to each other in gene evolution. Three predicted sulfotransferase cDNAs (XM-236811, XM-346008, and XM-346007), which are included in the SULT1C2 gene group based on the similarity of their deduced amino acid sequence, have been found in the rat database, although their enzymatic properties have not been characterized. Therefore it may be better to have these genes classified into different SULT1 subfamilies to avoid complication.

In addition, several genes predicted to encode cytosol sulfotransferase were found in the genome database of humans, rats, and mice (data not shown). However, most of them are predicted to be pseudogenes, judging from their deduced cDNA sequences. In recent cytochrome P450 nomenclature, the pseudogene was also named by a similar rule. Therefore the final SULT nomenclature can be expected when the human, rat, and mouse genome analyses are completed.

ARYL SULFOTRANSFERASE

ENZYME PROPERTY IN SULFO-CONJUGATION

Aryl sulfotransferase activity is found in all mammals (De Meio, 1975). The activity is usually detected in extrahepatic tissues such as brain, small intestine, lung, and blood platelet, although liver shows the highest activity in most substrates. Phenol sulfation is also detected in birds, mollusks, amphibians, insects, and araneids, but not in fishes (De Meio, 1975). In humans, β2-adrenalgic drugs such as salbutamol, terbutaline, and isoprenaline (Bonham Carter et al., 1983; Pacifici et al., 1993, 1996; Walle and Walle, 1992) et al., are mainly metabolized to sulfated conjugates. These metabolisms are catalyzed by aryl sulfotransferase. Moreover, minoxidil, an anti-hypertensine drug, is biotransformed into the pharmacologically active metabolite by aryl sulfotransferase (Buhl et al., 1990; Hamamoto and Mori, 1989). Aryl sulfotransferase is not only involved in the metabolism of exogenous compounds but also in biotransformation of endogenous substances. Dopamine, epinephrine, L-thyrosine, serotonin, norepinephrine, and thyroid hormones are known as their endogenous substrates (Fujita et al., 1997; Hidaka et al., 1969; Kuchel et al., 1982; Saeki et al., 1998). Sulfation of estrogens that have a phenolic hydroxyl residue is performed by aryl sulfotransferases (De Meio, 1975; Nose and Lipmann, 1958). Moreover, arylhydroxamic acids such as N-hydroxy-2-acetylaminofluorene (N-OH-AAF) inflict DNA damage through sulfotransferase-mediated activation (DeBaun et al., 1968; King and Phillips, 1968; Mulder, 1984). These enzyme activities are inhibited by pentachlorophenol and 2,6-dichloro-4-nitrophenol (Jakoby et al., 1984). In humans, large individual differences are observed in the expression level of liver aryl sulfotransferase (Nagata and Yamazoe, 2000; Weinshilboum, 1986). Moreover, a sex-related difference and development changes are observed in experimental

TABLE 12.3
Amino Acid Similarities among ST1C Forms

	ST1C1	ST1C2	ST1C3	St1c4	ST1C5	ST1C6	ST1C7	ST1C8	St1c9	ST1C10
ST1C1	100	63.7	63.1	94.4	62.3	60.8	59.7	68.4	59.7	56.6
ST1C2	63.7	100	62.9	61.6	85.1	84.5	80.4	59.0	83.1	58.8
ST1C3	63.1	62.9	100	63.1	62.5	59.9	60.2	62.9	61.6	55.3
St1c4	94.4	61.6	63.1	100	62.0	59.9	59.2	67.4	59.2	55.6
ST1C5	62.3	85.1	62.5	62.0	100	81.8	78.4	58.4	79.4	58.3
ST1C6	60.8	84.5	59.9	59.9	81.8	100	94.3	60.7	92.6	59.0
ST1C7	59.7	80.4	60.2	59.2	78.4	94.3	100	59.0	89.5	57.6
ST1C8	68.4	59.0	62.9	67.4	58.4	60.7	59.0	100	60.0	55.4
St1c9	59.7	83.1	61.6	59.2	79.4	92.6	89.5	60.0	100	58.3
ST1C10	56.6	58.8	55.3	55.6	58.3	59.0	57.6	55.4	58.3	100

Shaded numbers indicate amino acid similarities among SULT1C2 and SULT1C3 genes.
Numbers in boxes indicate amino acid similarities among SULT1C1 genes.

animals (Gong et al., 1991; Jackson and Irving, 1972; Okuda et al., 1989). Aryl sulfotransferase is classified into five SULT1 subfamilies at present.

SULT1A Subfamily

Rat aryl sulfotransferase cDNA was first isolated by Ozawa et al. (1990) by using the antibody of PST-1 from the male rat liver library. Thereafter, several novel aryl ST cDNAs have been isolated and their enzymatic properties characterized (Nagata and Yamazoe, 2000). Aryl sulfotransferase encoded by PST-1 cDNA was designated ST1A1 by our group (Nagata and Yamazoe, 2000). Subsequently, the same cDNA was also isolated by the groups of Falany, Ringer, and Jakoby (Chen et al., 1992; Hirshey et al., 1992; Yerokun et al., 1992). Single ST1A cDNA has been isolated from experimental animal species but not humans. Recombinant ST1A1 shows sulfation activity to phenols like p-aminophenol as well as N-oxide of minoxidil (Hirshey et al., 1992; Ozawa et al., 1993). ST1A1 does not mediate sulfation of dehydroepiandrosterone, but does mediate sulfation of β-estradiol and dopamine. Thereafter, two human ST1A forms, ST1A3 and ST1A2, corresponding to ST1A1 have been isolated from human liver cDNA libraries (Nagata and Yamazoe, 2000). ST1A3 and ST1A2, show 77% and 78% of homology to ST1A1, respectively, at amino acid level. The substitution of 12 amino acids was observed between both human forms. Furthermore, two forms of ST1A cDNAs were isolated from human brain cDNA libraries (Zhu et al., 1993). One of them was ST1A3 cDNA. Another was a cDNA encoding M-PST/ST1A5 that catalyzes sulfation of monoamines such as dopamine (Dooley et al., 1994). ST1A5 shows higher amino acid similarities to ST1A3 and ST1A2 than animal ST1A forms reported. Expressed amounts of ST1A forms in liver of various species are shown in Table 12.4 (Honma et al., 2001). ST1A

TABLE 12.4
Cytosolic Content of ST1A Forms in Livers of Various Species

Name	Content (pmol/mg Cytosolic Protein) [μg/mg Cytosolic Protein]	
ST1A3	120 ± 38.0 (n=21) [4.1 ± 1.3]	
ST1A5	6.4 ± 2.6 (n=21) [0.22 ± 0.09]	
ST1A1	270 ± 5.9 (male, $n=3$) [9.1 ± 0.2]	190 ± 15.0 (female, $n=3$) [6.3 ± 0.5]
St1a4	66 ± 2.9 (male, $n=3$) [2.3 ± 0.1]	130 ± 5.8 (female, $n=3$) [4.4 ± 0.2]
ST1A8	250 ± 10.0 (male, $n=3$) [8.3 ± 0.4]	

forms were detected in all five species. Protein level of ST1A forms is relatively similar among species, except for ST1A5. The level of ST1A5 is 20 times lower than that of ST1A3 in liver. Protein level is, however, 10 times higher with ST1A5 than with ST1A3 in small intestine.

Clear sex-related differences are detected on the level of ST1A1 and St1a4. ST1A1 is 1.5 times higher in adult male rats than in females. On the other hand, St1a4 is two times higher in adult female mice than in males. Catalytic activities of recombinant ST1A form are compared among animal species and shown in Table 12.5 and Table 12.6. ST1A5 prefers endogenous compounds such as tyramine, norepinephrine, and dopamine. These chemicals include amino moiety together with phenolic moiety. Conversely, all other ST1A forms display low activities to these chemicals. These results suggest that ST1A5 is a human-specific form playing an inevitable role for metabolism of neurotransmitters. ST1A forms listed in Table 12.5 and Table 12.6 display similar catalytic properties. Rates of sulfation for troglitazone and 4'-OH-PhIP are higher with ST1A3 than with ST1A forms of experimental animals. Sulfation rate of hydroxyindole is nearly equivalent for all ST1A forms. Mefenamic acid has been reported to be an inhibitor for p-NP sulfation in humans (Vietri et al., 2000). Mefenamic acid has no obvious effects on ST1A5. In fact, ST1A3, ST1A1, and St1a4 are strongly inhibited by mefenamic acid although ST1A8 was inhibited to lesser extents. The apparent kinetic parameters for p-NP, dopamine, and 6-hydroxymelatonin (6-HM) are compared among ST1A forms (Table 12.7). These STIA forms show a similar K_m 6-hydroxymelatonin (6-HM) value for p-NP but clearly differ for dopamine. The K_m values of ST1A3, ST1A1, and St1a4 for dopamine are ten times higher than that of ST1A5.

SULT1B Subfamily

ST1B1 cDNA was isolated together with ST1C1 cDNA from a rat liver cDNA library by a combination of antibodies against STIV and ST1A1 cDNA as probes (Fujita et al., 1997). This form is characterized as thyroid hormone sulfotransferase. Another group also purified ST1B1as dopa/tyrosine SULT and isolated its cDNA from rat livers (Sakakibara et al., 1995). Thereafter, ST1B2, St1b3, and ST1B4 have been identified from human, mouse and dog, respectively (Nagata and Yamazoe, 2000; Tsoi et al., 2001a). Among them, ST1B2 and ST1B1 have been characterized and their enzymatic properties compared by use of recombinant enzymes (Fujita et al., 1997). ST1B2 and ST1B1 are more thermostable enzymes as compared to ST1A forms and more resistant to inhibition by 2,6-chloro-4-nitrophenol, which is known as a typical inhibitor for ST1A forms (Fujita et al., 1999). ST1B1 and ST1B2 show similar affinities for PAPS and T_3 but differ for p-NP and dopamine, as shown in Table 12.8.

SULT1C Subfamily

A carcinogenic amine, N-OH-AAF, is activated by sulfation (DeBaun et al., 1968; King and Phillips, 1968). The sulfation was believed to be mediated by STIV from which it was purified from rat livers (Yerokun et al., 1992). However, HAST purified

TABLE 12.5
Sulfotransferase Activities to Catecholamines and Their Derivatives

Substrate (10 μM)	Sulfating Activities (nmol/nmol/min) [% of ST1A3 Activity]				
	ST1A5	ST1A3	ST1A1	St1a4	ST1A8
Tyramine	4.34 ± 0.02 [6199]	0.07 ± 0.01 [100]	0.08 ± 0.04 [118]	0.26 ± 0.02 [367]	0.03 ± 0.01 [46]
Norepinephrine	3.98 ± 0.16 [7955]	0.05 ± 0.00 [100]	0.18 ± 0.01 [366]	0.18 ± 0.00 [367]	0.06 ± 0.00 [115]
Normetanephrine	3.33 ± 0.07 [2559]	0.13 ± 0.00 [100]	0.04 ± 0.01 [33]	0.05 ± 0.01 [40]	0.03 ± 0.00 [22]
Dopamine	3.07 ± 0.18 [1025]	0.30 ± 0.01 [100]	0.39 ± 0.02 [132]	0.59 ± 0.07 [196]	0.13 ± 0.01 [42]
Epinephrine	2.93 ± 0.05 [3254]	0.09 ± 0.00 [100]	0.10 ± 0.00 [112]	0.12 ± 0.00 [130]	0.03 ± 0.00 [28]
HMPG	0.69 ± 0.03 [95]	0.72 ± 0.02 [100]	1.08 ± 0.03 [150]	1.58 ± 0.01 [219]	2.65 ± 0.06 [368]
DOPAC	0.06 ± 0.00 [23]	0.27 ± 0.01 [100]	0.54 ± 0.01 [199]	0.77 ± 0.01 [285]	0.30 ± 0.00 [110]
Phenylephrine	3.44 ± 0.15 [-]	n.d. [-]	n.d. [-]	n.d. [-]	n.d. [-]
Metaraminol	1.78 ± 0.05 [-]	n.d. [-]	n.d. [-]	n.d. [-]	n.d. [-]
Serotonin	0.36 ± 0.06 [1205]	0.03 ± 0.00 [100]	0.09 ± 0.00 [293]	0.10 ± 0.01 [323]	0.07 ± 0.01 [241]
5-Hydroxytryptophol	0.36 ± 0.04 [13]	2.79 ± 0.20 [100]	5.59 ± 0.09 [200]	2.27 ± 0.17 [81]	2.59 ± 0.08 [93]
5-Hydroxyindole	0.59 ± 0.03 [6]	10.24 ± 0.15 [100]	10.39 ± 0.19 [101]	9.33 ± 0.23 [91]	10.82 ± 0.35 [106]
6-HM	1.84 ± 0.07 [39]	4.70 ± 0.22 [100]	3.95 ± 0.15 [84]	3.41 ± 0.34 [73]	3.94 ± 0.10 [84]
Harmol	3.93 ± 0.13 [78]	5.07 ± 0.03 [100]	5.02 ± 0.18 [99]	6.24 ± 0.24 [123]	1.79 ± 0.09 [35]
5-HIAA	n.d. [-]	n.d. [-]	n.d. [-]	n.d. [-]	n.d. [-]

Assays were performed with 10 mM substrates at pH 7.4. After 20-min incubations, metabolites were separated by thin layer plates. The radioactive spots were quantified by BAS 1000. Each value represents mean ± S.D. ($n = 3$).

n.d., not detected (less than 10 pmol/mg/min).

6-HM: 6-hydroxymelatonin; 5-HIAA: 5-hydroxyindoleacetic acid.

HMPG: 4-hydroxy-3-methoxyphenylglycol; DOPAC: 3,4-dihydroxyphenylacetic acid.

in our laboratory as a sulfotransferase mediating the metabolic activation of N-OH-AAF was not in agreement with STIV in enzymatic properties (Gong et al., 1991). To clarify the correspondence of these enzymes, the antibodies against HAST and PST-1 were used for isolation of their cDNA from a rat liver λgt11 cDNA library

TABLE 12.6
Sulfotransferase Activities towards *p*-nitrophenol (*p*-NP) and Other Small Phenolic Compounds

Substrate (10 µM)	Sulfating Activities (nmol/nmol/min) [% of ST1A3 Activity]				
	ST1A5	ST1A3	ST1A1	St1a4	ST1A8
p-NP	0.09 ± 0.00 [2]	5.79 ± 0.22 [100]	6.10 ± 0.23 [105]	4.77 ± 0.17 [82]	3.94 ± 0.14 [68]
HMC	0.17 ± 0.01 [2]	7.24 ± 0.22 [100]	7.53 ± 0.19 [104]	8.07 ± 0.17 [111]	7.52 ± 0.39 [104]
Umbelliferone	0.22 ± 0.00 [3]	6.87 ± 0.51 [100]	6.46 ± 0.28 [94]	6.60 ± 0.12 [96]	6.803 ± 0.22 [99]
Salicylamide	0.34 ± 0.01 [9]	3.98 ± 0.15 [100]	3.59 ± 0.16 [90]	4.04 ± 0.25 [101]	3.58 ± 0.18 [90]
Naringenin	3.62 ± 0.15 [50]	7.24 ± 0.27 [100]	7.17 ± 0.31 [99]	7.70 ± 0.51 [106]	5.01 ± 0.11 [69]
DHF	2.52 ± 0.08 [60]	4.22 ± 0.09 [100]	4.86 ± 0.03 [115]	4.64 ± 0.16 [110]	3.02 ± 0.05 [71]
Minoxidil	0.04 ± 0.00 [22]	0.19 ± 0.01 [100]	0.25 ± 0.02 [131]	0.36 ± 0.03 [190]	0.06 ± 0.01 [31]
HPTH	0.02 ± 0.00 [6]	0.31 ± 0.01 [100]	0.36 ± 0.02 [115]	0.44 ± 0.01 [142]	0.39 ± 0.02 [127]
3-OH-B[a]P	0.13 ± 0.01 [37]	0.36 ± 0.10 [100]	0.35 ± 0.06 [98]	0.55 ± 0.03 [152]	0.35 ± 0.03 [96]
Phentolamine	0.09 ± 0.01 [11]	0.83 ± 0.04 [100]	0.19 ± 0.00 [23]	0.28 ± 0.01 [34]	0.14 ± 0.05 [17]
Estradiol	0.08 ± 0.01 [7]	1.09 ± 0.01 [100]	0.50 ± 0.01 [46]	0.32 ± 0.01 [29]	0.14 ± 0.00 [13]
Troglitazone	n.d. [6]	2.53 ± 0.09 [100]	0.50 ± 0.12 [101]	0.81 ± 0.03 [91]	0.28 ± 0.03 [106]
6-HM	1.84 ± 0.07 [–]	4.70 ± 0.22 [100]	3.95 ± 0.15 [20]	3.41 ± 0.34 [32]	3.94 ± 0.10 [11]
4′-OH-PhIP	3.08 ± 0.00 [1]	9.41 ± 0.23 [100]	0.47 ± 0.02 [5]	1.25 ± 0.08 [13]	1.36 ± 0.08 [14]

Assays were performed as described in Table 12.3. Each value represents mean ± S.D. of three samples. *n.d.*, not detected (less than 10 pmol/mg/min).
p-NP, *p*-nitrophenol, HMC, 7-hydroxy-4-methylcoumarin; DHF, 5, 7-dihydroxyflavonone.
HPTH, 5-(*p*-hydroxyphenyl)-5-(*p*-tolyl)-hydantoin; 3-OH-B[*a*]P, 3-hydroxybenzo[a]pyrene.
4′-OH-PhIP, 2-amino-4′-hydroxy-1-methyl-6-phenylimidazo[4, 5-*b*]pyridine.

(Nagata et al., 1993). Consequently, three related cDNA clones encoding ST1A1, ST1B1, and ST1C1 were obtained. Newly isolated ST1B1 and ST1C1 cDNAs have translation domains that are different in PST-1 cDNA. All the recombinant enzymes expressed in COS-I cells showed high sulfating activities to *p*-NP and T$_3$, whereas ST1C1 expressed in COS-1 cells showed the high sulfating activity of N-OH-AAF

TABLE 12.7
Apparent Kinetic Parameters for p-NP, Dopamine, and 6-HM Sulfations by Recombinant ST1A Forms

| | [Km (μM) V_{max} (nmol/nmol/min)] | | | | | |
| | p-NP | | Dopamine | | 6-HM | |
Enzymes	K_m	V_{max}	K_m	V_{max}	K_m	V_{max}
ST1A5	>500	—	14	5.06	65	18.08
ST1A3	3.0	3.98	130	3.98	18	11.94
ST1A1	3.8	4.66	130	5.02	30	17.93
St1a4	3.2	4.04	98	6.24	40	18.34
ST1A8	3.3	3.94	>500	—	29	17.90

Each parameter is derived from analyses of Lineweaver-Burk plots and shown as the mean of three separated experiments. Assays were performed at pH 7.4 with various concentrations of substrates (p-NP: p-nitrophenol; 0.5-1000 mM, dopamine; 1-1000 mM and 6-HM, 6-hydroxymelatonin; 0.5-100 mM).

—: not determined.

and supported the DNA binding. The enzyme activities were about 250 and 60 times higher than those of recombinant ST1A1, respectively. Recombinant ST1B1 did not activate N-OH-AAF. There are large age and sex differences in hepatic levels of ST1C1 mRNA in the rat. These expression profiles are quite consistent with those of cytosol sulfating activity to N-OH-AAF and the level of HAST. Furthermore, the N-terminal amino acid sequence deduced from ST1C1 cDNA was completely in agreement with that of HAST. Anti-HAST antibodies selectively recognized recombinant ST1C1, but not ST1A1. Thus, ST1C1 is judged to be the main enzyme mediating N-OH-AAF activation in rat livers although low activity was detected in ST1A1 corresponding to STIV (Nagata et al., 1993).

Mouse St1c4 cDNA was also isolated from an olfactory cDNA library (Tamura et al., 1998). The deduced amino acid sequence showed high similarity (94%) with that of ST1C1 cDNA. St1c4 mRNA was only detected as olfactory. Several ST1C cDNAs have been isolated from humans, rats, rabbits, and chickens. Their deduced amino acid sequences, however, show about only 60% similarity with those of the ST1C cDNAs mentioned above. Based on similarity of amino acid sequence, the SULT1C subfamily consists of three groups as shown in Figure 12.1 and Table 12.3. These groups share about 60% similarity to each other. These enzymatic properties, especially in catalytic activity, have not yet been well characterized. Recombinant ST1C2, ST1C3, and ST1C5 have shown sulfating activities for p-NP but not for dopamine (Hehonah et al., 1999; Her et al., 1997; Sakakibara et al., 1998a). Sulfation of N-OH-AAF was also measured in human ST1C2 and ST1C3. Recombinant ST1C2 and ST1C3 purified from *Escherichia coli* showed limited amounts of

TABLE 12.8
Apparent Kinetic Parameters for
Typical Substrates of
Sulfotransferase by Recombinant
ST1B1 and ST1B2

Substrate		ST1B2	ST1B1
p-NP	K_m	21.23	281.9
	K_{cat}	1.21	0.74
	K_{cat}/K_m	56.84	2.6
Dopamine	K_m	4880.95	552.3
	K_{cat}	0.344	0.13
	K_{cat}/K_m	0.07	0.23
T3	K_m	46.23	44.4
	K_{cat}	0.42	0.46
	K_{cat}/K_m	9.00	10.3
E2	K_m	n.d.	n.d.
	K_{cat}	n.d.	n.d.
	K_{cat}/K_m	n.d.	n.d.
DHEA	K_m	n.d.	n.d.
	K_{cat}	n.d.	n.d.
	K_{cat}/K_m	n.d.	n.d.

p-NP: p-nitrophenol; T3: triiodethyronine; E2:
estradiol; DHEA: dehydroepiandrosterone.

n.d.: not detected.

sulfating activities compared to that of rat and mouse ST1C forms as shown in Table 12.9. Metabolic activations of N-OH-AAF by ST1C1 and ST1C2 expressed in *E. coli* are shown in Table 12.9 as DNA binding. Large DNA binding was observed on ST1C1 but not on ST1C2. Specific mRNA of ST1C2 is detected in kidney and stomach in humans and rats as well as kidney, stomach, intestine, colon, and liver in rabbits (Hehonah et al., 1999; Her et al., 1997; Xiangrong et al., 2000). We previously reported on parts of the human SULT1C1 gene and the lack of expression of the mRNA in livers (Nagata et al., 1997). Searching the draft sequence of the human genome, the human SULT1C1 gene of which exon 7 and 8 shared high similarity in nucleotide sequence with ST1C1 cDNA is actually found in 2q12.2 of chromosome 2 but is likely to be a pseudogene (Nagata et al., 1997). Therefore, ST1A2 and ST1A3, instead of the ST1C form, are likely to mediate the metabolic activation of N-OH-AAF in human tissues.

TABLE 12.9
Sulfotransferase Activities and
PAPS-Dependent DNA Binding of
N-OH-AAF Recombinant Purified
ST1C Forms

	Rat Liver Cytosol	ST1C1	ST1C2
(nmol/min /mg protein)			
p-NP			
pH5.6	3.57	331	n.d.
pH7.4	0.94	12.5	8.31
(pmol/min /mg protein /mg DNA)			
N-OH-AAF (DNA Adduct)			
+PAPS	106	6521	109
−PAPS	47	179	116
Net	59	6342	<0

Net values are calculated from the means of complete minus the means of −PAPS (3-phosphoadenosine phosphosulfarte).

p-NP: p-nitrophenol; N-OH-AAF: N-Hydroxy-2-acetylaminofluorene.
n.d.: not detected.

SULT1D Subfamily

By searching with BLAST, several novel cDNA fragments sharing high similarity in nucleotide sequence with ST cDNAs were found as expression-sequenced tag (EST) in GenBank. One of them encoded a new sulfotransferase belonging to the SULT1 family. The similarity of amino acid sequence, however, shows less than 60% to sulfotransferases belonging to SULT1A, 1B, 1C, and 1E subfamilies. This novel form consists of a new SULT1 subfamily and was named St1d1 (Sakakibara et al., 1998b, Liu et al., 1999). Thereafter, ST1D cDNAs have been identified from rat and dog (Tsoi et al., 2001). However, rat ST1D2 has not been characterized. Mouse St1d1 is detected mainly in kidney and dog ST1D3 in kidney, small intestine, and colon (Liu et al., 1999). Both ST1D forms catalyze sulfation of dopamine as well as p-NP. Especially among catecholamines and their metabolites, St1d1 showed markedly high activities to 3,4-dihydroxyphenylacetic acid (230.2 ± 9.21 nmol/mg/min), 3,4-dihydroxymandelic acid (20.3 ± 2.69 nmol/mg/min), and 3,4-dihydrosyphenyletylene glycol (209.3 ± 13.7 nmol/mg/min) as shown in Table 12.10. These metabolites are formed from their parent amines by monoamine oxidase, followed by aldehyde oxidase or aldehyde reductase. St1d1 also catalyzed sulfation of homovanilic acid (116.9 ± 7.04 nmol/mg/min) and 4-hydroxy-3-methoxyphenyl-ethylene glycol (72.4 ± 7.00 nmol/mg/min), except for vanillylmandelic acid (Shimada et al., 2004). Kinetic parameters were examined for dopamine and DOPAC sulfations. Apparent K_m and V_{max} values of St1d1 were 713.2 ± 14.2 µM and 1474 ± 9.03 nmol/mg protein/min, respectively, for dopamine sulfation, while they were 35.02 ± 1.13 µM and 1678 ±159 nmol/mg protein/min for 3,4-dihydroxyphenylacetic acid sulfation. The values of V_{max}/K_m for dopamine and 3,4-dihydroxyphenylacetic acid sulfations were 2.07 and 47.92, respectively (Shimada et al., 2004). Under a high substrate concentration (200 µM), recombinant St1d1 also mediates sulfation of eicosanoids (Liu et al., 1999). The ST1D form is not expressed in humans although the human SULT1D1 gene was found between SULT1B1 and SULT1E1 genes on chromosome 4 (Meinl and Glatt, 2001). The nucleotide mutation at II, IV, and VI exon–intron junctions produced the putative pseudogene.

TABLE 12.10
Sulfation Activities to Catecholamines and Their Metabolites by Recombinant St1d1

Substrates	nmol/mg protein/min
L-DOPA	n.d.
Dopamine	22.8 ± 3.14
3-Methoxytyramine	n.d.
3,4-Dihydroxyphenylacetic acid	230.2 ± 9.21
Homovanilic acid	116.9 ± 7.04
Norepinephrine	6.04 ± 1.59
Norethanephrine	n.d.
3,4-dihydroxymandelic acid	20.3 ± 2.69
Vanillylmandelic acid	n.d.
Epinephrine	1.40 ± 0.07
Metanphrine	n.d.
3,4-Dihydrosyphenyletylene glycol	209.3 ± 13.7
4-Hydroxy-3-methoxyphenylethylene glycol	72.4 ± 7.00

Each value represents the mean ± S.D. of three different experiments. The assays were performed at pH 7.4 in the presence of 10 µM substrate and 50 ng of recombinant St1d1 proteins. n.d.: not detected. (<0.15 nmol/mg protein/min). (−)-Norepinephrine and (−)-epinephrine were used as norepinephrine and epinephrine. Normethanephrine, vanillylmandelic acid, metanephrine, and 3,4-dihydroxyphenyletylene glycol were used as racemic forms. Other catecholamines and their metabolites were used as kimeric L-forms.

SULT1E Subfamily

Sulfation of estrogen has a physiological role and is also thought to be associated with estrogen-mediated cancer by estrogen. Existence of this enzyme was reported for the first time in rabbit liver in 1958 and is now found in many animals (De Meio, 1975). In particular, high catalytic activity was detected in the adrenal gland, placenta, and liver. A sex difference of estrogen sulfation is also found in rat liver. The activity is lower in the embryo and infant than in the mature rat. The primary structure of the SULT 1E1 gene was reported for the first time through the isolation of the bovine cDNA in 1988 (Nash et al., 1988). This deduced amino acid sequence is the first report among all SULT genes. Thereafter, from the liver of rat and man and the adrenal gland of guinea pig, ST1E cDNAs were isolated and their primary structures clarified (Aksoy et al., 1994; Bernier et al., 1994; Demyan et al., 1992; Oeda et al., 1992). Rat ST1E4 cDNA consists of 295 amino acids, which is the same number as that of a cow. The ST1E3 isolated from the adrenal gland of a guinea pig has 296 amino acids and tandem methionines at the N terminus. ST1E4 and ST1E3 expressed in COS-7 cells were reported to have shown sulfating activity to estradiol. Other ST1 forms included in the SULT1A, 1C, and 1D subfamilies also mediated sulfation of estrogen (Fujita et al., 1999). Kinetic studies of their recombinant forms indicated that ST1E4 displays the highest activity of estrogen sulfation (Fujita et al., 1999). Under the physiological condition, therefore, ST1E4 is mainly contributing to sulfation of estrogen in human.

HYDROXYSTEROID SULT

Enzymes catalyzing sulfation of primary or secondary alcohols were classified into hydroxysteroid (alcohol), bile acid, and cholesterol sulfotransferases in the past (De Meio, 1975; Jakoby et al., 1984). At present, these sulfotransferases are known to consist of the SULT2 family, which includes the SULT2A and SULT2B subfamilies (Nagata and Yamazoe, 2000).

SULT2A Subfamily

ST2A1 cDNA was isolated for the first time by Ogura et al. (1989) from a rat liver cDNA library and the primary structure clarified by analysis of the nucleotide sequence. ST2A1 cDNA (ST-20) consists of 284 amino acids, and high homology is observed on SMP-2/ST2A11 reported as a senescence marker protein (SMP; Chatterjee et al., 1987; Ogura et al., 1990a). The level of ST2A1 mRNA was higher in male than in female rat livers. These data are consistent with expression profiles of ST-a purified from rat livers (Ogura et al., 1990b). However, differences are observed in two residues in 20 amino acid N-terminal sequences of ST2A1 and ST-20a. Further investigation resulted in the isolation of ST2A2 cDNA (ST-40) from a female rat liver cDNA library, which matched the N-terminal amino acid sequence of ST-a (Watabe et al., 1994). ST2A2 also consists of 284 amino acids, but ST2A1 and ST2A2 were different in 29 amino acids and shared 89.8% homology. Otterness et al. (1992) and we (Yoshinari et al., 1998a) have isolated human ST2A3 cDNA from human liver cDNA libraries. ST2A3 consists of 285 amino acids in which one amino acid is longer than in those of ST2A1 and ST2A2. Two related SULT2A forms have been identified within one species from rats, guinea pigs, and mice. Only a single ST2A form has been found in human, monkey, and rabbit livers (Nagata and Yamazoe, 2000).

To verify profiles of species differences in hydroxysteroid sulfation, ST2A forms from humans, rats, and rabbits were expressed in *E. coli* and purified from the bacterial cytosol (Yoshinari et al., 1998a). Sulfating activities for hydroxysteroids are shown as the ratio to that for prognenolone (PRG) in Table 12.11. Rabbit ST2A8 shows the highest activity to DHEA, and comparatively high activity to PRG and testosterone (TEST). ST2A8 also catalyzes sulfation of 3β-hydroxyl-5-cholenoic acid (3HCA) but not sulfation of lithocholic acid. Recombinant rat ST2A1 shows activity to DHEA, PRG, androsterone (AND), and 3HCA. Rat ST2A2 mediates sulfation of DHEA, PRG, TEST, AND, and 3HCA. The activities of those SULT2A forms to corticosterone and cortisol are low however. Conversely, rat ST2A1 mediates sulfation of conticosterone at rates similar to those of DHEA. Although high activity is found in 3HCA, activities to TEST and AND are also very low. These results suggest that the sulfation of 3HCA is a common property in the SULT2A subfamily. It has been reported that excretion of lithocholic acid sulfate in rabbit urine is very low or is hardly detected. This is in agreement with data from recombinant ST2A8 showing no sulfation of lithocholic acid. Mouse St2a4 catalyzes sulfation of lithocholic acid as well as DHEA (Shimada, Nagata, and Yamazoe: unpublished data).

TABLE 12.11
Comparison of Hydroxyasteroid Sulfation Rates among ST2A Forms in Human and Experimental Animals

Enzyme	Substrate							
	PRG	DHEA	CRC	CRS	TEST	AND	LCA	3HCA
ST2A3	100	177	4	0.4	15	44	2.1	45
ST2A1	100	43	54	3.9	3	5	2.3	112
ST2A2	100	201	11	3.9	50	61	9.4	61
ST2A8	100	440	2	0.3	58	11	n.d.	63

Substrates are abbreviated as follows: PRG: pregnenolone; DHEA: dehydroepiandrosterone; CRC: corticosterone; CRS: cortisol; TEST: testosterone; AND: androsterone; LCA: lithocholic acid; 3HCA: 3β-hydroxy-5-cholenoic acid. Sulfating activities are shown as relative activities (% of those toward pregnenolone) for each enzyme. The activities to pregnenolone of ST2A8, ST2A1, ST2A2, and ST2A3 were 5,380, 96, 711, and 12,700 pmol/mg of cytosolic protein/min, respectively. n.d. represents activities below the detection limit (6 pmol/mg of cytosolic protein/min [0.001%]).

SULT2B SUBFAMILY

Based on information of EST database from mouse embryo and human placenta, a novel SULT has been identified and termed SULT2B1 (Her et al., 1998; Sakakibara et al., 1998). These recombinant enzymes mediate sulfation of DHEA. From human placenta, two different cDNAs, ST2B1a and ST2B1b, produced by the differential splicing from a single gene, are isolated (Her et al., 1998). The amino acid residue at the N terminus of ST2B1b is longer than that of ST2B1a, and the difference is found in the PAPS affinity when DHEA is used as a substrate. Recently, ST2B1 was reported to have an activity not only to DHEA but also to cholesterol and oxysterol. Based on study of mRNA and protein expression, formation of cholesterol sulfate in skin was reported to be catalyzed by SULT2B1. St2b2 cDNA isolated from mouse brain was expressed in *E. coli* (Shimada et al., 2002). The recombinant form mediates sulfation of cholesterol and DHEA but not that of litocholic acid and amines (Shimada et al., 2002).

AMINE SULFOTRANSFERASE

Formation of 2-naphthylsulfamate was first reported in 1957 (Boyland et al., 1957). Higher amounts of the metabolite are found in rabbit than in rat urine after 2-NA administrations. Two forms of sulfotransferase, AST-RB1 and AST-RB2, mediating sulfation of DETR are purified from rabbit livers (Shiraga et al., 1999). Corresponding cDNAs have been isolated from a rabbit liver cDNA library by use of antibodies against AST-RB1 and AST-RB2 (Yoshinari et al., 1998b, 1998a). Judging from their primary structures, a cDNA isolated with an AST-RB2 antibody encodes a hydroxysteroid sulfotransferase, ST2A8 (Yoshinari et al., 1998a), and another cDNA isolated with an AST-RB1 antibody encodes a novel sulfotransferase consisting of a new family (Yoshinari et al., 1998b). ST3A1 mediated sulfamate formation of desipramine and 4-phenyl-1,2,3,6-tetrahydropyridine (PTHP) in Table 12.12, although the K_m value for PTHP is very high. Interestingly, rat ST2A2, but not ST2A1, also mediates sulfamate reaction. The K_m value and V_{max} of ST2A2 for PTHP are lower (about 30 times) and higher (5 times) than those of rabbit ST2A8, respectively. Mouse St3a2 cDNA was also isolated but not characterized until now. By searching the gene database in the NCBI homepage, the human SULT3A1 gene is found at 14q12 in chromosome 14, but it is a putative pseudogene. Rat SULT3A1 gene also could not be found in the database at present. Therefore, sulfamate reaction is likely to be mediated by ST2A forms in humans and rats and ST3A1 as well as ST2A8 in rabbits.

OTHER SULFOTRANSFERASE

Recently, based on the database of ESTs, novel sulfotransferases have been identified. Among them, brain-specific forms, the SULT4A1 gene, have been found in humans, rats, rabbits, and mice (Liyou et al., 2003; Sakakibara et al., 2002). The amino acid sequence of SULT4A1 is highly conserved at more than 97.9% among them. In particular, rat and mouse SULT4A1s show the same amino acid sequence, although nucleotide sequence in the open reading frame of both SULT4A1s shared

TABLE 12.12
Apparent Kinetic Parameters for
DMI and PTHP Sulfations by ST3A1
and ST2A Forms

Enzymes	[K_m (μM) V_{max} (nmol/mg protein/min)]			
	DMI		PTHP	
	K_m	V_{max}	K_m	V_{max}
ST3A1	12	60	76	74
ST2A8	135	79	8600	124
Rabbit cytosol	16	0.60	86	1.3
ST2A1	n.d.	n.d.	n.d.	n.d.
ST2A2	67	8.6	2.7	369
Rat cytosol	19	0.3	1.6	5.8

Each parameter is derived from analysis of
Lineweaver-Burk analysis and is as the mean of
three separated experiments. Assays were performed at pH 10 with various concentrations of
substrates (DIM; 2.5-400 μM, and PTHP; 1.25-
800 μM).

DIM: desipramine; PTHP: 4-phenyl-1,2,3,6-
tetrahydropyridine.
n.d.: not detected.

95.6% similarity. Judging from the highly conserved sequence among animals, the
very important role of SULT4A1 in brain function is predicted; however, the substrate is still unknown. Other unique sulfotransferases identified from mice and
chickens are SULT5A1 and SULT6A1, which comprise a new family. Substrates
for those forms are also unknown.

CONCLUSION

Species differences of sulfation among experimental animals have been thought to
depend on both the kinetic properties and distinct expression profiles in tissues.
However, recent gene analyses of humans and experimental animals have indicated
that the species difference also depends on the occurrence of the specific gene
destruction during animal evolution.

   More than 50 cDNAs of cytosol sulfotransferase have been isolated from vertebrates at present. In the Genbank database, a number of unknown genes predicted
to encode cytosol sulfotransferase are also found. Most of them seem to be a putative
pseudogene. As shown in Table 12.13, ten cytosolic sulfotransferases have been
identified in humans. On the other hand, 15 genes of the sulfotransferase are found

TABLE 12.13

Sulfotransferase Genes in Human and Experimental Animals

	Human (h)	Rat (r)	Mouse (m)	Rabbit (rb)	Guinea Pig (gp)	Dog (d)	Monkey (mf)	Bovine (b)	Pig (p)	Chicken (ck)
SULT1A1	ST1A3	ST1A1	St1a4	ST1A8		ST1A7	ST1A9	ST1A6		
SULT1A2	ST1A2									
SULT1A3	ST1A5									
SULT1B1	ST1B2	ST1B1	St1b3			ST1B4				ST1B6
SULT1C1	★	ST1C1	St1c4							ST1C8
SULT1D1	★	ST1D2	St1d1			ST1D3				
SULT1E1	ST1E4	ST1E4	St1e5		ST1E3			ST1E1	ST1E7	
SULT1E2		ST1E6								
SULT1F1	ST1C2	ST1C6	St1c9	ST1C5						
SULT1F2		ST1C7								
SULT1G1	ST1C3									
SULT2A1	ST2A3	ST2A1	St2a4	ST2A8	ST2A6		ST2A10			
SULT2A2		ST2A11	St2a9	ST2A7						
SULT2A3		ST2A2								
SULT2A4		ST2A5								
SULT2B1	ST2B1		St2b2							
SULT2C1	★		St2c1							
SULT3A1	★		St3a2	ST3A1						
SULT4A1	ST5A1	ST5A3	St5a2							
SULT5A1	★		St4a1							
SULT6A1	nf									ST6A1

★: Pseudogene, nf: not found in human genome.

in the human genomic database. Five genes are putative pseudogenes. In other animals, similar situations may be found after analyses of their genomic sequence.

To avoid confusion in the future nomenclature, we propose a new nomenclature for cytosol sulfotransferase as shown in Table 12.13. This system is an adopted version of the sulfotransferase homepage (http://www.fccc.edu/research/labs/rafto-gianis/sult/index.html) except for SULT1C. As described above, SULT1C forms are classified into three groups from their similarities of amino acid sequence. In animal evolution, the divergence of ST1C1 and ST1C6 or ST1C7 from a single gene occurred at an earlier time than appearance of those animals as shown. Furthermore, unknown genes as mentioned above sharing a high similarity with ST1C6 and ST1C7 were identified. Therefore, in this nomenclature system, a paralogous gene is classified into a different gene subfamily even if both genes share nearly 60% similarity. Similarly, ST2C1 is termed as a novel SULT2 gene. A symbol for each animal species should be used in the fort of the familiar name as shown in Table 12.13.

We hope that our nomenclature system for cytosol sulfotransferase (visit the following home page: http://www.pharm.tohoku.ac.jp/~doutai/doutai-n.html/index.html) is commonly acceptable and provides for improved understanding of the research in this field.

REFERENCES

Aksoy, I.A., Wood, T.C., and Weinshilboum, R. (1994). Human liver estrogen sulfotransferase: identification by cDNA cloning and expression. *Biochem. Biophys. Res. Commun.* 200: 1621–1629.

Bernier, F., Lopez Solache, I., Labrie, F., and Luu-The, V. (1994). Cloning and expression of cDNA encoding human placental estrogen sulfotransferase. *Mol. Cell Endocrinol.* 99: R11–R15.

Bonham Carter, S.M., Rein, G., Glover, V., Sandler, M., and Caldwell, J. (1983). Human platelet phenolsulphotransferase M and P: substrate specificities and correlation with in vivo sulphoconjugation of paracetamol and salicylamide. *Br. J. Clin. Pharmacol.* 15: 323–330.

Boyland, E., Manson, D., and Orr, S.F.D. (1957). The biochemistry of aromatic amines 2: the conversion of arylamines into arylsulphamic acids and arylamine-N-glucosiduronic acids. *Biochem. J.* 65: 417–423.

Buhl, A.E., Waldon, D.J., Baker, C.A., and Johnson, G.A. (1990). Minoxidil sulfate is the active metabolite that stimulates hair follicles. *J. Invest. Dermatol.* 95: 553–557.

Chatterjee, B., Majumdar, D., Ozbilen, O., Murty, C.V., and Roy, A.K. (1987). Molecular cloning and characterization of cDNA for androgen-repressible rat liver protein, SMP-2. *J. Biol. Chem.* 262: 822–825.

Chen, X., Yang, Y.S., Zheng, Y., Martin, B.M., Duffel, M.W., and Jakoby, W.B. (1992). Tyrosine-ester sulfotransferase from rat liver: bacterial expression and identification. *Protein Expr. Purif.* 3: 421–426.

Corpechot, C., Robel, P., Axelson, M., Sjovall, J., and Baulieu, E.E. (1981). Characterization and measurement of dehydroepiandrosterone sulfate in rat brain. *Proc. Natl. Acad. Sci. U S A* 78: 4704–4707.

De Meio, R.H. (1975). Sulfate activation and transfer, in *Metabolism Pathways*. Greenberg, D.M., Ed., Academic, NY, 7: 287–357.

DeBaun, J.R., Rowley, J.Y., Miller, E.C., and Miller, J.A. (1968). Sulfotransferase activation of N-hydroxy-2-acetylaminofluorene in rodent livers susceptible and resistant to this carcinogen. *Proc. Soc. Exp. Biol. Med.* 129: 268–273.

Demyan, W.F., Song, C.S., Kim, D.S., Her, S., Gallwitz, W., Rao, T.R., Slomczynska, M., Chatterjee, B., and Roy, A.K. (1992). Estrogen sulfotransferase of the rat liver: complementary DNA cloning and age- and sex-specific regulation of messenger RNA. *Mol. Endocrinol.* 6: 589–597.

Dooley, T.P., Probst, P., Munroe, P.B., Mole, S.E., Liu, Z., and Doggett, N.A. (1994). Genomic organization and DNA sequence of the human catecholamine-sulfating phenol sulfotransferase gene (STM). *Biochem. Biophys. Res. Commun.* 205: 1325–1332.

Elias, P.M., Williams, M.L., Maloney, M.E., Bonifas, J.A., Brown, B.E., Grayson, S., and Epstein, E.H., Jr. (1984). Stratum corneum lipids in disorders of cornification: steroid sulfatase and cholesterol sulfate in normal desquamation and the pathogenesis of recessive X-linked ichthyosis. *J. Clin. Invest.* 74: 1414–1421.

Fujita, K., Nagata, K., Ozawa, S., Sasano, H., and Yamazoe, Y. (1997). Molecular cloning and characterization of rat ST1B1 and human ST1B2 cDNAs, encoding thyroid hormone sulfotransferases. *J. Biochem. (Tokyo)* 122: 1052–1061.

Fujita, K., Nagata, K., Yamazaki, T., Watanabe, E., Shimada, M., and Yamazoe, Y. (1999). Enzymatic characterization of human cytosolic sulfotransferases; identification of ST1B2 as a thyroid hormone sulfotransferase. *Biol. Pharm. Bull.* 22: 446–452.

Gong, D.W., Ozawa, S., Yamazoe, Y., and Kato, R. (1991). Purification of hepatic N-hydroxyarylamine sulfotransferases and their regulation by growth hormone and thyroid hormone in rats. *J. Biochem. (Tokyo)* 110: 226–231.

Hamamoto, T. and Mori, Y. (1989). Sulfation of minoxidil in keratinocytes and hair follicles. *Res. Commun. Chem. Pathol. Pharmacol.* 66: 33–44.

Hehonah, N., Zhu, X., Brix, L., Bolton-Grob, R., Barnett, A., Windmill, K., and McManus, M. (1999). Molecular cloning, expression, localisation and functional characterisation of a rabbit SULT1C2 sulfotransferase. *Int. J. Biochem. Cell. Biol.* 31: 869–882.

Her, C., Kaur, G.P., Athwal, R.S., and Weinshilboum, R.M. (1997). Human sulfotransferase SULT1C1: cDNA cloning, tissue-specific expression, and chromosomal localization. *Genomics* 41: 467–470.

Her, C., Wood, T.C., Eichler, E.E., Mohrenweiser, H.W., Ramagli, L.S., Siciliano, M.J., and Weinshilboum, R.M. (1998). Human hydroxysteroid sulfotransferase SULT2B1: two enzymes encoded by a single chromosome 19 gene. *Genomics* 53: 284–295.

Hidaka, H., Nagatsu, T., and Yagi, K. (1969). Formation of serotonin O-sulfate by sulfotransferase of rabbit liver. *Biochim. Biophys. Acta* 177: 354–357.

Hirshey, S.J., Dooley, T.P., Reardon, I.M., Heinrikson, R.L., and Falany, C.N. (1992). Sequence analysis, in vitro translation, and expression of the cDNA for rat liver minoxidil sulfotransferase. *Mol. Pharmacol.* 42: 257–264.

Honma, W., Kamiyama, Y., Yoshinari, K., Sasano, H., Shimada, M., Nagata, K., and Yamazoe, Y. (2001). Enzymatic characterization and interspecies difference of phenol sulfotransferases, ST1A forms. *Drug Metab. Dispos.* 29: 274–281.

Jackson, C.D. and Irving, C.C. (1972). Sex differences in cell proliferation and N-hydroxy-2-acetylaminofluorene sulfotransferase levels in rat liver during 2-acetylaminofluorene administration. *Cancer Res* 32: 1590–1594.

Jakoby, W.B., Duffel, M.W., Lyon, E.S., and Ramaswamy, S. (1984). Sulfotransferases active with xenobiotics — comments on mechanism, in *Progress in Drug Metabolism*. Bridges, J.W., Ed., Taylor & Francis Ltd., London 8: 11–33.

King, C.M. and Phillips, B. (1968). Enzyme-catalyzed reactions of the carciongen N-hydroxy-2-fluorenylacetamide with nucleic acid. *Science* 159: 1351.

Kuchel, O., Buu, N.T., and Serri, O. (1982). Sulfoconjugation of catecholamines, nutrition, and hypertension. *Hypertension* 4: III93–III98.

Liu, M.C., Sakakibara, Y., and Liu, C.C. (1999). Bacterial expression, purification, and characterization of a novel mouse sulfotransferase that catalyzes the sulfation of eicosanoids. *Biochem. Biophys. Res. Commun.* 254: 65–69.

Liyou, N.E., Buller, K.M., Tresillian, M.J., Elvin, C.M., Scott, H.L., Dodd, P.R., Tannenberg, A.E., and McManus, M.E. (2003). Localization of a brain sulfotransferase, SULT4A1, in the human and rat brain: an immunohistochemical study. *J. Histochem. Cytochem.* 51: 1655–1664.

Marsolais, F. and Varin, L. (1998). Recent developments in the study of the structure-function relationship of flavonol sulfotransferases. *Chem. Biol. Interact.* 109: 117–122.

Meinl, W. and Glatt, H. (2001). Structure and localization of the human SULT1B1 gene: neighborhood to SULT1E1 and a SULT1D pseudogene. *Biochem. Biophys. Res. Commun.* 288: 855–862.

Mulder, G.J. (1984). Sulfation-metabolic aspects, in *Progress in Drug Metabolism*. Bridges, J.W. and Chasseaud, L.F., Eds., Taylor & Francis Ltd, London, pp. 35–100.

Nagata, K., Ozawa, S., Miyata, M., Shimada, M., Gong, D.W., Yamazoe, Y., and Kato, R. (1993). Isolation and expression of a cDNA encoding a male-specific rat sulfotransferase that catalyzes activation of N-hydroxy-2-acetylaminofluorene. *J. Biol. Chem.* 268: 24720–24725.

Nagata, K. and Yamazoe, Y. (2000). Pharmacogenetics of sulfotransferase. *Annu. Rev. Pharmacol. Toxicol.* 40: 159–176.

Nagata, K., Yoshinari, K., Ozawa, S., and Yamazoe, Y. (1997). Arylamine activating sulfotransferase in liver. *Mutat. Res.* 376: 267–272.

Nash, A.R., Glenn, W.K., Moore, S.S., Kerr, J., Thompson, A.R., and Thompson, E.O. (1988). Oestrogen sulfotransferase: molecular cloning and sequencing of cDNA for the bovine placental enzyme. *Aust. J. Biol. Sci.* 41: 507–516.

Nose, Y. and Lipmann, F. (1958). Separation of steroid sulfokinases. *J. Biol. Chem.* 233: 1348.

Oeda, T., Lee, Y.C., Driscoll, W.J., Chen, H.C., and Strott, C.A. (1992). Molecular cloning and expression of a full-length complementary DNA encoding the guinea pig adrenocortical estrogen sulfotransferase. *Mol. Endocrinol.* 6: 1216–1226.

Ogura, K., Kajita, J., Narihata, H., Watabe, T., Ozawa, S., Nagata, K., Yamazoe, Y., and Kato, R. (1989). Cloning and sequence analysis of a rat liver cDNA encoding hydroxysteroid sulfotransferase. Biochem. *Biophys. Res. Commun.* 165: 168–174.

Ogura, K., Kajita, J., Narihata, H., Watabe, T., Ozawa, S., Nagata, K., Yamazoe, Y., and Kato, R. (1990a). cDNA cloning of the hydroxysteroid sulfotransferase STa sharing a strong homology in amino acid sequence with the senescence marker protein SMP-2 in rat livers. *Biochem. Biophys. Res. Commun.* 166: 1494–1500.

Ogura, K., Sohtome, T., Sugiyama, A., Okuda, H., Hiratsuka, A., and Watabe, T. (1990b). Rat liver cytosolic hydroxysteroid sulfotransferase (sulfotransferase a) catalyzing the formation of reactive sulfate esters from carcinogenic polycyclic hydroxymethylarenes. *Mol. Pharmacol.* 37: 848–854.

Okuda, H., Nojima, H., Watanabe, N., and Watabe, T. (1989). Sulphotransferase-mediated activation of the carcinogen 5-hydroxymethyl-chrysene: species and sex differences in tissue distribution of the enzyme activity and a possible participation of hydroxysteroid sulphotransferases. *Biochem. Pharmacol.* 38: 3003–3009.

Otterness, D.M., Wieben, E.D., Wood, T.C., Watson, W.G., Madden, B.J., McCormick, D.J., and Weinshilboum, R.M. (1992). Human liver dehydroepiandrosterone sulfotransferase: molecular cloning and expression of cDNA. *Mol. Pharmacol.* 41: 865–872.

Ozawa, S., Nagata, K., Gong, D.W., Yamazoe, Y., and Kato, R. (1990). Nucleotide sequence of a full-length cDNA (PST-1) for aryl sulfotransferase from rat liver. *Nucleic Acids Res.* 18: 4001.

Ozawa, S., Nagata, K., Gong, D.W., Yamazoe, Y., and Kato, R. (1993). Expression and functional characterization of a rat sulfotransferase (ST1A1) cDNA for sulfations of phenolic substrates in COS-1 cells. *Jpn. J. Pharmacol.* 61: 153–156.

Pacifici, G.M., De Santi, C., Mussi, A., and Ageletti, C.A. (1996). Interindividual variability in the rate of salbutamol sulphation in the human lung. *Eur. J. Clin. Pharmacol.* 49: 299–303.

Pacifici, G.M., Eligi, M., and Giuliani, L. (1993). (+) and (−) terbutaline are sulphated at a higher rate in human intestine than in liver. *Eur. J. Clin. Pharmacol.* 45: 483–487.

Robbins, P.W. and Lipmann, F. (1957). Isolation and identification of active sulfate. *J. Biol. Chem.* 229: 837.

Saeki, Y., Sakakibara, Y., Araki, Y., Yanagisawa, K., Suiko, M., Nakajima, H., and Liu, M.C. (1998). Molecular cloning, expression, and characterization of a novel mouse liver SULT1B1 sulfotransferase. *J. Biochem. (Tokyo)* 124: 55–64.

Sakakibara, Y., Suiko, M., Pai, T.G., Nakayama, T., Takami, Y., Katafuchi, J., and Liu, M.C. (2002). Highly conserved mouse and human brain sulfotransferases: molecular cloning, expression, and functional characterization. *Gene* 285: 39–47.

Sakakibara, Y., Takami, Y., Zwieb, C., Nakayama, T., Suiko, M., Nakajima, H., and Liu, M.C. (1995). Purification, characterization, and molecular cloning of a novel rat liver dopa/tyrosine sulfotransferase. *J. Biol. Chem.* 270: 30470–30478.

Sakakibara, Y., Yanagisawa, K., Katafuchi, J., Ringer, D.P., Takami, Y., Nakayama, T., Suiko, M., and Liu, M.C. (1998a). Molecular cloning, expression, and characterization of novel human SULT1C sulfotransferases that catalyze the sulfonation of N-hydroxy-2-acetylaminofluorene. *J. Biol. Chem.* 273: 33929–33935.

Sakakibara, Y., Yanagisawa, K., Takami, Y., Nakayama, T., Suiko, M., and Liu, M.C. (1998b). Molecular cloning, expression, and functional characterization of novel mouse sulfotransferases. *Biochem. Biophys. Res. Commun.* 247: 681–686.

Shimada, M., Kamiyama, Y., Sato, A., Honma, W., Nagata, K., and Yamazoe, Y. (2002). A hydroxysteroid sulfotransferase, st2b2, is a skin cholesterol sulfotransferase in mice. *J. Biochem. (Tokyo)* 131: 167–169.

Shimada, M., Terazawa, R., Kamiyama, Y., Honma, W., Nagata, K., and Yamazoe, Y. (2004). Unique properties of a renal sulfotransferase, St1d1 in dopamine metabolism. *J. Pharmacol. Exp. Ther.,* 310: 808–814.

Shiraga, T., Iwasaki, K., Hata, T., Yoshinari, K., Nagata, K., Yamazoe, Y., and Ohno, Y. (1999). Purification and characterization of two amine N-sulfotransferases, AST-RB1 (ST3A1) and AST-RB2 (ST2A8), from liver cytosols of male rabbits. *Arch. Biochem. Biophys.* 362: 265–274.

Shiraga, T., Iwasaki, K., Takeshita, K., Matsuda, H., Niwa, T., Tozuka, Z., Hata, T., and Guengerich, F.P. (1995). Species- and gender-related differences in amine, alcohol and phenol sulphoconjugations. *Xenobiotica* 25: 1063–1071.

Strott, C.A. (1996). Steroid sulfotransferases. *Endocr. Rev.* 17: 670–697.

Tamura, H.O., Harada, Y., Miyawaki, A., Mikoshiba, K., and Matsui, M. (1998). Molecular cloning and expression of a cDNA encoding an olfactory-specific mouse phenol sulphotransferase. *Biochem. J.* 331 (Pt 3): 953–958.

Tsoi, C., Falany, C.N., Morgenstern, R., and Swedmark, S. (2001a). Molecular cloning, expression, and characterization of a canine sulfotransferase that is a human ST1B2 ortholog. *Arch. Biochem. Biophys.* 390: 87–92.

Tsoi, C., Falany, C.N., Morgenstern, R., and Swedmark, S. (2001b). Identification of a new subfamily of sulphotransferases: cloning and characterization of canine SULT1D1. *Biochem. J.* 356: 891–897.

Vietri, M., De Santi, C., Pietrabissa, A., Mosca, F., and Pacifici, G.M. (2000). Fenamates and the potent inhibition of human liver phenol sulphotransferase. *Xenobiotica* 30: 111–116.

Walle, T. and Walle, U.K. (1992). Stereoselective sulfate conjugation of 4-hydroxypropranolol and terbutaline by the human liver phenolsulfotransferases. *Drug Metab. Dispos.* 20: 333–336.

Watabe, T., Ogura, K., Satsukawa, M., Okuda, H., and Hiratsuka, A. (1994). Molecular cloning and functions of rat liver hydroxysteroid sulfotransferases catalysing covalent binding of carcinogenic polycyclic arylmethanols to DNA. *Chem.-Biol. Interact.* 92: 87–105.

Weinshilboum, R.M. (1986). Phenol sulfotransferase in humans: properties, regulation, and function. *Fed. Proc.* 45: 2223–2228.

Xiangrong, L., Johnk, C., Hartmann, D., Schestag, F., Kromer, W., and Gieselmann, V. (2000). Enzymatic properties, tissue-specific expression, and lysosomal location of two highly homologous rat SULT1C2 sulfotransferases. *Biochem. Biophys. Res. Commun.* 272: 242–250.

Yerokun, T., Etheredge, J.L., Norton, T.R., Carter, H.A., Chung, K.H., Birckbichler, P.J., and Ringer, D.P. (1992). Characterization of a complementary DNA for rat liver aryl sulfotransferase IV and use in evaluating the hepatic gene transcript levels of rats at various stages of 2-acetylaminofluorene-induced hepatocarcinogenesis. *Cancer Res.* 52: 4779–4786.

Yoshinari, K., Nagata, K., Shiraga, T., Iwasaki, K., Hata, T., Ogino, M., Ueda, R., Fujita, K., Shimada, M., and Yamazoe, Y. (1998a). Molecular cloning, expression, and enzymatic characterization of rabbit hydroxysteroid sulfotransferase AST-RB2. *J. Biochem. (Tokyo)* 123: 740–746.

Yoshinari, K., Nagata, K., Ogino, M., Fujita, K., Shiraga, T., Iwasaki, K., Hata, T., and Yamazoe, Y. (1998b). Molecular cloning and expression of an amine sulfotransferase cDNA: a new gene family of cytosolic sulfotransferases in mammals. *J. Biochem. (Tokyo)* 123: 479–486.

Zhu, X., Veronese, M.E., Bernard, C.C., Sansom, L.N., and McManus, M.E. (1993). Identification of two human brain aryl sulfotransferase cDNAs. *Biochem. Biophys. Res. Commun.* 195: 120–127.

13 Activation and Inactivation of Carcinogens and Mutagens by Human Sulfotransferases

Hansruedi Glatt

CONTENTS

METABOLIC ACTIVATION OF CARCINOGENS

The induction of mutations in tumor-suppresser genes and proto-oncogenes is a major mechanism of chemical carcinogenesis. In general, the carcinogen binds covalently to the DNA and disturbs its template function at replication or leads to errors during repair processes. Most carcinogens to which the human is exposed do not directly react with DNA but require metabolic activation to chemically reactive, usually short-lived intermediates (termed ultimate carcinogens/mutagens). Activation may occur during phase-I metabolism (functionalization), whereas phase-II metabolism (conjugation) is primarily associated with detoxification by many toxicologists and pharmacologists. However, this view is oversimplified, as numerous examples are known for activation via various conjugation reactions (Glatt, 2000a; King et al., 1997; Ritter, 2000; van Bladeren, 2000). It is probable that the role of phase-II reactions in the activation is underestimated for technical reasons (different state of knowledge of the various enzyme classes; differences in membrane penetration between phase-I and phase-II metabolites). Indeed, using bacterial and eukaryotic target cells engineered for expression of individual human sulfotransferase (SULT) forms, we demonstrated an activation to mutagens by these enzymes for more than 100 chemicals, as outlined in this article. A list of further compounds activated by human SULTs is contained in a previous review article (Glatt, 2000a).

SULT-DEPENDENT ACTIVATION AND INACTIVATION: CHEMICAL BACKGROUND

SULTs transfer the sulfo/sulfuryl group from the cofactor 3'-phosphoadenosine-5'-phosphosulfate (PAPS) to nucleophilic groups of their substrates (Glatt, 2002). The resulting sulfate, thiosulfate, and sulfamate groups are electron-withdrawing and are therefore good leaving groups in certain chemical linkages. A trivial example is H_2SO_4, which is a strong bivalent acid. Likewise, the sulfate group is readily cleaved off spontaneously in sulfuric acid esters of many benzylic alcohols, allylic alcohols, aromatic hydroxylamines, and aromatic hydroxamic acids as the resulting carbonium or carbonium/nitrenium ions are resonance-stabilized. Other mechanisms, such as inductive effects, may also facilitate heterolytic cleavage. Reactive sulfuric acid esters will react, usually via an SN1 mechanism, with cellular nucleophiles, such as the DNA, but also with water, regenerating the metabolic precursor of the sulfoconjugate (Landsiedel et al., 1996), as illustrated in Figure 13.1 for 1-hydroxymethylpyrene (**1**; underlined bold numbers indicate structural formulas presented in numerical order in Figure 13.1, Figure 13.2, Figure 13.5, Figure 13.6, and Figure 13.10).

Most ultimate mutagens are electrophiles, which find reaction partners in the numerous nucleophilic sites present in DNA (and other macromolecules; Miller, 1970). The sulfo acceptor sites of the substrates of SULTs are nucleophilic and, therefore, are not reactive toward DNA, which lacks significant electrophilic sites. In this regard, SULTs are not suited to detoxify ultimate mutagens. However, SULTs may compete with other enzymatic and spontaneous (e.g., autoxidative) reaction pathways, including those generating electrophilic products, for the same substrates. Therefore, sulfoconjugation can be beneficial via the sequestration of potential proximate toxicants. In fact, this situation is rather common. With various classes of chemicals, for example aromatic amines, sulfoconjugation is involved in both the activation as well as the sequestration of progenotoxicants (Figure 13.2).

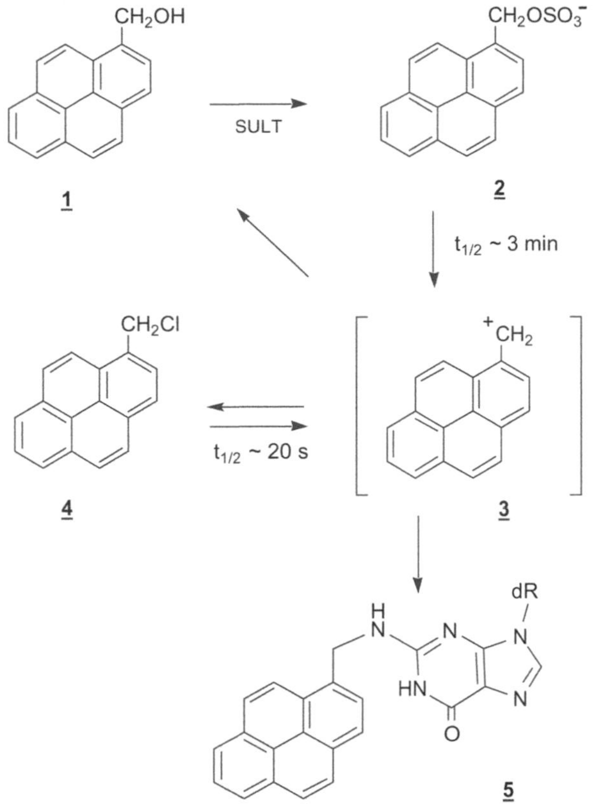

FIGURE 13.1 Metabolic pathways of 1-hydroxymethylpyrene (**1**) leading to the formation of DNA adducts. The conversion of **1** to 1-sulfooxymethylpyrene (**2**) is catalyzed by SULTs. All other reactions occur spontaneously in the presence of the appropriate reactants (H_2O, Cl^-, and DNA, respectively). Cl^- represents here numerous small nucleophilic molecules that may lead to the formation of secondary reactive species (such as 1-chloromethylpyrene, **4**; Landsiedel et al., 1996). In addition to the desoxyguanosine adduct (**5**), several minor DNA adducts are formed.

SULT-DEPENDENT ACTIVATION OF CARCINOGENS: HISTORICAL BACKGROUND

It is worth mentioning that a sulfuric acid ester, rather than a phase-I metabolite, was the first electrophilic metabolite of a carcinogen to be discovered. This compound was 2-acetylaminofluorene N-sulfate (DeBaun et al., 1968; King and Phillips, 1968). Later it was shown that the SULT inhibitor pentachlorophenol drastically decreases the formation of DNA adducts and the induction of tumors by 2-acetyl-aminofluorene and N-hydroxy-2-acetylaminofluorene (**29**) in rats and mice (Lai et al., 1985, 1987; Meerman et al., 1981; Ringer et al., 1988). A prominent role of sulfonation was also detected in the activation of other carcinogenic aromatic amines, alkenyl benzenes, and polycyclic benzylic alcohols, as deduced, for example, from

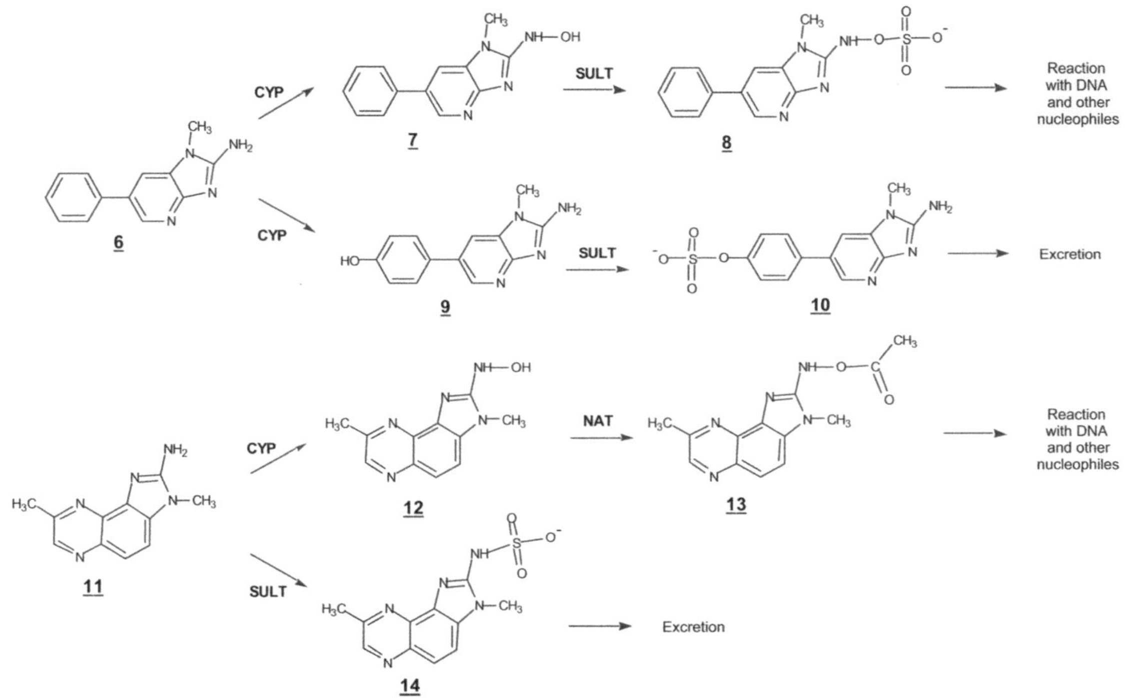

FIGURE 13.2 Role of SULTs in the activation and inactivation of aromatic amines, exemplified for the heterocyclic amines 2-amino-1-methyl-6-phenylimidazo[4,5-b]pyridine (PhIP, **6**) and 2-amino-3,8-dimethylimidazo[4,5-f]quinoxaline (MeIQx, **11**). The most common activation pathway involves N-hydroxylation by a cytochrome P450 (CYP) followed by the introduction of a leaving group by SULT or acetyltransferase (NAT). The predominant human conjugating enzymes involved in the activation of PhIP are SULT1A1 and SULT1A2, whereas the activation of MeIQx and 2-amino-3-methylimidazo[4,5-f]quinoline (IQ) appears to be mediated exclusively by NATs (Muckel et al., 2002). Sulfo conjugation of the aromatic amines, rather than the hydroxylamines, leads to the formation of stable metabolites (e.g., **14**), which can be readily excreted. This pathway has been reported to occur in humans with MeIQx (Stillwell et al., 1994) but has not been observed with PhIP (e.g., Malfatti et al., 1999). CYPs also form ring-hydroxylated metabolites, such as 4′-OH-PhIP. Although these phenolic metabolites still could be activated at the exocyclic amino group, they usually are conjugated and excreted. PhIP-4-sulfate (**10**) has been detected in human urine (Malfatti et al., 1999).

protection by SULT inhibitors or by a genetic defect in the PAPS synthesis (in brachymorphic mice; Miller, 1994). These findings were made in animal models, and it remains to be shown whether a particular chemical is activated via the same metabolic pathway in the human. Answering this question requires metabolism studies in humans *in vivo* or in model systems containing active human SULT enzymes. Andreas Czich et al. (1994) were the first to express a heterologous SULT in a target cell of a mutagenicity assay, in Chinese hamster V79 cells (which do not express any endogenous SULTs). Shortly afterwards, Ingrid Bartsch expressed the first mammalian SULT in Ames's *his⁻ Salmonella typhimurium* strains, the most widely used bacterial mutagenicity assay (Glatt et al., 1995); the vector was obtained from Charles Falany. In the meantime, my work group has expressed a total of 18 individual human SULTs (including allelic variants) and 20 additional forms from laboratory animals in these target cells of standard mutagenicity assays. This approach was particularly useful as externally generated reactive sulfuric esters may not readily penetrate into the target cell (next section).

SULT-DEPENDENT MUTAGENICITY *IN VITRO*: PROBLEMS BY CELL MEMBRANES

Mutagenicity tests in bacteria and mammalian cells in culture are the most important tools in the detection of mutagens/carcinogens and the elucidation of activation pathways. Since standard target cells lack most of the bioactivating enzymes, the ultimate mutagen has to be added directly or be generated *in situ* by an appropriate external metabolizing system from its metabolic precursors. However, the ultimate carcinogen, 2-acetylaminofluorene *N*-sulfate (sulfoconjugate of **29**), showed negligible mutagenic effects to *S. typhimurium* (Smith et al., 1986). We suspected that the short-lived, anionic sulfuric acid ester may not penetrate the bacteria, and we therefore expressed mammalian SULTs in the bacteria to generate 2-acetylaminofluorene *N*-sulfate from its precursor, *N*-hydroxy-2-acetylaminofluorene (**29**), directly within the target cells. Indeed, *N*-hydroxy-2-acetylaminofluorene showed strong mutagenic activity in an *S. typhimurium* strain expressing human sulfotransferase 1A2 (SULT1A2) but not in the SULT-deficient parental bacterial strain (Figure 13.7A). We made similar findings with a number of other sulfoconjugates. For example, 1-thiosulfooxymethylpyrene (sulfoconjugate of **26**) was not mutagenic when added to *S. typhimurium* (even at a dose of 1000 nmol) but showed clear mutagenic effects when it was generated by a cDNA-expressed human SULT within the bacteria from 1-thiomethylpyrene (**26**), a dose of 0.01 nmol of this metabolic precursor being sufficient (Figure 13.3). These results corroborate the notion that membranes may form barriers for reactive sulfuric acid esters.

Nevertheless, a number of other externally added sulfuric acid esters, such as 1-sulfooxymethylpyrene (**2**) and 6-sulfooxymethylbenzo[*a*]pyrene (sulfoconjugate of **33**), showed high mutagenic activity in *S. typhimurium* (Enders et al., 1993; Glatt, 2000a; Glatt et al., 1994a, 1994b, 1995; Rogan et al., 1986; Watabe et al., 1982). We then found that the mutagenic activity of these compounds strongly depends on the composition of the exposure medium. For example, the activity of 1-sulfooxymethylpyrene was

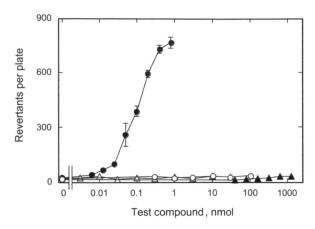

FIGURE 13.3 Mutagenicity of exogenously added and intracellularly generated 1-thiosulfooxymethylpyrene (sulfoconjugate of **26**) to *S. typhimurium*. Triangles: 1-thiosulfooxymethylpyrene tested in strain TA1538 (two experiments covering different dose ranges); solid circles: 1-thiomethylpyrene (**26**) tested in a TA1538-derived strain expressing human SULT1A1; open circles: 1-thiomethylpyrene tested in strain TA1538. The construction of SULT-expressing strains and the mutagenicity assay used have been described by Meinl et al. (2002). Values are means and SE of 3 plates.

400-fold higher when the medium contained chloride ions than when only phosphate buffer or water was used (Figure 13.4). We could demonstrate the transient formation of a secondary, membrane-penetrating reactive species, 1-chloromethylpyrene (**3**), in the presence of chloride anions (Landsiedel et al., 1996). Although chloride strongly enhanced the mutagenicity and DNA adduct formation by 1-sulfooxymethylpyrene (**2**) in bacteria, it had only minimal effects on the formation of DNA adducts by the same compound in a cell-free system. Thus, charged leaving groups (such as sulfate) may impede the transmembrane transfer of a reactive species but may be spontaneously replaced in some cases by another leaving group that facilitates the transfer (Figure 13.1).

SUBSTRATE SPECIFICITY OF HUMAN SULT FORMS TOWARD PROMUTAGENS

Sulfo conjugation of xenobiotics in vertebrates is mediated by soluble enzymes, which are members of a single superfamily abbreviated SULT (or ST in the older literature) (Falany, 1997; Glatt, 2002; Nagata and Yamazoe, 2000; Weinshilboum et al., 1997). A total of 11 *SULT* genes encoding 12 enzyme forms have been detected in the human.

Figure 13.5 and Figure 13.6 show the structural formulas of some compounds that are activated to mutagens by SULTs expressed in *S. typhimurium* strains or V79 cells. Each of these compounds is activated by certain SULT forms, but not by others, as summarized in Table 13.1 and illustrated by the dose–response curves of representative compounds in Figure 13.7. 1-hydroxymethylpyrene (**1**) is unusual as it is activated by many different human SULTs (Figure 13.7C), although with varying

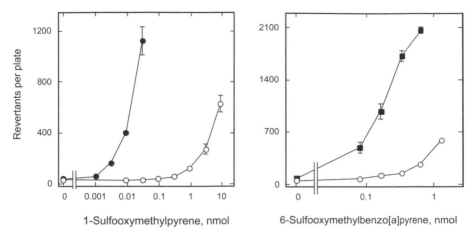

FIGURE 13.4 Influence of the exposure medium on the mutagenicity of exogenously added 1-sulfooxymethylpyrene (**2**) and 6-sulfooxymethylbenzo[*a*]pyrene (sulfoconjugate of **33**) to *S. typhimurium* TA98. The exposure medium was water containing no additional components (open circles), 135 mM KCl (solid circles), or 135 glutamine (solid squares). Values are means and SE of 3 plates. Data from Enders et al. (1993) and Landsiedel et al. (1996).

efficiency. *N*-Hydroxy-2-acetylaminofluorene (**29**; Figure 13.7A) and 4*H*-cyclopenta[*def*]chrysen-4-ol (**32**; Figure 13.7B) are more typical examples in that they are activated exceptionally well by a single human SULT form (SULT1A2 and SULT1B1, respectively), considerably less by a second form (SULT1A1 and SULT2A1, respectively), and very little or not at all by the other forms. Some human SULT forms, in particular SULT1A1, activate a large number of promutagens (Figure 13.5, Table 13.1). Others (such as SULT1B1 and 1E1) are much more selective but activate the appropriate compounds with very high efficiency.

SULT1A1 is characterized by a very broad substrate tolerance (especially, but not only, toward phenols) and by a high expression in numerous tissues (Glatt et al., 2001). We found more than 80 compounds that are activated to genotoxicants by SULT1A1 expressed in *S. typhimurium* strains or V79 cells. Representative examples are the secondary nitroalkane 2-nitropropane (**15a**) after spontaneous tautomerization to propane 2-nitronate (**15b**; Kreis et al., 2000); the *N*-hydroxy derivatives of the heterocyclic amines 2-amino-1-methyl-6-phenylimidazo[4,5-*b*]pyridine (PhIP; **7**; Muckel et al., 2002), 2-amino-5-phenylpyridine (Phe-P-1; **23**; Meinl et al., 2002), and 2-amino-3-methyl-9*H*-pyrido[2,3-*b*]indole (MeAαC; **24**; Glatt et al., 2004); various nitroarenes (following reduction to the corresponding hydroxylamines), such as the chemical intermediates 2-nitrotoluene, 3-nitrotoluene, 4-nitrotoluene (**20**), and 2,4-dinitrotoluene (**19**; manuscript in preparation), the pesticide nitrofen (**21**; Glatt and Meinl, 2004a), and the diesel pyrolysis product 3-nitrobenzanthrone (**22**; Arlt et al., 2002); the allylic alcohol 5-hydroxymethylfurfural (**16**; an abundant Maillard reaction product; Sommer et al., 2003); various benzylic alcohols derived from polycyclic aromatic hydrocarbons, such as 1-hydroxymethylpyrene (**1**; Figure 13.7C) and 7-hydroxy-7,8,9,10-tetrahydrobenzo[*a*]pyrene (**28**; manuscript in preparation); as well as benzylic thiols and amines, such as

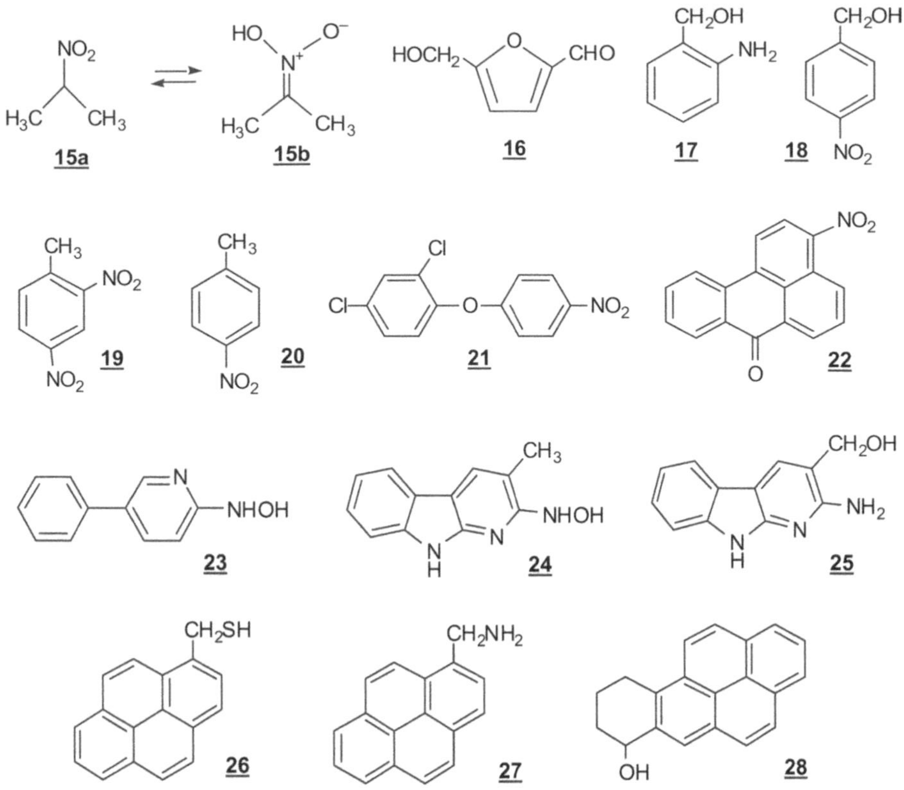

FIGURE 13.5 Structural formulas of representative compounds activated to mutagens by human SULT1A1 expressed in target cells: 2-nitropropane (**15a**; Kreis et al., 2000), 5-hydroxymethylfurfural (**16**; Sommer et al., 2003), 2-aminobenzylalcohol (**17**), 4-nitrobenzyl-alcohol (**18**), 2,4-dinitrotoluene (**19**), 4-nitrotoluene (**20**), nitrofen (**21**), 3-nitrobenzanthrone (**22**), 2-hydroxylamino-5-phenylpyridine (**23**; N-OH-Phe-P-1; Meinl et al., 2002), 2-hydroxyl-amino-3-methyl-9H-pyrido[2,3-b]indole (**24**; N-OH-MeAαC; Glatt et al., 2004), 2-amino-3-hydroxymethyl-9H-pyrido[2,3-b]indole (**25**; benzylic alcohol of MeAαC; Glatt et al., 2004), 1-thiomethylpyrene (**26**, sulfur analogue of **1**), 1-aminomethylpyrene (**27**, nitrogen analogue of **1**), and (±)-7-hydroxy-7,8,9,10-tetrahydrobenzo[a]pyrene (**28**). Mutagenicity was detected in SULT1A1-expressing target cells derived from S. typhimurium stains TA1538, TA1538-1,8-DNP (a derivative of TA1538 deficient in endogenous acetyltransferase) or TA100 (reversion to histidine prototrophy). Selected compounds were additionally investigated in Chinese hamster V79 cells (forward mutation at the hprt locus) expressing SULTs; in all cases the result in both test systems was consistent. 2-nitropropane (**15a**) is sulfoconjugated after nonenzymatic tautomerization to propane 2-nitronate (**15b**). Nitroarenes have to be reduced by the target cells to the corresponding hydroxylamines before they can be sulfoconjugated.

1-thiomethylpyrene (**26**; Figure 13.2) and 1-aminomethylpyrene (**27**; manuscript in preparation). Thus, relatively small to relatively large molecules, containing various functional groups, are activated by human SULT1A1. However, the list of activated compounds does not yet include any phenols, although SULT1A1 is the prototypic phenol SULT. The reason is the lack of electrophilic reactivity of phenolic

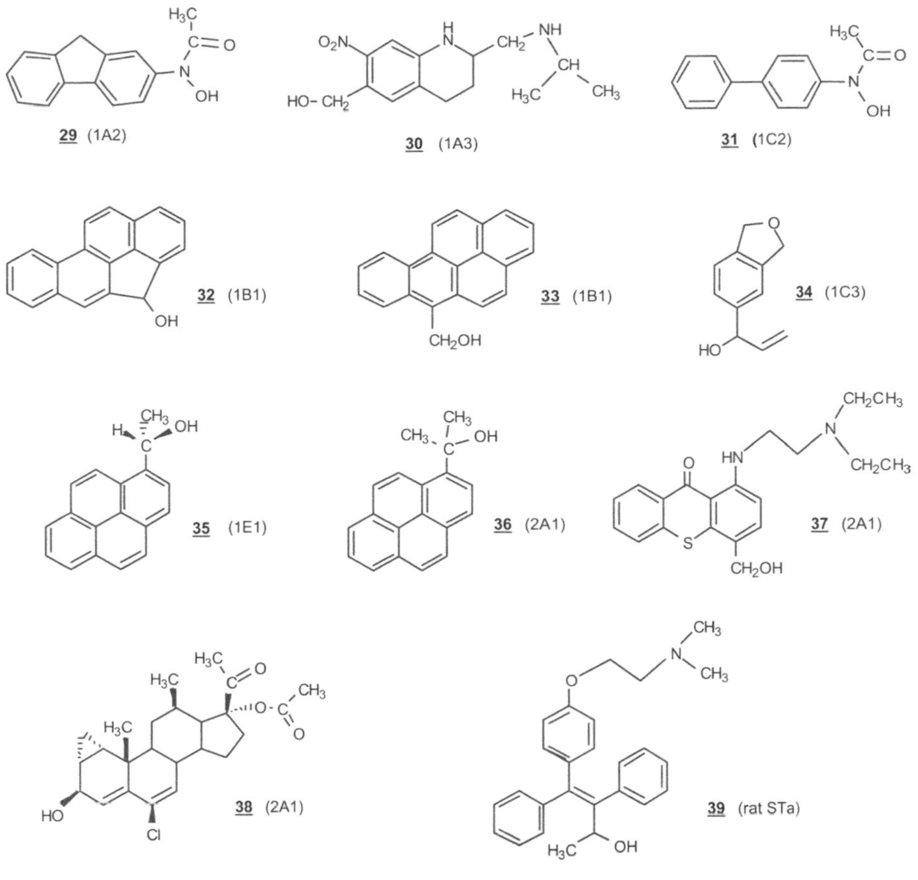

FIGURE 13.6 Structural formulas of promutagens activated with high preference by SULTs other than human SULT1A1: *N*-hydroxy-2-acetylaminofluorene (**29**; Figure 13.7A), oxamniquine (**30**), *N*-hydroxy-4-acetylaminobiphenyl (**31**), 4*H*-cyclopenta[*def*]chrysen-4-ol (**32**; Figure 13.7B), 6-hydroxymethylbenzo[*a*]pyrene (**33**), 1'-hydroxysafrole (**34**), (*S*)-1-(α-hydroxyethyl)pyrene (**35**), 1-(α-hydroxyisopropyl)pyrene (**36**), hycanthone (**37**; Glatt, 1997), 3β-hydroxycyproterone acetate (**38**), and α-hydroxytamoxifen (**39**; Glatt et al., 1998). These compounds are activated with high efficiency by the indicated human SULTs. α-Hydroxytamoxifen did not show any mutagenic effects in target cells expressing the various human SULTs but was mutagenic when rat hydroxysteroid sulfotransferase a (rSTa, member of the SULT2A subfamily) was expressed. See also legend to Figure 13.5.

sulfoconjugates. However, some phenols can be activated to mutagens via other metabolic pathways. For example, phenol was strongly mutagenic to a V79-derived cell line coexpressing human cytochrome (CYP) 2E1 and SULT1A1 (Liu et al., 2003). Its mutagenicity was abolished in the presence of a CYP2E1 inhibitor but enhanced in the presence of a SULT1A1 inhibitor (manuscript in preparation).

SULT1A2 differs in only twelve amino acid residues from SULT1A1. Its tissue distribution is narrower, and its expression levels and its catalytic efficiency toward various phenols are lower than those of SULT1A1. Many promutagens listed above

TABLE 13.1
Activation of Promutagens by Human SULTs Expressed in *S. typhimurium* TA1538 and/or TA100[a]

SULT Form	1-HMP (1)	(−)-1-HEP (35)	(+)-1-HEP	4-NBzA (18)	2,4-DNBzA	1-HIPP (36)	1-TMP (26)	6-HMBP (33)	4-OH-CPC (32)	N-OH-AAF (29)	N-OH-Phe-P1 (23)	N-OH-PhIP (7)
1A1	+++	+	+	+++	++	0	+++	0	0	++	++	+++
1A2	+	+	−	+++	+	−	0	0	0	+++	+++	+++
1A3	+	+	+	0	0	0	0	0	0	0	0	0
1B1	+	+	+	0	0	0	+	+++	+++	+	+	0
1C1[b]	0	+	0	0	−	−	−	0	0	0[c]	0	0
1C2[b]	+	+	+	+	−	−	0	0	0	0[c]	0	0
1C3	+	+	+	+	−	−	−	−	−	0	0	0
1E1	+++	++	++	++	++	+	+++	0	++	0	0	0[d]
2A1	++	+	++	++	+	++	0	++	++	0[c]	0	0
2B1a	0	0	0	0	−	−	−	0	0	0	0	0
2B1b	0	0	0	0	−	−	0	0	0	0	0	0
4A1	0	0	−	−	−	−	0	0	0	0	0	0

SULT Form	N-OH-MeAαC (24)	4-Nitrotoluene (20)	2,4-DNBzA	Nitrofen (21)	Hycanthone (37)	Oxamniquine (30)	3α-OH-CPA	3β-OH-CPA (38)	OH-TAM (39)
1A1	+++	+++	++	++	0	0	0	0	0
1A2	+	++	+	0	0	−	−	−	−
1A3	0	−	0	+	0	+	0	0	0
1B1	+	0	0	0	0	−	−	−	0

1C1[b]	0	+	0	0	0	+	-	-	-	-
1C2[b]	+	+	0	0	0	0	-	-	-	-
1C3	0	0	0	-	0	0	-	-	-	-
1E1	0	0	0	0	+	0	0	0	0	0
2A1	0	0	0	0	+++	+++	+	++	0[e]	
2B1a	0	0	0	0	0	0	0	0	-	
2B1b	0	0	0	0	0	0	0	0	0	
4A1	0	0	0	0	0	0	-	-	-	

[a] Data from references cited in the main text, updated with unpublished results. 1-HMP, 1-hydroxymethylpyrene; 1-(−)-HEP, S-(−)-1-(α-hydroxyethyl)pyrene; 1-(+)-HEP, R-(+)-1-(α-hydroxyethyl)pyrene; 1-HIIP, 1-(α-hydroxyisoproyl)pyrene; 1-TMP, 1-thiomethylpyrene; 6-HMBP, 6-hydroxymethyl-benzo[a]pyrene; 4-OH-CPC, 4H-cyclopenta[def]chrysen-4-ol; N-OH-AAF, N-hydroxy-2-acetylaminofluorene; N-OH-Phe-P-1, 2-hydroxylamino-5-phenylpyridine; N-OH-PhIP, 2-hydroxylamino-1-methyl-6-phenylimidazo[4,5-b]pyridine; N-OH-MeAαC, 2-hydroxylamino-3-methyl-9H-pyrido[2,3-b]indole; 4-NBzA, 4-nitrobenzylalcohol; 2,4-DNBzA, 2,4-dinitrobenzylalcohol; 3α-OH-CPA, 3α-hydroxycyproterone acetate; 3β-OH-CPA, 3β-hydroxycyproterone acetate; OH-TAM, α-hydroxytamoxifen; +++, very strong activation; ++, strong activation; +, moderate activation; 0, no or negligible activation; −, not tested. The table is primarily designed for comparing mutagenic potencies within columns, rather than within lines.

[b] Using the nomenclature of Freimuth et al. (2000).

[c] Lewis et al. (2000) reported that N-OH-AAF is sulfonated by cDNA-expressed human SULT2A1. The apparent K_m for N-OH-AAF was 166 μM. Likewise, N-OH-AAF, at a substrate concentration of 50 μM, was sulfonated by purified, cDNA-expressed SULT1C1 and 1C2 (Sakakibara et al., 1998). However, these concentrations are far above those used in the mutagenicity experiment with the recombinant bacterial strains. For example, a concentration of 4 nM of N-OH-AAF was sufficient to double the number of revertants in strain TA1538-SULT1A2 (Figure 13.7A).

[d] Lewis et al. (1998) reported that recombinant SULT1E1 increased the covalent binding of N-OH-PhIP, used at concentrations of 1 to 100 μM, to calf thymus DNA. Much lower concentrations were used in our recombinant bacterial strains. A concentration of 0.05 nM of N-OH-PhIP was sufficient to double the number of revertants in strain TA1538-SULT1A1.

[e] An orthologous rat SULT (rSTa, a member of the SULT2A subfamily), expressed in S. typhimurium TA1538 or Chinese hamster V79 cells, efficiently activated α-hydroxytamoxifen to a mutagen (Glatt et al., 1998).

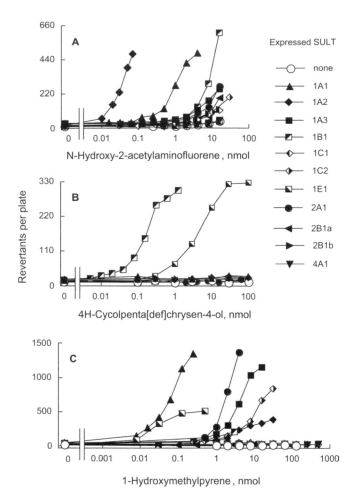

FIGURE 13.7 Mutagenicity of *N*-hydroxy-2-acetylaminofluorene (**29**; A), 4*H*-cyclopenta[*def*]chrysen-4-ol (**32**; B), and 1-hydroxymethylpyrene (**1**; C) to *S. typhimurium* TA1538-derived strains expressing the indicated human SULTs. Values are means of 3 plates; SE usually was <10% of the mean. Data for *N*-hydroxy-2-acetylaminofluorene and 1-hydroxy-methylpyrene are from Meinl et al. (2002). Data for 4*H*-cyclopenta[*def*]chrysen-4-ol were obtained using the same experimental methods.

with SULT1A1 are also activated by SULT1A2, usually with lower efficiency. However, certain aromatic hydroxylamines (such as *N*-hydroxy-Phe-P-1, **23**) and in particular the hydroxamic acid, *N*-hydroxy-2-acetylaminofluorene (**29**), are activated most efficiently by SULT1A2 among all human SULT forms (Meinl et al., 2002).

The *SULT1A3* gene, like *SULT1A2*, represents a phylogenetically young duplication of *SULT1A1*. The SULT1A3 enzyme is characterized by an enhanced affinity for dopamine and related catecholamines and a decreased affinity for numerous xenobiotic phenols compared to SULT1A1. Likewise, its substrate specificity toward promutagens is substantially narrower than that of SULT1A1. 2-nitropropane (**15a**;

Kreis et al., 2000), 1-hydroxymethylpyrene (**1**; Meinl et al., 2002), and nitrofen (**21**; Glatt and Meinl, 2004a) are activated by SULT1A3, although with much lower catalytic efficiency than by SULT1A1. To date, the schistosomacidal drug oxamniquine (**30**) was the only promutagen that was activated by SULT1A3 more efficiently than by any other human SULT form (manuscript in preparation).

SULT1B1 activated various aromatic hydroxylamines and hydroxamic acids as well as benzylic alcohols to mutagens, but in most cases its catalytic efficiency was much lower than that of other, individual human SULT forms. However, two benzylic alcohols derived from polycyclic aromatic hydrocarbons, 4*H*-cyclopenta[*def*]chrysen-4-ol (**32**; Figure 13.7B) and 6-hydroxymethylbenzo[*a*]pyrene (**33**), were activated with exceptional efficiency by SULT1B1. Furthermore, the allylic alcohol 5-hydroxymethylfurfural (**16**) is also a relatively good substrate of SULT1B1 (Sommer et al., 2003).

The physiological function of SULT1C1 (using the nomenclature of Freimuth et al., 2000) is not understood, as it only shows extremely low affinities and turnover rates with all conventional substrates studied. Likewise, nitrofen (**21**), 4-nitrotoluene (**20**), and (−)-1-(α-hydroxyethyl)pyrene (**35**) were the only promutagens activated by SULT1C1 in our recombinant test system until now (Glatt and Meinl, 2004a). *N*-hydroxy-2-acetylaminofluorene (**29**) may also be a substrate but has to be used at extremely high concentrations (Sakakibara et al., 1998).

SULT1C2 activated a number of benzylic alcohols (including 1-hydroxymethylpyrene, **1**) and aromatic hydroxylamines. In most cases, its efficiency was markedly lower than that of some other human SULT forms. However, *N*-hydroxy-4-acetylaminobiphenyl (**31**) demonstrated higher mutagenic activity in an *S. typhimurium* strain expressing SULT1C2 than in the strains engineered for expression of the other human SULT forms (manuscript in preparation).

SULT1C3 cDNA was recently constructed on the basis of human genomic sequences (manuscript in preparation). So far we have studied only a handful of promutagenic substrates with this enzyme. 1-hydroxymethylpyrene (**1**) was activated by SULT1C3, although with a much lower efficiency than by several other human SULTs. Both enantiomers of 1-(α-hydroxyethyl)pyrene (*S*-enantiomer; see **35**) were activated with reasonable efficiency by SULT1C3. 1′-hydroxysafrole (**34**), the proximate carcinogen of the secondary plant metabolite safrole (Miller, 1994), was more efficiently activated to a mutagen by SULT1C3 than by other human SULTs (manuscript in preparation).

SULT1D1 is a pseudogene in the human (Meinl and Glatt, 2001). Mouse Sult1d1 expressed in *S. typhimurium* activates a number of hydroxylamino- and nitroarenes, such as *N*-OH-PhIP (**7**) and 4-nitrotoluene (**20**), with high efficiency (manuscript in preparation).

SULT1E1 shows much higher affinity for the physiological estrogens 17β-estradiol and 17β-estrone than does any other human SULT (Falany, 1997) and, therefore, is often named estrogen SULT. Nevertheless, a number of drugs and drug metabolites, such as 4-hydroxylonazolac, are also excellent substrates for SULT1E1 (Engst et al., 2002). The *S*-enantiomer of the secondary benzylic alcohol 1-hydroxyethylpyrene (**35**) is an exceptionally potent mutagen in an *S. typhimurium* strain and in a V79-derived cell line expressing human SULT1E1 (Glatt et al., 2002). A few other polycyclic benzylic alcohols, such as 1-hydroxymethylpyrene (**1**; Figure 13.7C) and 4*H*-cyclopenta[*def*]chrysen-4-ol (**32**; Figure 13.7B), are also activated with relatively high efficiency.

SULT2A1 activates numerous benzylic alcohols of polycyclic aromatic hydrocarbons to mutagens (e.g., **1**, **33**, and **36**; Glatt, 1997). Furthermore, the schistosomacidal drug hycanthone (**37**) and some metabolites of steroidal drugs, such as the 3β-hydroxy derivatives of cyproterone acetate (**38**), megestrole acetate, and chlormadinone acetate, demonstrated substantial mutagenic activity in *S. typhimurium* strains expressing human SULT2A1. Other human SULT forms showed negligible activity toward these drugs and drug metabolites.

SULT2B1a and SULT2B1b are encoded by splice variants of the *SULT2B1* gene. Expression of these enzymes enhanced the mutagenicity of some nitroarenes, but expression of other human SULT forms led to much stronger effects.

Although SULT4A1 genetically is a member of the SULT superfamily, no conventional substrate has been found for this form. Therefore, it may not be surprising that its expression in target cells of mutagenicity tests has not yet led to the activation of any compounds studied. On the contrary, certain compounds that are somewhat mutagenic even in SULT-deficient control strains, such as *N*-OH-PhIP (**7**) and *N*-hydroxy-4-acetylaminobiphenyl (**31**; Meinl et al., 2002), showed reduced effects after the expression of SULT4A1 (whereas certain other human SULTs drastically enhanced the mutagenicity). Thus, it may be speculated that SULT4A1 acted as a binding protein in these cases.

INFLUENCE OF GENETIC POLYMORPHISMS ON THE ACTIVATION AND INACTIVATION OF PROMUTAGENS

SULT1A1 and *SULT1A2* show common genetic polymorphisms that affect the amino acid sequence (Raftogianis et al., 1999). The Arg213His exchange in SULT1A1 is associated with decreased SULT1A1 activity (Raftogianis et al., 1999) and enzyme protein level (R. Kollock, H. Schneider, W. Meinl, and H. R. Glatt, submitted) *in vivo*, at least in platelets, which are readily accessible for analysis, whereas the polymorphism has little effect on the kinetic properties of the cDNA-expressed enzyme using PAPS and 4-nitrophenol as sulfo donor and acceptor (Raftogianis et al., 1999). It is probable that such a genotype-dependent expression level, if it occurs in target tissues, affects the susceptibility toward SULT1A1-activated (or -sequestered) toxicants, as this conjugation reaction is usually competed by other metabolic pathways. It is difficult to model this situation in *in vitro* test systems. Increasing the expression level of SULT1A1 in *S. typhimurium* or V79 cells usually enhanced the mutagenic response although a saturation level was obtained with some promutagens. When SULT1A1*1 (Arg[213]) and *2 (His[213]) alloenzymes were expressed at an equal level, the mutagenic effects were similar in both systems or higher in the SULT1A1*1-expressing cells, depending on the promutagen studied and the recipient cells used. The molecular basis of these differences requires further investigation.

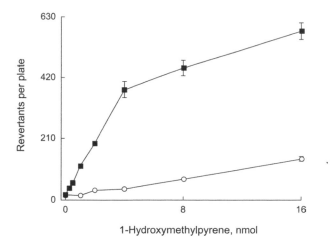

FIGURE 13.8 Mutagenicity of 1-hydroxymethylpyrene (**1**) in *S. typhimurium* TA1538-derived strains expressing the human allelic variants SULT1A2*1 (squares) and SULT1A2*2 (circles) at equal levels (strains TA1538-SULT1A1*1Z and TA1538-SULT1A1*2Y). Values are means ± SE from 3 plates. Data from Meinl et al. (2002).

It is not known whether the Asn235Thr exchange and the genetically linked Ile7Thr exchange in SULT1A2 affect the expression level in tissues *in vivo*, although they lead to reduced expression in recombinant *S. typhimurium* strains (Meinl et al., 2002). In any case, these exchanges drastically reduce the affinity of the enzyme for the standard substrate 4-nitrophenol (Meinl et al., 2002; Raftogianis et al., 1999). When these alloenzymes were expressed at equal levels in *S. typhimurium* (using synonymous nucleotide exchanges), alloenzyme SULT1A2*1 activated various promutagens more efficiently than did alloenzyme SULT1A2*2 (Figure 13.8). Similar results were obtained when these alloenzymes were expressed in V79 cells (manuscript in preparation).

SPECIES-DEPENDENT DIFFERENCES IN SULT-MEDIATED ACTIVATION OF PROMUTAGENS/PROCARCINOGENS

The antiestrogenic drug tamoxifen is a potent inducer of DNA adducts and tumors in the liver of the rat (da Costa et al., 2002; Greaves et al., 1993; White et al., 1992). α-hydroxytamoxifen (**39**), which is a major metabolite of tamoxifen, is strongly mutagenic to *S. typhimurium* strains and V79 cells engineered for expression of rat hydroxysteroid sulfotransferase a (rSTa, a member of the SULT2A subfamily; Glatt et al., 1998). This enzyme is not taken into account in standard *in vitro* mutagenicity test systems, explaining why these systems consistently gave negative results with tamoxifen and its metabolites. No human SULT form was capable of activating α-hydroxytamoxifen to a mutagen, explaining the apparent lack of hepatocarcinogenicity and DNA adduct formation in the human (Glatt et al., 1998).

TABLE 13.2
SULT1B1 Enzymes from Different Species: Expression in Colon Mucosa and the Activation of 4*H*-Cyclopenta[*def*]chrysen-4-ol 4-OH-CPC and 6-Hydroxymethylbenzo[*a*]pyrene (6-HMBP) to Mutagens[a]

SULT	Expression Level in Colon (% of Level in Recombinant Bacteria)	Mutagenicity in *S. typhimurium* Strain Expressing the Corresponding SULT (Revertants per nmol)		
		4-OH-CPC (32), Enantiomer 1	4-OH-CPC (32), Enantiomer 2	HMBP (33)
Human 1B1	6	2100	4400	13000
Rat 1B1	0.4	860	2700	700
Mouse 1B1	0.4	410	170	350
Dog 1B1	0.3	4	120	500

[a] SULT1B1 enzymes from different species were expressed in *S. typhimurium* TA1538. The levels of expression in the resulting strains were similar (approximately 1% of the cytosolic protein). All species express SULT1B1 in colon mucosa but to varying levels (as estimated from comparative western-blot analysis with the recombinant *S. typhimurium* strains, indicated in the second column). Details will be published elsewhere (H. Schneider, W. Teubner, W. Meinl, C. Tsoi, S. Swedmark, C. N. Falany, and H. R. Glatt, manuscript in preparation).

As a counterpart to α-hydroxytamoxifen, we detected many compounds that were activated by human SULTs but not (or much less) by corresponding rodent SULTs (with the limitation that we have not yet expressed all rat and mouse SULTs). For example, nitrofen (21) exhibited much stronger mutagenic activity when human, rather than rat or mouse, SULT1A1 was expressed in *S. typhimurium* (Glatt and Meinl, 2004a).

As mentioned in a previous section, 4*H*-cyclopenta[*def*]chrysen-4-ol (32) and 6-hydroxymethylbenzo[*a*]pyrene (33) are exceptionally potent mutagens in *S. typhimurium* strains engineered for expression of human SULT1B1. However, they were substantially less active when the orthologous enzymes of rat, mouse, or dog were expressed at similar absolute levels (Table 13.2). The species-dependent differences in the activation potential are even larger when they are related to expression levels in colon mucosa, where the human SULT1B1 enzyme is much more highly expressed than the SULT1B1 of the other species.

TISSUE DISTRIBUTION OF SULT-DEPENDENT DNA DAMAGE AND CARCINOGENESIS

The various human SULT forms strongly differ in their tissue distribution (Glatt, 2002). Therefore, it may be hypothesized that the tissues expressing a SULT required for the activation of a given compound are at particular risk. This hypothesis has not yet been tested in the human. However, some findings in the rat are significant with regard to this aspect, with the limitations that the expression of most SULT forms in the rat, unlike in the human, is focused in the liver.

Tamoxifen exerts its carcinogenic, mutagenic, and DNA-adduct-forming effects with high specificity in the liver but not in other investigated tissues of the rat (da Costa et al., 2002; Greaves et al., 1993; White et al., 1992). These findings correlate with the observation that the activation of tamoxifen is specifically dependent on rSTa, an enzyme expressed with high specificity in the liver (Dunn and Klaassen, 1998). Interestingly, rSTa is constitutively expressed in the liver of adult females, whereas it is induced by tamoxifen in male liver (Davis et al., 2000).

Cyproterone acetate is another rat hepatocarcinogen (Schuppler and Günzel, 1979). It forms DNA adducts exclusively in the liver (Topinka et al., 1993). Among the rat SULTs expressed in *S. typhimurium*, only rSTa activated 3α- and 3β-hydroxy-cyproterone acetate (**38**) to mutagens (manuscript in preparation).

Several other SULT-dependent carcinogens, such as 2-acetylaminofluorene and safrole, are primarily carcinogenic to the liver in the rat (Miller, 1994). Again, the SULTs involved in their activation are highly expressed in liver but low or absent in other tissues.

The benzylic alcohol 1-hydroxymethylpyrene (**1**) is activated to a mutagen by many different rat SULTs. Nevertheless, all these forms are mainly expressed in liver. 1-hydroxymethylpyrene SULT activity in each of the 20 extrahepatic rat cytosolic preparations studied amounted to <1% of the hepatic activity in females (and <2% in males; Glatt et al., 2003). We then treated rats with 1-hydroxymethylpyrene. Taking into account the above-described findings with tamoxifen and cyproterone acetate, we would have predicted a liver-specific effect. Surprisingly, kidney showed clearly the highest adduct levels under all conditions, followed by liver and lung (Glatt et al., 2003). Only low adduct levels were detected in other tissues investigated. Even within the liver, levels of adducts were rather low in Kupffer and endothelial cells (showing low 1-hydroxymethylpyrene SULT activity) compared to hepatocytes (showing high enzyme activity; Monnerjahn et al., 1993). Although the active metabolite of 1-hydroxymethylpyrene, 1-sulfooxymethylpyrene (**2**) is short-lived in water ($t_{1/2}$ ~ 3 min at 37°C), it was detected in blood plasma of 1-hydroxymethylpyrene-treated rats (Ma et al., 2003). We then found that $t_{1/2}$ of 1-sulfooxymethylpyrene was strongly prolonged in rat blood plasma *in vitro* (to 990 min). Thus, it appears, that 1-hydroxymethylpyrene was converted in the liver to 1-sulfooxymethylpyrene, which was rapidly exported into the blood, where it was stabilized by reversible binding to albumin and transported to the kidney. Metabolites that extensively bind to serum albumin cannot be filtrated in the glomerulus but require active transport through the kidney cell for their excretion. The much smaller size of kidney compared to liver is a possible reason for the exceptionally high levels of adducts formed in that tissue, although the (unknown) passage times in both tissues may also be important.

These examples corroborate the hypothesis that the site of formation of a reactive sulfuric ester is at a particular risk for corresponding compounds. With some SULT-dependent genotoxicants, damage was practically limited to that tissue. With other compounds, substantial DNA damage may occur in specific additional tissues as a result of an efficient vectorial transfer of the reactive species. The life span of the reactive species, the expression and substrate specificity of transmembrane transporters, the blood flow, the interaction with serum proteins, and the distribution of detoxification systems are some potential factors that may affect the formation of DNA adducts in tissues other than the site of formation of the sulfuric acid ester.

CHEMICAL MODULATION OF SULT-DEPENDENT GENOTOXIC AND CARCINOGENIC EFFECTS *IN VIVO*

The dependence of a mutagenic effect in recombinant test systems is usually demonstrated by conducting control experiments in the parental bacterial strain or cell line lacking the specific enzyme. As an alternative, an appropriate agent, such as pentachlorophenol with SULT1A1, can inhibit the heterologous enzyme. The test system can be used to study defined chemicals or complex samples for chemopreventive activity. For example, dealcoholized red wine completely suppressed the mutagenicity of N-OH-PhIP (**7**; Figure 13.9) and 1-hydroxymethylpyrene (**1**; Glatt, 2000b) to a TA1538-derived *S. typhimurium* strain expressing human SULT1A1. In this case, the effect was due to inhibition of the enzyme. It would be important to know whether this chemoprotection would also occur *in vivo* in the human. This situation is typically characterized by a low exposure to procarcinogens, a high total enzyme capacity, and limited (rather than total) inhibition by food-borne factors. With the limitation that this situation is not met in the animal models used, pretreatment of rats and mice with SULT inhibitors led to a strong reduction of the genotoxic and tumorigenic effects of various carcinogens (reviewed by Glatt, 2002). It is probable that inhibition of SULT shifted the metabolism into alternative, less harmful

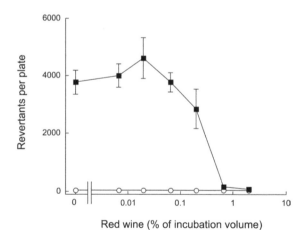

Red wine (% of incubation volume)

FIGURE 13.9 Prevention of the mutagenicity of N-OH-PhIP (**7**) in an *S. typhimurium* TA1538-derived strain expressing human SULT1A1 in the presence of dealcoholized red wine. The experiment was conducted as described elsewhere (Meinl et al., 2002) with the modification that dealcoholized red wine was added to the incubation mixture in the indicated concentration. N-OH-PhIP was used at a dose of 10 and 0 pmol per plate (squares and circles respectively). Values are means ± SE of 3 plates.

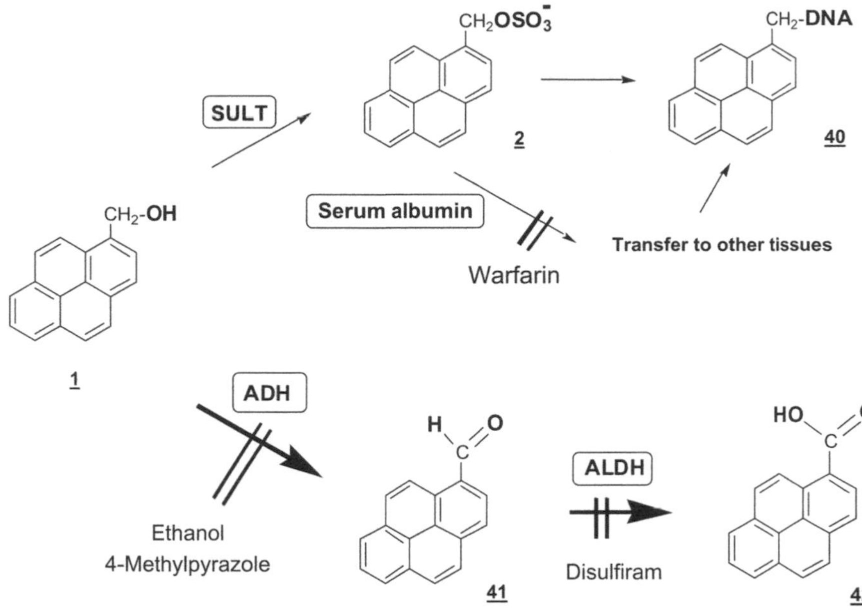

FIGURE 13.10 Major metabolic pathways 1-hydroxymethylpyrene (**1**) in the rat and chemicals that modulate its metabolism and the distribution of its DNA-reactive metabolite 1-sulfooxymethylpyrene (**2**). ADH, alcohol dehydrogenase; ALDH, aldehyde dehydrogenase. Data published by Ma et al. (2002, 2003).

pathways. Indeed, SULTs usually compete with other enzymes (such as UDP-glucuronosyltransferases, acetyltransferases, and alcohol dehydrogenases) for their substrates. Therefore, inhibition of these alternative pathways may enhance SULT-mediated activation. For example, we observed that oxidation of the side-chain, leading to 1-pyrenyl-carboxylic acid (**42**), is the major metabolic pathway of 1-hydroxymethylpyrene (**1**) in the rat (scheme in Figure 13.10); this pathway is inhibited in the presence of 4-methylpyrazole (an inhibitor of alcohol dehydrogenases), disulfiram (an inhibitor of aldehyde dehydrogenases), or ethanol (a substrate of alcohol dehydrogenases). Administration of these compounds to rats, prior to treatment with 1-hydroxymethylpyrene, dramatically enhanced the levels of DNA adducts formed in tissues (Table 13.3; for further data see Ma et al., 2002).

Warfarin competes with 1-sulfooxymethylpyrene for binding to serum albumin (Ma et al., 2003). Treatment of rats with warfarin shortly before the intraperitoneal administration of 1-sulfooxymethylpyrene markedly reduced the plasma concentration of 1-sulfooxymethylpyrene as well as the formation of DNA adducts (Ma et al., 2003).

EPIDEMIOLOGY

We have recently reviewed epidemiological studies on the association of SULT genotypes or phenotypes and certain forms of neoplasias (Glatt and Meinl, 2004b).

TABLE 13.3
Influence of Ethanol (a Competing Substrate of Alcohol Dehydrogenase, ADH) and 4-Methylpyrazole (an Inhibitor of ADH) on the Formation of Renal DNA Adducts by 1-Hydroxymethylpyrene (HMP; <u>1</u>) in the Rat *In Vivo*[a]

Treatment	1-Methylpyrenenyl-DNA Adducts in Kidney, per 10^8 Nucleotides[c]
Only HMP	270 ± 220
HMP and ethanol	20000 ± 6000
HMP and 4-methylpyrazole	35000 ± 15000

[a] HMP (250 µmol/kg body mass) was administered intraperitoneally to male adult Shoe:Wistar rats 3 h before killing. Ethanol (8 g/kg body mass) and 4-methyl-pyrazole (1 mmol/kg body mass) were given intragastrally 10 min prior to the administration of HMP. Adducts were determined using ^{32}P-postlabeling. Values are means ± SD of 5 animals. A scheme of the influence of the modulators on the biotransformation of HMP is given in Figure 13.10. Data from L. Ma, Ph.D. thesis; manuscript in preparation.

In general, these studies did not take into account exposures to specific chemicals. This is a serious shortcoming, as SULTs are double-sided swords, which can activate certain procarcinogens and sequester others. Furthermore, SULT1A1 and SULT1A2 are the only SULT forms for which common genetic polymorphisms are known and have been investigated in epidemiological studies. The major polymorphisms of both genes are closely linked together. In Caucasians, the frequencies of *1 and *2 alleles occur in a ratio of approximately 1:2, leading to genotype frequencies near 45% for *1/*1 (putative fast sulfonators), 45% for *1/*2 (rather fast sulfonators), and 10% for *2/*2 (slow sulfonators). In Asian populations, the *2 allele is less common than in Caucasians and therefore the frequency of putative slow sulfonators is very low. Under these conditions, it is not surprising that various case-control studies did not detect any association with the SULT genotype. For details, we refer the reader to our review article (Glatt and Meinl, 2004b). Here, we only present selected positive results. (a) Several studies found statistically significant associations between SULT1A1 genotype and breast cancer at least in certain subgroups. The observation that different genotypes were associated with an increased risk in different subgroups/populations may not be surprising, taking into account that SULT1A1 may be involved in the activation and sequestration of procarcinogens, in the regulation of 17β-estradiol, as well as in the elimination of drugs used in the treatment of breast cancer (e.g., tamoxifen); (b) the fast-sulfonator SULT1A1 genotypes (at least one *1 allele) enhanced the association between cigarette smoking and colorectal adenomas (Tiemersma et al., 2004); (c) the presence of at least one SULT1A1*2 allele was associated with a statistically increased risk of esophageal cancer in Taiwan (odds ratio of 3.53 [confidence interval 2.12 to 5.87], compared to the *1/*1 genotype; Wu et al., 2003).

CONCLUSIONS AND PERSPECTIVES

Many structurally diverse xenobiotics can be activated to genotoxicants by SULTs, as demonstrated, for example, by using recombinant test systems and specific SULT inhibitors. In many other cases, sulfoconjugation leads to inactive, excretable products and should be protective via competition with toxifying pathways. Although such a protective role is plausible, it has rarely been demonstrated directly in carcinogenicity and genotoxicity studies.

Results from recombinant test systems demonstrate a high selectivity of the individual SULT forms toward promutagenic substrates, at least at low substrate concentrations. At high substrate concentrations, SULTs often become promiscuous toward substrates giving stable sulfoconjugates (e.g., Dajani et al., 1999; Falany, 1997) as well as progenotoxicants (e.g., footnotes c and d to Table 13.1). However, human exposure to environmental carcinogens and their phase-I metabolites usually is extremely low. Therefore, we expect that the high-affinity forms primarily are of toxicological relevance to the human.

Human SULTs and corresponding forms from laboratory animals can substantially differ in their regulation (e.g., tissue distribution) and their substrate specificity. These differences substantially limit the value of results from animal studies for predicting and characterizing the risk for the human. To tackle this problem, we are constructing humanized mouse models for SULTs. In these mice, individual murine *SULT* genes are replaced by human *SULT* genes.

Each human SULT form appears to have unique expression characteristics with regard to tissues, cell types, ontogenesis, physiological states, and other parameters. In combination with the fact that sulfoconjugates — as anionic species — do not freely permeate cell membranes, this may lead to specific sites of action of SULT-dependent genotoxicants. However, the situation is complicated by the fact that physiological sulfoconjugates (such as terminal metabolites of xenobiotics or storage forms of hormones) are subjected to vectorial transfer processes. Some reactive sulfoconjugates may also be distributed in the organism by these systems, depending on an appropriate life span of the conjugate and the substrate specificity of the transporters of the host. The life span of the conjugate may be determined by its chemical reactivity and the presence of detoxifying enzymes, such as certain forms of glutathione transferases (Hiratsuka et al., 1994). The tissue dependence of SULT-mediated genotoxic effects in humans is an unexplored but challenging area.

Large variation of SULT activity has been observed in tissue samples from different human subjects (discussed by Glatt, 2002). Genetic polymorphisms appear to be one among many factors leading to this variation. It is probable that this variation will affect the individual susceptibility. In addition to the level of enzyme, the presence of SULT inhibitors and the modulation of competing biotransformation pathways and transport processes have strongly affected SULT-mediated genotoxic and carcinogenic effects in animal models. It is probable that these factors are also pivotal in the human. It will be an important but demanding task to specify important interactions and promising chemoprotective strategies in humans. Humanized animal models may be very useful in this work.

ACKNOWLEDGMENTS

Our work on SULTs was financially supported by Deutsche Forschungsgemeinschaft (INK 16), European Union (BMH1-CT92-0097, QLK1-CT99-01197, FP6-506820), and Bundesministerium für Forschung und Technologie (0311243 and Bioprofile grant PTJ-BIO/0313028A).

REFERENCES

Arlt, V.M., Glatt, H.R., Muckel, E., Pabel, U., Sorg, B.L., Schmeiser, H.H., and Phillips, D.H. (2002) Metabolic activation of the environmental contaminant 3-nitrobenzanthrone by human acetyltransferases and sulfotransferase. *Carcinogenesis* 23: 1937–1945.

Czich, A., Bartsch, I., Dogra, S., Hornhardt, S., and Glatt, H.R. (1994) Stable heterologous expression of hydroxysteroid sulphotransferase in Chinese hamster V79 cells and their use for toxicological investigations. *Chem.-Biol. Interact.* 92: 119–128.

da Costa, G.G., Manjanatha, M.G., Marques, M.M., and Beland, F.A. (2002) Induction of *Lac*I mutations in Big Blue rats treated with tamoxifen and α-hydroxytamoxifen. *Cancer Lett.* 176: 37–45.

Dajani, R., Cleasby, A., Neu, M., Wonacott, A.J., Jhoti, H., Hood, A.M., Modi, S., Hersey, A., Taskinen, J., Cooke, R.M., Manchee, G.R., and Coughtrie, M.W.H. (1999) X-ray crystal structure of human dopamine sulfotransferase, SULT1A3: molecular modeling and quantitative structure-activity relationship analysis demonstrate a molecular basis for sulfotransferase substrate specificity. *J. Biol. Chem.* 274: 37862–37868.

Davis, W., Hewer, A., Rajkowski, K.M., Meinl, W., Glatt, H.R., and Phillips, D.H. (2000) Sex differences in the activation of tamoxifen to DNA binding species in rat liver *in vivo* and in rat hepatocytes *in vitro*: role of sulfotransferase induction. *Cancer Res.* 60: 2887–2891.

DeBaun, J.R., Rowley, J.Y., Miller, E.C., and Miller, J.A. (1968) Sulfotransferase activation of *N*-hydroxy-2-acetylaminofluorene in rodent livers susceptible and resistant to this carcinogen. *Proc. Soc. Exptl. Biol. Med.* 129: 268–273.

Dunn, R.T. and Klaassen, C.D. (1998) Tissue-specific expression of rat sulfotransferase messenger RNAs. *Drug Metab. Dispos.* 26: 598–604.

Enders, N., Seidel, A., Monnerjahn, S., and Glatt, H.R. (1993) Synthesis of 11 benzylic sulfate esters, their bacterial mutagenicity and its modulation by chloride, bromide and acetate anions. *Polycyclic Aromat. Compds.* 3: 887s–894s.

Engst, W., Pabel, U., and Glatt, H.R. (2002) Conjugation of 4-nitrophenol and 4-hydroxylonazolac in V79-derived cells expressing individual forms of human sulphotransferases. *Environ. Toxicol. Appl. Pharmacol.* 11: 243–250.

Falany, C.N. (1997) Sulfation and sulfotransferases: 3. Enzymology of human cytosolic sulfotransferases. *FASEB J.* 11: 206–216.

Freimuth, R.R., Raftogianis, R.B., Wood, T.C., Moon, E., Kim, U.J., Xu, J., Siciliano, M.J., and Weinshilboum, R.M. (2000) Human sulfotransferases SULT1C1 and SULT1C2: cDNA characterization, gene cloning, and chromosomal localization. *Genomics* 65: 157–165.

Glatt, H.R. (1997) Sulfation and sulfotransferases: 4. Bioactivation of mutagens via sulfation. *FASEB J.* 11: 314–321.

Glatt, H.R. (2000a) Sulfotransferases in the bioactivation of xenobiotics. *Chem.-Biol. Interact.* 129: 141–170.

Glatt, H.R. (2000b) An overview of bioactivation of chemical carcinogens. *Biochem. Soc. Trans.* 28: 1–6.

Glatt, H.R. (2002) Sulphotransferases, in *Handbook of Enzyme Systems that Metabolise Drugs and Other Xenobiotics.* Ioannides, C., Ed., John Wiley & Sons, Sussex, UK, pp. 353–439.

Glatt, H.R. and Meinl, W. (2004a) Use of genetically manipulated *Salmonella typhimurium* strains to evaluate the role of sulfotransferases and acetyltransferases in nitrofen mutagenicity. *Carcinogenesis* 25: 779–786.

Glatt, H.R. and Meinl, W. (2004b) Pharmacogenetics of soluble sulfotransferases (SULTs). *Naunyn-Schmiedeberg's Arch. Pharmacol.* 369: 55–68.

Glatt, H.R., Seidel, A., Harvey, R.G., and Coughtrie, M.W.H. (1994a) Activation of benzylic alcohols to mutagens by human hepatic sulphotransferases. *Mutagenesis* 9: 553–557.

Glatt, H.R., Pauly, K., Frank, H., Seidel, A., Oesch, F., Harvey, R.G., and Werle-Schneider, G. (1994b) Substance-dependent sex differences in the activation of benzylic alcohols to mutagens by hepatic sulfotransferases of the rat. *Carcinogenesis* 15: 2605–2611.

Glatt, H.R., Bartsch, I., Czich, A., Seidel, A., and Falany, C.N. (1995) *Salmonella* strains and mammalian cells genetically engineered for expression of sulfotransferases. *Toxicol. Lett.* 82–83: 829–834.

Glatt, H.R., Davis, W., Meinl, W., Hermersdörfer, H., Venitt, S., and Phillips, D.H. (1998) Rat, but not human, sulfotransferase activates a tamoxifen metabolite to produce DNA adducts and gene mutations in bacteria and mammalian cells in culture. *Carcinogenesis* 19: 1709–1713.

Glatt, H.R., Boeing, H., Engelke, C.E.H., Kuhlow, L.M.A., Pabel, U., Pomplun, D., Teubner, W., and Meinl, W. (2001) Human cytosolic sulphotransferases: genetics, characteristics, toxicological aspects. *Mutation Res.* 482: 27–40.

Glatt, H.R., Pabel, U., Muckel, E., and Meinl, W. (2002) Activation of polycyclic aromatic compounds by cDNA-expressed phase I and phase II enzymes. *Polycyclic Aromat. Compds.* 22: 955–967.

Glatt, H.R., Meinl, W., Kuhlow, A., and Ma, L. (2003) Metabolic formation, distribution and toxicological effects of reactive sulphuric acid esters. *Nova Acta Leopoldina NF87* 329: 151–161.

Glatt, H.R., Pabel, U., Meinl, W., Frederiksen, H., Frandsen, H., and Muckel, E. (2004) Bioactivation of the heterocyclic aromatic amine 2-amino-3-methyl-9*H*-pyrido[2,3-*b*]indole (MeAαC) in recombinant test system expressing human xenobiotic-metabolizing enzymes. *Carcinogenesis* 25: 801–807.

Greaves, P., Goonetilleke, R., Nunn, G., Topham, J., and Orton, T. (1993) Two-year carcinogenicity study of tamoxifen in Alderley Park Wistar-derived rats. *Cancer Res.* 53: 3919–3924.

Hiratsuka, A., Okada, T., Nishiyama, T., Fujikawa, M., Ogura, K., Okuda, H., Watabe, T., and Watabe, T. (1994) Novel *theta* class glutathione *S*-transferases Yrs-Yrs' and Yrs'-Yrs' in rat liver cytosol: their potent activity toward 5-sulfoxymethylchrysene, a reactive metabolite of the carcinogen 5-hydroxymethylchrysene. *Biochem. Biophys. Res. Commun.* 202: 278–284.

King, C.M. and Phillips, B. (1968) Enzyme-catalyzed reactions of the carcinogen *N*-hydroxy-2-fluorenylacetamide with nucleic acid. *Science* 159: 1351–1353.

King, C.M., Land, S.J., Jones, R.F., Debiec-Rychter, M., Lee, M.-S., and Wang, C.Y. (1997) Role of acetyltransferases in the metabolism and carcinogenicity of aromatic amines. *Mutation Res.* 376: 123–128.

Kreis, P., Brandner, S., Coughtrie, M.W.H., Pabel, U., Meinl, W., Glatt, H.R., and Andrae, U. (2000) Human phenol sulfotransferases hP-PST and hM-PST activate propane 2-nitronate to a genotoxicant. *Carcinogenesis* 21: 295–299.

Lai, C.-C., Miller, J.A., Miller, E.C., and Liem, A. (1985) *N*-Sulfoöxy-2-aminofluorene is the major ultimate electrophilic and carcinogenic metabolite of *N*-hydroxy-2-acetyl-aminofluorene in the livers of infant male C57BL/6J x C3H/HeJ F₁ (B6C3F₁) mice. *Carcinogenesis* 6: 1037–1045.

Lai, C.-C., Miller, E.C., Miller, J.A., and Liem, A. (1987) Initiation of hepatocarcinogenesis in infant male B6C3F₁ mice by *N*-hydroxy-2-aminofluorene or *N*-hydroxy-2-acetyl-aminofluorene depends primarily on metabolism to *N*-sulfooxy-2-aminofluorene and formation of DNA-(deoxyguanosin-8-yl)-2-aminofluorene adducts. *Carcinogenesis* 8: 471–478.

Landsiedel, R., Engst, W., Scholtyssek, M., Seidel, A., and Glatt, H.R. (1996) Benzylic sulphuric acid esters react with diverse functional groups and often form secondary reactive species. *Polycyclic Aromat. Compds.* 11: 341–348.

Lewis, A.J., Walle, U.K., King, R.S., Kadlubar, F.F., Falany, C.N., and Walle, T. (1998) Bioactivation of the cooked food mutagen *N*-hydroxy-2-amino-1-methyl-6-phenylim-idazo[4,5-*b*]pyridine by estrogen sulfotransferase in cultured human mammary epi-thelial cells. *Carcinogenesis* 19: 2049–2053.

Lewis, A.J., Otake, Y., Walle, U.K., and Walle, T. (2000) Sulphonation of *N*-hydroxy-2-acetylaminofluorene by human dehydroepiandrosterone sulphotransferase. *Xenobiot-ica* 30: 253–261.

Liu, Y.-G., Muckel, E., Doehmer, J., and Glatt, H.R. (2003) Phenol and hydroquinone induce gene mutations in V79-derived cells expressing human xenobiotic-metabolising enzymes. *Nova Acta Leopoldina NF87* 329: 231–237.

Ma, L., Kuhlow, A., and Glatt, H.R. (2002) Ethanol enhances the activation of 1-hydroxy-methylpyrene to DNA adduct-forming species in the rat. *Polycyclic Aromat. Compds.* 22: 933–946.

Ma, L., Kuhlow, A., and Glatt, H.R. (2003) Albumin strongly prolongs the lifetime of chemically reactive sulphuric acid esters and affects their biological activities in the rat. *Nova Acta Leopoldina NF87* 329: 265–272.

Malfatti, M.A., Kulp, K.S., Knize, M.G., Davis, C., Massengill, J.P., Williams, S., Nowell, S., MacLeod, S., Dingley, K.H., Turteltaub, K.W., Lang, N.P., and Felton, J.S. (1999) The identification of [2-¹⁴C]2-amino-1-methyl-6-phenylimidazo[4,5-*b*]pyridine metabolites in humans. *Carcinogenesis* 20: 705–713.

Meerman, J.H.N., Beland, F.A., and Mulder, G.J. (1981) Role of sulfation in the formation of DNA adducts from *N*-hydroxy-2-acetylaminofluorene in rat liver *in vivo*: inhibition of *N*-acetylated aminofluorene adduct formation by pentachlorophenol. *Carcinogen-esis* 2: 413–416.

Meinl, W. and Glatt, H.R. (2001) Structure and localization of the human *SULT1B1* gene: neighborhood to *SULT1E1* and a *SULT1D* pseudogene. *Biochem. Biophys. Res. Commun.* 288: 855–862.

Meinl, W., Meerman, J.H., and Glatt, H.R. (2002) Differential activation of promutagens by alloenzymes of human sulfotransferase 1A2 expressed in *Salmonella typhimurium*. *Pharmacogenetics* 12: 677–689.

Miller, J.A. (1970) Carcinogenesis by chemicals: an overview. *Cancer Res.* 30: 559–576.

Miller, J.A. (1994) Sulfonation in chemical carcinogenesis: history and present status. *Chem.-Biol. Interact.* 92: 329–341.

Monnerjahn, S., Seidel, A., Steinberg, P., Oesch, F., Hinz, M., Stezowsky, J.J., Hewer, A., Phillips, D.H., and Glatt, H.R. (1993) Formation of DNA adducts from 1-hydroxy-methylpyrene in liver cells *in vivo* and *in vitro*, in *Postlabelling Methods for Detection of DNA Adducts*. Phillips, D.H., Castegnaro, M., and Bartsch, H., Eds., IARC, Lyon, France, pp. 189–193.

Muckel, E., Frandsen, H., and Glatt, H.R. (2002) Heterologous expression of human *N*-acetyltransferases 1 and 2 and sulfotransferase 1A1 in *Salmonella typhimurium* for mutagenicity testing of heterocyclic amines. *Food Chem. Toxicol.* 40: 1063–1068.

Nagata, K. and Yamazoe, Y. (2000) Pharmacogenetics of sulfotransferase. *Annu. Rev. Pharmacol. Toxicol.* 40: 159–176.

Raftogianis, R.B., Wood, T.C., and Weinshilboum, R.M. (1999) Human phenol sulfotransferases SULT1A2 and SULT1A1: genetic polymorphisms, allozyme properties, and human liver genotype-phenotype correlations. *Biochem. Pharmacol.* 58: 605–616.

Ringer, D.P., Norton, T.R., Cox, B., and Howell, B.A. (1988) Changes in rat liver *N*-hydroxy-2-acetylaminofluorene aryl sulfotransferase activity at early and late stages of hepatocarcinogenesis resulting from dietary administration of 2-acetylaminofluorene. *Cancer Lett.* 40: 247–255.

Ritter, J.K. (2000) Roles of glucuronidation and UDP-glucuronosyltransferases in xenobiotic bioactivation reactions. *Chem.-Biol. Interact.* 129: 171–193.

Rogan, E.G., Cavalieri, E.L., Walker, B.A., Balasubramanian, R., Wislocki, P.G., Roth, R.W., and Saugier, R.K. (1986) Mutagenicity of benzylic acetates, sulfates and bromides of polycyclic aromatic hydrocarbons. *Chem.-Biol. Interact.* 58: 253–275.

Sakakibara, Y., Yanagisawa, K., Katafuchi, J., Ringer, D.P., Takami, Y., Nakayama, T., Suiko, M., and Liu, M.C. (1998) Molecular cloning, expression, and characterization of novel human SULT1C sulfotransferases that catalyze the sulfonation of *N*-hydroxy-2-acetylaminofluorene. *J. Biol. Chem.* 273: 33929–33935.

Schuppler, J. and Günzel, P. (1979) Liver tumors and steroid hormones in rats and mice. *Arch. Toxicol. Suppl.* 2: 181–195.

Smith, B.A., Springfield, J.R., and Gutmann, H.R. (1986) Interaction of the synthetic ultimate carcinogens, *N*-sulfonoxy- and *N*-acetoxy-2-acetylaminofluorene, and of enzymatically activated *N*-hydroxy-2-acetylaminofluorene with nucleophiles. *Carcinogenesis* 7: 405–411.

Sommer, Y., Hollnagel, H., Schneider, H., and Glatt, H.R. (2003) Metabolism of 5-hydroxymethyl-2-furfural (HMF) to the mutagen, 5-sulfoxymethyl-2-furfural (SMF) by individual human sulfotransferases. *Naunyn-Schmiedeberg's Arch. Pharmacol.* 367: R166.

Stillwell, W.G., Turesky, R.J., Gross, G.A., Skipper, P.L., and Tannenbaum, S.R. (1994) Human urinary excretion of sulfamate and glucuronide conjugates of 2-amino-3,8-dimethylimidazo[4,5-*f*]quinoxaline (MeIQX). *Cancer Epidemiol. Biomarkers Prevent.* 3: 399–405.

Tiemersma, E.W., Bunschoten, A., Kok, F.J., Glatt, H.R., de Boer, S.Y., and Kampman, E. (2004) Effect of SULT1A1 and NAT2 genetic polymorphism on the association between cigarette smoking and colorectal adenomas. *Int. J. Cancer* 108: 97–103.

Topinka, J., Andrae, U., Schwarz, L.R., and Wolff, T. (1993) Cyproterone acetate generates DNA adducts in rat liver and in primary rat hepatocyte cultures. *Carcinogenesis* 14: 423–427.

van Bladeren, P.J. (2000) Glutathione conjugation as a bioactivation reaction. *Chem.-Biol. Interact.* 129: 61–76.

Watabe, T., Ishizuka, T., Isobe, M., and Ozawa, N. (1982) A 7-hydroxymethyl sulfate ester as an active metabolite of 7,12-dimethylbenz[*a*]anthracene. *Science* 215: 403–405.

Weinshilboum, R.M., Otterness, D.M., Aksoy, I.A., Wood, T.C., Her, C., and Raftogianis, R.B. (1997) Sulfotransferase molecular biology: 1. cDNAs and genes. *FASEB J.* 11: 3–14.

White, I.N., de Matteis, F., Davies, A., Smith, L.L., Crofton-Sleigh, C., Venitt, S., Hewer, A., and Phillips, D.H. (1992) Genotoxic potential of tamoxifen and analogues in female Fischer F344/n rats, DBA/2 and C57BL/6 mice and in human MCL-5 cells. *Carcinogenesis* 13: 2197–2203.

Wu, M.T., Wang, Y.T., Ho, C.K., Wu, D.C., Lee, Y.C., Hsu, H.K., Kao, E.L., and Lee, J.M. (2003) SULT1A1 polymorphism and esophageal cancer in males. *Int. J. Cancer* 103: 101–104.

Index

A

Access gate, substrate binding, 36–37, 196, 200
Acetaminophen, 1, 11
2-Acetylaminofluorene, 212, 295
2-Acetylaminofluorene *N*-sulfate, 281
Acetyltransferases, 145, 147, 181
Achondrogenesis type 1B, 45
Adenosine 5'-phosphosulfate (APS), 44, 47
Adrenal gland
    DHEA sulfate, 80, 243
    hydroxysteroids and, 80, 243
    SULT2A1 expression, 85, 232, 234, 241
        fetal, 108–109, 115, 243
Adrenaline, 109, 181, 214–215
Adrenarche, 80–81
Aging
    and androgen responsiveness, 243
    and DHEA levels, 71
Albumin, steroid sulfate binding, 232
Alkenyl benzenes, 281, 283
Allopregnenelone, 14
Allozymes
    activation/inactivation of promutagens,
        292–293
    nomenclature, 10–11
Allylic alcohols, sulfonated, 181
Alzheimer's disease, 215
Amaranth, 170
Amides, SULT1A2 allozymes and, 192
Amine sulfotransferases, 271; *see also* SULT3
Amines, dietary, SULT1A3 sulfation of, 91
Amniotic fluid, iodothyronine levels, 124, 129
Amphibians, SULTs, 106
Androgens
    aromatase pathway, 136
    SULT2A1 sulfation, 242
    synthesis, 81
Androsterone, SULT2 sulfation, 14, 234, 270
Anhydroretinol, 15
Aniline, 15
Anthocyanins, 170
Apomorphine, 38, 170–171
APS-kinase, 44
    Penicillium chrysogenum, 48
    reaction mechanism, 45
    structure/function, 47–51

Aromatic amines, 192, 280–283
Aromatic hydroxylamines, 290
Arthritis, degenerative, *PAPSS2* mutations and, 73
Arylamines, 181
Arylhydroxamic acids, 259–260, 262–266
Arylsulfatase (ARS) enzyme family, 106–109,
    129–130
Arylsulfotransferases, 259–269; *see also*
        subfamilies and individual enzymes
    rat, 35
Asn235, 190
Asp86, 201
Atelosteogenesis type II, 45
ATP-sulfurylase, 44
    Penicillium chrysogenum, 48
    reaction mechanism, 45
    structure/function, 47–51

B

Banana, 170
Bartsch, Ingrid, 283
Benzylic alcohols, 291
β-adrenergic agonists
    metabolism, 215
    sulfation of, 180, 259
Bi Bi mechanism, 197–198
Bile acids sulfation, 81, 86, 179
    SULT2A1, 234, 242–243
Binding pocket loop 1–3, 32
Bipolar disorder, 215
Blanchard, Rebecca, 255
BPL1, 32
BPL2, 32
BPL3, 32
Brachymorphism, 54
Brain
    fetal, SULT1A expression, 114, 127, 207
    PAPS synthetase demand, 55
    rat, SULT2A1 expression, 243
    SULT1A1 expression, 95, 206–208
    SULT1A3 expression, 95
    SULT2B1a expression, 244
Brain sulfotransferases, 94; *see also*
        SULT4A1
Brassinosteroids, 16

T - #0376 - 071024 - C4 - 234/156/15 - PB - 9780367392567 - Gloss Lamination